T0184523

Machine Learning Support for Fault Diagnosis
of System-on-Chip

Patrick Girard • Shawn Blanton • Li-C. Wang

Editors

Machine Learning Support for Fault Diagnosis of System-on-Chip

 Springer

Editors
Patrick Girard
LIRMM - CNRS
Montpellier, France

Shawn Blanton
Carnegie Mellon University
Pittsburgh, PA, USA

Li-C. Wang
University of California, Santa Barbara
Santa Barbara, CA, USA

ISBN 978-3-031-19641-6   ISBN 978-3-031-19639-3   (eBook)
https://doi.org/10.1007/978-3-031-19639-3

© The Editor(s) (if applicable) and The Author(s), under exclusive license to Springer Nature Switzerland AG 2023
This work is subject to copyright. All rights are solely and exclusively licensed by the Publisher, whether the whole or part of the material is concerned, specifically the rights of translation, reprinting, reuse of illustrations, recitation, broadcasting, reproduction on microfilms or in any other physical way, and transmission or information storage and retrieval, electronic adaptation, computer software, or by similar or dissimilar methodology now known or hereafter developed.
The use of general descriptive names, registered names, trademarks, service marks, etc. in this publication does not imply, even in the absence of a specific statement, that such names are exempt from the relevant protective laws and regulations and therefore free for general use.
The publisher, the authors, and the editors are safe to assume that the advice and information in this book are believed to be true and accurate at the date of publication. Neither the publisher nor the authors or the editors give a warranty, expressed or implied, with respect to the material contained herein or for any errors or omissions that may have been made. The publisher remains neutral with regard to jurisdictional claims in published maps and institutional affiliations.

This Springer imprint is published by the registered company Springer Nature Switzerland AG
The registered company address is: Gewerbestrasse 11, 6330 Cham, Switzerland

General Introduction

Today's electronic systems are composed of complex Systems on a Chip (SoCs) made of heterogeneous blocks that comprise memories, digital circuits, analog and mixed-signal circuits, etc. To fit a critical application standard requirement, SoCs pass through a comprehensive test flow (functional, structural, parametric, etc.) at the end of the manufacturing process. The goal is to achieve near-zero Defective Parts per Million (DPPM) so as to ensure the quality level required by the standard. Unfortunately, imperfections in the manufacturing process may introduce systematic defects, especially when the first devices are produced while the process is not yet mature. Identification of these systematic defects and correction of the related manufacturing process call for efficient diagnosis techniques. Hence, the goal of diagnosis is to extract information from test data in order to identify the nature and the causes of defects that have occurred in a SoC. Note that additional data can also be produced and used to improve the diagnosis process, such as distinguishing test patterns used only during the diagnosis phase.

Failure isolation is critical to identify root-cause of manufacturing issues. The scaling of manufacturing process technology and shrinking of device sizes and interconnects in nano-scale geometries make defect isolation more and more challenging. With the introduction of new transistor devices, lithography, and fabrication technologies, the demand for faster and precise defect isolation will continue to grow. Along with manufacturing process complexity, design complexity has also increased as functionality and computing needs have grown manyfold. Increasing design complexity makes the defect isolation in billon transistor chip ever more challenging. Therefore, fast diagnosis and root-cause identification of failures are essential to maintain high production yields and keep the Moore's Law in meeting the time to market demand.

Chips that pass manufacturing test are then shipped to the customer and integrated in their host system. However, despite the quality level of the manufacturing test procedures, SoCs may fail in the field, either due to the occurrence of a defect not covered during the manufacturing test phase or due to early-life failures or failures caused by various wear-out mechanisms. Early-life failures are caused by latent defects that are not exposed during manufacturing tests, but that are degraded due to electrical and thermal stress during in-field use, and lead to a failure in functionality. Wear-out or aging manifesting as progressive performance degradation, is induced by various mechanisms such as negative-bias temperature instability or hot-carrier injection. To avoid catastrophic consequences, many systems include in-the-field test techniques, which allow the detection of such problems. After detection, the defective SoC must be diagnosed to identify any possible systematic degradation patterns and avoid their re-occurrence in next-generation products. In this context, the first step during the failure identification process is to reproduce the failure mechanism with any original test and test conditions. Next, a diagnosis program made of several routines is used to identify, step by step, the failing part of the defective SoC and, finally, the suspected defects. Each routine corresponds to the application of a diagnosis algorithm at a given hierarchy level (system, core and cell levels).

Irrespective of the phase during which a SoC fails (after manufacturing, or after online test if the defective SoC is sent back to the manufacturer), defective SoCs undergo logic diagnosis to locate the fault, and then physical failure analysis (PFA) to characterize the fault. Diagnosis is a software-based method that analyzes the applied tests, the tester responses, and the netlist (possibly with layout information) to produce a list of candidates that represent the possible locations and types of defects (or faults) within the defective circuit. The quality of a diagnosis outcome is usually evaluated owing to two metrics: accuracy and resolution. A diagnosis is accurate if the actual defect is included in the reported list of candidates. Resolution refers to the total number of candidates reported for each actual defect. An accurate diagnosis with perfect resolution (i.e., one) is the ideal case. Diagnosis is usually followed by physical failure analysis (PFA), a time-consuming process for exposing the defect physically in order to characterize the failure mechanism. Due to the high cost and destructive nature of PFA, diagnosis resolution is of critical importance. In practice, it is very uncommon to perform PFA on any defect with more than five candidates. This ensures that the likelihood for uncovering the root-cause of failure is maximized when performing PFA.

Historically, conventional approaches based on cause-effect (i.e., fault simulation) and/or effect-cause (i.e., critical path tracing) analysis were used in industry for defect and fault diagnosis. However, with the fast development and vast application of machine learning (ML) in recent years, ML-based techniques have been shown to be highly valuable for diagnosis. They can be used for volume diagnosis after manufacturing to improve production yield or for diagnosis of customer returns to identify any possible systematic degradation patterns. The main advantage of ML-based diagnosis techniques is that they can deal with huge amount of insightful test data that could not be efficiently exploited otherwise in a reasonable amount of time.

A wide range of solutions based on supervised, unsupervised and reinforcement learning have been proposed in the last 10 years. They can be used for failure isolation in logic or analog parts of SoCs, board-level fault diagnosis, or even wafer-level failure cluster identification. A plethora of ML algorithms have been experimented and implemented in new diagnosis tools used today in industry. Benefits can be measured in terms of diagnosis accuracy, resolution and duration.

This book identifies the key challenges in fault diagnosis of system-on-chip and presents the solutions and corresponding results that have emerged from leading-edge research in this domain. In a comprehensive form, it provides necessary background to the reader and proposes a compendium of solutions existing in this field. The book explains and applies optimized techniques from the machine learning domain to solve the fault diagnosis problem in the realm of electronic system design and manufacturing. It demonstrates techniques based on industrial data and feedback from actual PFA analysis. It also discusses practical problems, including test sequence quality, diagnosis resolution, accuracy, and time cost.

The first chapter gives some prerequisites on fault diagnosis. Basic terms, such as defect, fault, and failure, are first enumerated. Then, basic concepts of test and fault simulation are described. After that, the basics of volume diagnosis for yield improvement and fault diagnosis of customer returns are given. Finally, basic information on yield and failure analysis is provided.

Chapter 2 is dedicated to the presentation of conventional methods for fault diagnosis. The chapter focuses on the automated tools and methods along with design features at the architectural, logic, circuit, and layout level that are needed to facilitate silicon debug and defect diagnosis of integrated circuits. These design features are generally referred to as design for debug and diagnosis (DFD). The chapter describes how these DFD features along with automated tools and methods are used effectively in a debug or diagnosis environment for applications ranging from design validation, low yield analysis, and all the way to field failure analysis. The chapter can serve as a steppingstone to understand further how conventional methods for fault diagnosis can be improved by using machine learning–based techniques.

The third chapter provides details of machine learning techniques proposed so far to solve various VLSI testing problems. It focuses on explaining scope of machine learning in VLSI testing. First, it gives a high-level overview of machine learning. After that, it describes the types of machine learning algorithms. Then, it explains some popular and commonly used machine learning algorithms. After that, this chapter discusses some recent machine learning based solutions proposed to solve VLSI testing problems. Finally, it discusses the strength and limitations of these methods.

Chapter 4 is dedicated to machine learning support for logic diagnosis and defect classification. After a preliminary discussion about attempts to distinguish maleficent defects from benign variations, the chapter presents machine learning techniques developed so far for distinguishing variations from reliability threats due to defects. Then, machine learning techniques for identifying different defect types during diagnosis are discussed. A neural network-based fault classifier is presented that can distinguish different fault models at gate level. Finally, the chapter concentrates on distinguishing between transient errors covered by hardening or masking techniques from intermittent faults. A solution based on Bayesian networks is presented for classifying intermittent, transient, and permanent faults.

The fifth chapter is dedicated to machine learning in logic circuit diagnosis. It is organized into three main sections that describe the use of ML for pre-diagnosis, during-diagnosis, and post-diagnosis, so as to characterize when and how a given methodology enhances the classic outcomes of diagnosis that include localization, failure behavior identification, and root cause of failure. The first section is dedicated to pre-diagnosis, which is concerned with any activities that are performed before diagnosis is deployed. Examples of pre-diagnosis activities include classic work such as diagnostic ATPG, and DFT for increasing testability. In the second section, the use of ML in during-diagnosis activities is described, which generally involves learning while diagnosis executes. For example, a k-nearest neighbor model can be created, evolved, and used during on-chip diagnosis to improve diagnosis outcomes. In the third and last section, post-diagnosis is discussed, which includes all activities that occur after diagnosis execution. These approaches usually involve volume diagnosis (i.e., using the outcome results of many diagnoses) to improve diagnostic resolution.

Chapter 6 gives an overview of the various machine learning approaches and techniques proposed to support cell-aware generation, test, and diagnosis. The chapter focuses on the generation of the cell-aware models and their usage for diagnosis. After some backgrounds on conventional approaches to generate and diagnose cell-aware defects, the chapter will present a learning-based solution to generate cell-aware models. Then, it presents a ML-based cell-aware diagnosis technique. Effectiveness of existing techniques will be shown through industrial case studies and corresponding diagnosis results in terms of accuracy and resolution. The chapter will conclude with a discussion on the future directions in this field.

Chapter 7 discusses the state of the art on fault diagnosis for analog circuits with a focus on techniques that leverage machine learning. For a chip that has failed either in post-manufacturing testing or in the field of operation, fault diagnosis is launched to identify the root-cause of failure at subblock level and transistor-level. In this context, machine learning can be used to build a smart system that predicts the fault that has occurred from diagnostic measurements extracted on the chip. The chapter discusses the different elements of a diagnosis flow for analog circuits, including fault modeling, fault simulation, diagnostic measurement extraction and selection, and the machine learning algorithms that compose the prediction system. A machine learning–based diagnosis flow experimented on an industrial case study is finally presented.

The eighth chapter discusses machine learning support for board-level functional fault diagnosis. First, the chapter presents an overview of board-level manufacturing tests and conventional fault-diagnosis models. Next, it discusses the motivation of utilizing machine learning techniques and presents the existing machine learning–based diagnosis models. To address the practical issues that arise in real testing data, the chapter next presents a diagnosis system based on online learning algorithms and incremental updates. In the following, it also presents a diagnosis system that utilizes domain-adaption algorithms to transfer the knowledge learned from mature boards to a new board.

Chapter 9 is dedicated to wafer-level failure pattern analytics. In the first section of the chapter, the application is about early detection of yield excursions with the goal to automatically recognize the existence of a systematic failure cluster when one occurs. In the second section, analytics is formulated as solving a multi-class classification problem and the discussion focuses on training a high-accuracy neural network classifier. In the next section, techniques to learn an individual recognizer for one pattern class are discussed with the goal to learn with very few training samples. Generative adversarial networks (GANs) and tensor computation–based techniques are used together to implement an unsupervised wafer pattern classification and recognition flow. The last section of the chapter introduces language-driven analytics and explains its use in the analytics context. The authors show how a pretrained language model like GPT-3 can play a role in solving the problem.

Finally, a conclusion summarizes the contribution of the book and some perspectives in the field of fault and defect diagnosis of circuits and systems by using machine learning techniques are given.

Contents

Prerequisites on Fault Diagnosis

Harry H. Chen, Xiaoqing Wen, and Wu-Tung Cheng

1 Defect, Fault, Error, Failure

The life of an *integrated circuit* (IC) consists of three phases, namely the *design* phase, the *manufacturing* phase, and the *operational* phase. Ideally, an IC should perform all required functions correctly at its required speed and within its required power limit. In practice, unavoidable imperfection in the three phases may cause adverse impact on an IC. As illustrated in Fig. 1, an *error* may occur at an IC (i.e., the IC outputs a wrong signal) and may eventually cause a *failure* of a system based on the IC (i.e., the system shows a wrong behavior). System failures may cause anything from inconvenience to catastrophe. Although techniques for robust system design exist that can mask the impact of errors to some extent, they are generally expensive because of larger circuit area and higher power consumption. Therefore, it is imperative to minimize the occurrence of IC errors by all means so as to minimize the possibility of system failures.

An IC error can be caused by incorrectness (i.e., *bugs*) introduced in the design phase. However, discussions on how to reduce design bugs are not the focus of this book; instead, this book assumes that IC design is correctly conducted. As illustrated in Fig. 1, for a correctly-designed IC, an IC error is the effect of either an internal cause (*defect*) or an external cause (*radiation*).

H. H. Chen
MediaTek Inc., Hsinchu, Taiwan

X. Wen
Kyushu Institute of Technology, Iizuka, Japan

W.-T. Cheng (✉)
Siemens Digital Industries Software, Plano, TX, USA
e-mail: wu-tung.cheng@siemens.com

© The Author(s), under exclusive license to Springer Nature Switzerland AG 2023
P. Girard et al. (eds.), *Machine Learning Support for Fault Diagnosis of System-on-Chip*, https://doi.org/10.1007/978-3-031-19639-3_1

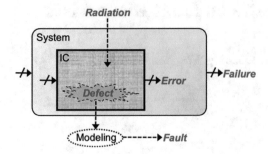

Fig. 1 Various terms

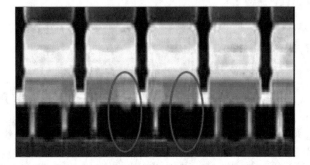

Fig. 2 Example of manufacturing defects (missing vias)

The internal cause of an IC error is a *defect*, which is the unintended difference between the implemented hardware and its intended design [5]. A defect physically and permanently exists since its occurrence. An IC with one or more defects is defective; otherwise, the IC is defect-free.

Defects may occur in the manufacturing phase. An example is shown in Fig. 2. Manufacturing defects are caused by random imperfection (such as particle contamination) or systematic reasons (such as process variations). Therefore, all manufactured ICs need to go through a check process, called *manufacturing test*. *Ideal manufacturing test* passes all defect-free ICs but fails all defective ICs. *Realistic manufacturing test*, however, is imperfect in that it may pass some defective ICs (i.e., *under-test*) and may fail some defect-free ICs (i.e., *over-test*). Passing ICs are shipped to customers for use in electronic systems and failing ICs are discarded. *Yield* is defined as the percentage of passing ICs among all manufactured ICs. Yield loss is *catastrophic* if it is caused by random defects, or *parametric* if it is caused by defects due to process variations. Low yield not only is the nightmare for IC manufacturers but also makes customers worry about IC's quality and reliability. There are two major design-based approaches to improving yield, namely *design for yield enhancement* (DFY) and *design for manufacturability* (DFM). DFY tries to reduce the effect of process variation while DFM tries to avoid

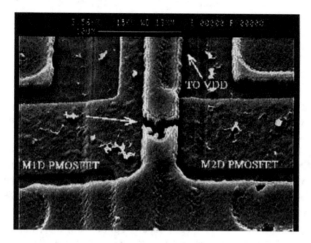

Fig. 3 Example of age defects (stress migration)

random defects. In addition, efforts can be made to reduce over-test, especially for low-power ICs [11].

Defects may also occur in the operational phase, in the forms of *latent defects* and *age defects,* which may cause IC errors and eventually system failures. Latent defects are manufacturing defects that escape manufacturing test. This is especially the case where the quality of manufacturing test is low, causing severe under-test. Age defects are caused by the wear-out of ICs under electrical, thermal, and mechanical stresses in the forms of metal fatigue, hot carriers, electromigration, dielectric breakdown, etc. An example is shown in Fig. 3. In the early operation stage of a system, latent IC defects are the dominant cause of system failures, called *early failures.* In the middle operation stage of a system, latent IC defects become rare and age defects are yet to show up, resulting in only sparse system failures, called *random failures.* In the end operation stage of a system, age defects are the dominant cause of system failures, called *wear-out failures.* Therefore, it is desirable that an IC used in a system can also go through a check process, called *field test.* Defective ICs found by field test can then be replaced before they cause any system failure.

The external cause of an IC error is *radiation*, which comes from the space as protons, electrons, and heavy ions or from IC package materials as alpha particles. Radiation may cause a *single event upset (SEU)*, which is usually a bit flip ($0 \rightarrow 1$ or $1 \rightarrow 0$) at a storage element (a memory cell, a flip-flop, or a latch) in an IC at a certain time [10]. Such a (local) SEU may in turn cause a (global) error at an output of the IC, eventually resulting in a failure of a system based on the IC. An SEU is also called a *soft error* since the affected IC can be free of any permanent defects and its impact is *transient* (i.e., the impact of bit flip at a storage cell disappears after a correct value is loaded). With ever-decreasing feature sizes and power supply voltages, ICs are increasingly becoming susceptible to soft errors.

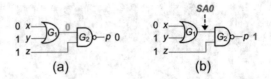

Fig. 4 Stuck-at fault model

However, due to this transient nature, soft errors cannot be tested for as defects that are permanent in nature. The common practice for mitigating soft errors is to adopt radiation-hardening design techniques for storage elements to reduce the chance of radiation to cause soft errors [28]. Such design techniques are referred to as *design for reliability* (DFR).

Generally, defects are the dominant cause of IC errors; thus, defects need to be targeted in manufacturing test as well as field test. However, directly dealing with defects in test-related tasks (e.g., quantifying test quality, generating test data, etc.) is often inefficient and sometimes even computationally infeasible. To solve this problem, one can model the behavior of a defect by a *fault* and only deals with faults explicitly in test-related tasks. Due to the complexity and variety of defects, their behaviors are widely diverse. As a result, no one fault model can be a good representative for all defects. Therefore, it is important to select or build a fault model by taking into consideration of process characteristics, transistor structures, anticipated defects, circuit abstraction levels, etc. Some typical fault models are introduced below.

- *Stuck-At Fault Model*: A gate-level logic circuit is modeled by gates and their interconnections. If the circuit is defect-free, any signal line should be able to take both logic values (0 and 1). The existence of some sort of defects in the circuit may make a signal line to take only one logic value (either 0 or 1). This defect behavior can be modeled by a *stuck-at fault* [9]. If the signal line can only take logic 0 (1), it is said to have a *stuck-at-0 (stuck-at-1) fault* or a *SA0 (SA1) fault*. Fig. 4a shows a fault-free circuit, whose output (p) is 0 for the input value combination $<x = 0, y = 1, z = 1>$. If the output of the OR gate G_1 has a SA0 (i.e., the output of G_1 is fixed at 0 due to the existence of some defect), the circuit output p will be 1 as shown in Fig. 4b. In this case, $<x = 0, y = 1, z = 1>$ is said to be a *test vector* for the SA0 fault.

- *Transistor Stuck Fault Model*: A transistor-level circuit is modeled by transistors and their interconnections. If the circuit is defect-free, any transistor in the circuit should be able to be turned on and off. The existence of some sort of defects in the circuit may make a transistor to be permanently turned on (off). This defect behavior can be modeled by a *transistor stuck-on (stuck-off) fault*. A transistor stuck-on (stuck-off) fault is also referred to as a *transistor stuck-short (stuck-open) fault*. As illustrated in Fig. 5, an inverter, whose input and output are x and z, respectively, consists of two transistors, P and N. Thus, this transistor has 4 transistor stuck faults ("*P stuck-on*", "*N stuck-on*", "*P stuck-off*", "*N stuck-off*").

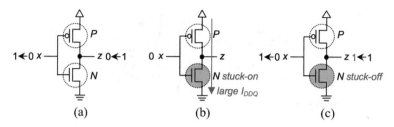

Fig. 5 Transistor stuck fault model

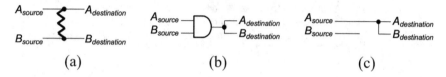

Fig. 6 Bridging fault model

Whether "*N stuck-on*" exists can be determined by applying 0 to x and measuring the quiescent power supply current, I_{DDQ} as shown in Fig. 5b. This is because, if "*N stuck-on*" exists, applying 0 to x will turn P on, resulting in a current path from the power supply to the ground [19]. Whether "*N stuck-off*" exists can be determined by first applying 0 and then 1 to x as shown in Fig. 5c. If the value of z changes from 1 to 0, "*N stuck-off*" does not exist. If the value of z remains at 1, "*N stuck-off*" exists. That is, determining whether "*N stuck-off*" exists needs two test vectors [25].

- *Bridging Fault Model*: Gates in a gate-level circuit and transistors in a transistor-level circuit are all interconnected by wires. Some defects may short two separate wires together, and its effect can be modeled by a *bridging fault* [21] Depending on how to determine the resultant logic value when the two involved wires have opposite logic values, bridging faults can be classified into a few types. For the bipolar technology, the resultant logic value can be assumed to be the AND (OR) result of the logic values of the two wires, resulting in a *wired-AND (wired-OR) bridging fault*. Fig. 6a shows that wires A and B are shorted together. Its effect can be modeled by a wired-AND bridging fault, whose gate-level representation is shown in Fig. 6b. For the *complementary metal oxide semiconductor* (CMOS) technology, it is more appropriate to model a pair of shorted wires as a *dominant bridging fault*, whose logic value is assumed to be determined by the stronger driver for the two shorted wires. For example, the shorted wires A and B as shown in Fig. 6a can be modeled by a *dominant bridging fault* (A dominates B), whose gate-level representation is shown in Fig. 6c.

- *Delay Fault Model*: Extra delay can be introduced to wires by resistive open and short defects, or to transistors or gates by parameter variations. Such extra delay, either alone or in an accumulative manner, may increase signal propagation delay so much as to break timing requirements, resulting in wrong circuit behaviors.

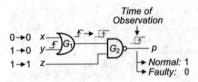

Fig. 7 Transition delay fault model

```
Cell "MUX2" {
    Fault "Z1" {
        test { StaticFault "Z'=1;Condition "D0'=0,"D1'=0,"S'=0;}
        test { StaticFault "Z'=1;Condition "D0'=0,"D1'=1,"S'=0;}
        test { StaticFault "Z'=1;Condition "D0'=0,"D1'=0,"S'=1;}
    }
}
```

Fig. 8 Cell-aware fault model

Extra delay can be modeled by a *delay fault* with several varieties. A *transition delay fault* is used to model an extra-delay-affected transition (*rise* $(0 \rightarrow 1)$ or *fall* $(1 \rightarrow 0)$) at a gate [18] and the extra delay is assumed to be large enough to prevent the transition from reaching any output of the circuit at the required time. Thus, each gate is associated with a *slow-to-rise transition delay fault* and a *slow-to-fall transition delay fault*. An example is shown in Fig. 7, where the output of G_1 has a slow-to-rise transition delay fault. Suppose that x remains at 1, z remains at 1, and y is applied with a transition of $1 \rightarrow 0$. In this case, at the time of observation, the output p of G_2 will be 0 instead of the correct value of 1, meaning that the slow-to-rise transition delay fault is detected. A *gate-delay fault* is used to model the extra delay of a gate by explicitly quantifying it [14, 15]. As a result, a gate-delay fault may not cause any wrong circuit behavior if the signal propagation path going through the gate is short enough to prevent the extra delay from breaking timing requirements. A *path-delay fault* is used to model the accumulative extra delay along a path comprising gates and wires [30].

- *Cell-Aware Fault Model*: Complex cells are widely used in ICs and the impact of intra-cell defects is increasingly becoming significant. In order to explicitly deal with intra-cell defects in test-related tasks for higher test quality, a defect-oriented fault model, called the *cell-aware fault model* [12], can be built as follows: First, all possible defects in a standard cell are extracted from its layout. After that, the behavior of each extracted defect is SPICE-simulated under all possible cell input combinations. Finally, all unexpected behaviors are modeled at the IO ports of the cell with a digitalized format for use in various test-related tasks, such as test generation and fault simulation. Fig. 8 shows an example, which shows that the fault Z1 in the cell MUX2 can be detected by any one of the three input value combinations. That is, when any of the input value combinations is applied to a MUX2 cell with the fault Z1, the output of the cell will show a value different from the one for a fault-free MUX2 cell.

2 Test Basics

Defects are the dominant cause of IC errors and they may lead to system failures. Therefore, *test*, a check process aimed at determining whether an IC is defective, is critically important. Since defects may occur both in the manufacturing phase and in the operational phase of ICs, both *manufacturing test* and *field test* are necessary. On the other hand, today's ICs usually come in the form of Systems on a Chip (SoCs) consisting of heterogeneous blocks that comprise logic circuits, memories, analog and mixed-signal circuits, etc. Therefore, test methods need to be developed for these different blocks and for the entire SoC as a whole. Furthermore, since ICs are often assembled on *printed circuit boards* (PCBs) to make electronic systems, boards also need to be tested. In this section, general information on manufacturing test and field test is provided first. After that, the basics for logic test, memory test, analog test, SoC test, and board test are briefly introduced.

2.1 Manufacturing Test and Field Test

The manufacturing of today's ICs involves highly complicated processes, materials, equipment, and operations. This makes it impossible to perfect every factor in manufacturing, rendering it inevitable that some manufactured ICs are defective. As a result, manufacturing test is indispensable in guaranteeing IC quality.

Manufacturing test usually consists of multiple rounds. The first round of test, namely *wafer test*, is conducted for bare dies on a wafer through direct electrical contact with the bonding pads of the dies. Its purpose is to identify defective dies so that only good dies are packaged. After packaging, the second round of test is conducted through the external pins of packaged ICs. This round of test is necessary because imperfect packaging may introduce new defects and wafer test is often not thorough due to various limitations. The passing ICs of this round of test often goes through a process called *burn-in*, which is conducted by applying electrical, thermal, mechanical and environmental stresses so as to accelerate the occurrence of potential defects to the manufacturing phase, instead of leaving them to occur in the operational phase. Burn-in is especially important for ICs intended for mission-critical applications. After burn-in, the third round of test is conducted to screen out defective ICs whose defects' occurence is accelerated by burn-in. Only ICs passing all three rounds of test are shipped to customers. Note that manufacturing test is usually conducted with powerful *automatic test equipment* (ATE). Each round of manufacturing test usually consists of open-short test, DC parametric test, high-speed I/O test, memory test, analog test, and logic test under various conditions related to voltages, temperatures, and operating speeds.

Field test is conducted for ICs already placed into a system, especially a mission-critical one. Different from manufacturing test, large and expensive ATE cannot be used for field test. Instead, self-contained test circuitry can be designed into a target IC so that it can test itself without using external ATE. This method is called *built-in self-test* (BIST). An alternative is to design self-contained test circuitry into the PCB board holding a target IC. This method is called *built-out self-test* (BOST).

2.2 Logic Test

The testing of a logic circuit is conducted by applying test stimuli to its inputs and observe test responses at its outputs. If all of the observed test responses match their corresponding expected test responses, the circuit is judged to be defect-free. Obviously, the confidence level of the judgement depends on the quality of the test stimuli used, which is measured by *fault coverage* achieved by the test stimuli. Fault coverage of a given set of test stimuli is obtained by *fault simulation*, while the process of generating a set of test stimuli to achieve high fault coverage is called *test generation*. Furthermore, it is important to note that, in order to improve test quality and test efficiency, the *circuit-under-test* (CUT) itself often needs to be modified or extended with some test-oriented circuitry. This concept is called *design for test* (DFT). In the following, DFT, fault simulation, and test generation are briefly introduced.

2.2.1 Design for Test (DFT)

Most logic circuits are sequential circuits that cannot be efficiently tested. This is because a sequential circuit contains *flip-flops* (FFs), whose inputs are hard to control and whose outputs are hard to observe. As a result, a common DFT methodology, called *scan design*, needs to be applied to sequential circuits [8]. As shown in Fig. 9a, scan design requires that all FFs replaced with *scan FFs*. A scan FF has an original *data input* and an added *scan input*, selectable by the *scan enable* (SE) signal. All scan FFs form one or more scan chains. *Scan test* is conducted as follows: In shift mode ($SE = 1$), a scan chain operates as a shift register, allowing test stimuli to be applied from the outside and previous test responses to be taken to the outside. In capture mode ($SE = 0$), all scan FFs operate as normal FFs to load test responses into FFs. This way, scan design makes all FFs both controllable and observable, thus greatly easing the test generation for sequential circuits. In order to reduce test data volume for scan test, *compressed scan design* can be applied. Fig. 9b illustrates such a technique, called *embedded deterministic test* (EDT), in which a ring-register-based decompressor restores compressed test stimuli on the input side and a compactor reduce test response data volume on the output

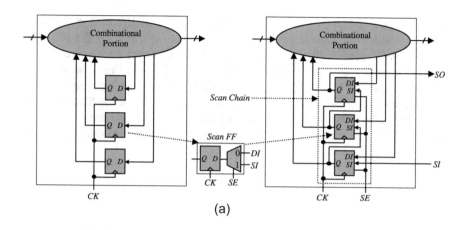

(a)

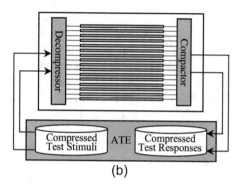

(b)

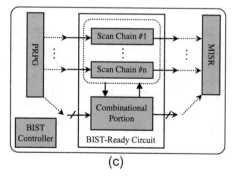

(c)

Fig. 9 Various scan-based DFT techniques

side [24]. Scan design is also the base for logic BIST as illustrated in Fig. 9c, in which test stimuli are generated by a *linear feedback shift register* (LFSR)-based *pseudo random pattern generator* (PRPG) and test responses are compressed into a single *signature* by a *multi-input shift register* (MISR) [4]. From the viewpoint of fault diagnosis, normal scan design provides the best test responses without any diagnostic information loss. Compressed scan design suffers from moderate diagnostic information loss due to the compaction of test responses while logic BIST suffers from the worst diagnostic information loss due to the compression of all test response into a single signature.

2.2.2 Fault Simulation

The most basic form of fault simulation for a combinational circuit (or the combinational portion of a scan circuit) c is conducted for a fault f and an input vector v to determine whether f is *defected* by v. Suppose that $r(c, v)$ is the output response to v by the fault-free circuit c and $r(c(f), v)$ is the output response to v by the circuit c with the fault f. If $r(c, v) \neq r(c(f), v)$, f is said to be *detected* by v and v is said to be a test vector for f. $r(c, v)$ can be obtained by logic simulation and $r(c(f), v)$ can be obtained by forcing the behavior of f into c. Note that a test vector v for two faults f_1 and f_2 may lead to identical output responses, i.e., $r(c, v) \neq r(c(f_1), v) = r(c(f_2), v)$. In this case, v can detect both f_1 and f_2 but cannot distinguish between them. A more general form of fault simulation is conducted for a set of faults F and a set of input vectors V. The purpose is to determine which faults in F can be detected by at least one vector in V. The percentage of defected faults is called the *fault coverage* of V. Compared with logic simulation in which each input vector only needs to be processed once, fault simulation is more time-consuming since each input vector needs to be processed once for each fault. To accelerate fault simulation, *parallel fault simulation* makes use of bit-parallelism of logical operations in a digital computer [29]. In the example shown in Fig. 10, a 4-bit word is used to store the signal values of each signal line. One bit is used to represent the fault-free value while the remaining three bits are used to represent the values corresponding to three faults. That is, three faults can be simulated simultaneously in a single pass. From the values obtained at the output q, the test vector $<x = 0$, $y = 0$, $z = 1>$ can detect two faults, "y SA1" and "p SA1". Other fault simulation approaches include *deductive fault simulation*, which deduces all signal values in

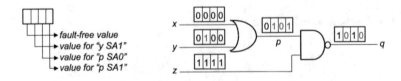

Fig. 10 Fault Simulation

each faulty circuit from the fault-free circuit values and the circuit structure in a single pass [2], and *concurrent fault simulation*, which emulates faults in a circuit in an event-driven manner to avoid unnecessary computation efforts [31].

2.2.3 Test Generation

The basic purpose of test generation for a circuit is to create a small as possible set of test vectors for achieving the highest possible fault coverage. Test generation is usually conducted with an algorithm in the form of *automatic test pattern generation* (ATPG). The target of generation is usually a combinational circuit or the combinational portion of a scan-based sequential circuit. Generally, there are two approaches to test generation, namely *non-fault-oriented* and *fault-oriented*.

Non-fault-oriented test generation for a circuit is conducted by considering the function of the circuit but ignoring its internal structure. Typical test generation based on this approach include *functional test generation, exhaustive test generation*, and *random test generation*. Although non-fault-oriented test generation is simple to implement, the number of resultant test vectors is usually large. In addition, although non-fault-oriented test generation itself does not need to consider faults, time-consuming fault simulation usually needs to be conducted to calculate fault coverage.

Fault-oriented test generation for a circuit is conducted by explicitly trying to create a test vector for each target fault under a fault model. Since structural information of the circuit is needed to define a fault model, fault-oriented test generation is also called *structural test generation*, which can be conducted with special algorithms, including D [27], PODEN (Goel 1981), FAN (Fujiwara 1983), and SOCRATES (Schulz 1988). A popular approach to structural ATPG, *path sensitization*, is illustrated in Fig. 11. First, in order to generate a test vector for the target fault as shown in Fig. 11a, "L_7 SA0", *fault activation* is conducted to make 1 to appear on L_7 as shown in Fig. 11b. As a result, the *fault effect D* (1 as the fault-free value and 0 as the value corresponding to "L_7 SA0") appears on L_7 as shown in Fig. 11c. Next, *fault propagation* is conducted to make the fault effect D or $\overline{D}$ to appear on at least on output of the circuit. For this purpose, 0 needs to be put on L_2 in order to make D to L_8 as shown in Fig. 11d and 1 needs to be put on L_3 in order to make $\overline{D}$ to the output x as shown in Fig. 11d and Fig. 11e. Note that fault activation and fault propagation lead to three value assignment requirements: $1 \rightarrow L_7, 0 \rightarrow L_2$, and $1 \rightarrow L_3$. Finally, *justification* is conducted to determine necessary values for the inputs of the circuit to satisfy the value assignment requirements, and the result is a *test cube* (containing a don't-care value X) as shown in Fig. 11f. Since X cannot be applied by ATE, a logic value needs to be assigned to X by *X-filling*, resulting in a test vector $<a = 0, b = 1, c = 1, d = 1>$.

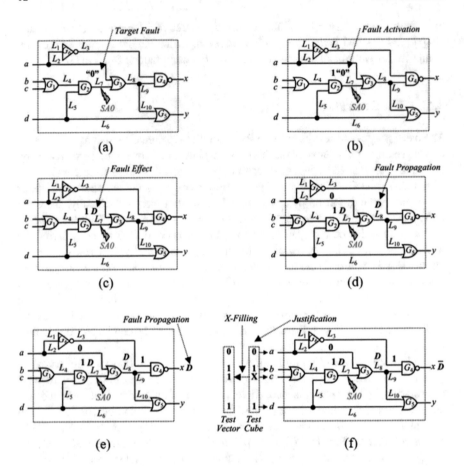

Fig. 11 Test generation basics

2.3   Memory Test

Different memories, such as *read-only memories* (ROMs) and *random access memories* (RAMs) require different approaches to testing. A ROM can be tested by reading out all stored values and checking if they are exactly what have been written into it. *Static RAMs* (SRAMs), *dynamic RAMs* (DRAMS), *electrically erasable ROM* (EEPROM) and flash memories need to be tested with test patterns targeted on various memory faults and generated by various memory test algorithms [1, 32].

Classical memory faults include *cell stuck-at faults* (i.e., the value of a memory cell is fixed at 0 or 1), *address decoder faults* (i.e., a memory cell corresponds to multiple addresses or one address corresponds to multiple memory cells), *data line faults* (i.e., defective input and output data registers prevent correct data from being written into or read from a memory cell), *read/write faults* (i.e., defective read/write control lines/logic prevent a read or write operation from being conducted), and *data retention faults* (i.e., a memory cell loses its content after a certain period of time). Modern high-density RAMs also suffer from *transition faults* (i.e., a memory cell cannot undergo a 0-to-1 or 1-to-0 transition), *destructive read faults* (i.e., a read operation changes the content of a memory cell), *coupling faults* (i.e., the content of a memory cell is affected by the operations on other memory cells), and *pattern sensitivity faults* (i.e., the content of a memory cell is affected by the contents of other memory cells).

In order to test for the faults mentioned above, many test pattern generation algorithms have been proposed over the years. These algorithms can be classified into different types, such as N, $N^{3/2}$, and N^2, where N is the number of address locations or words. Generally, test patterns generated by an N^2 algorithm can detect more faults than those generated by an N al algorithm but result in longer test time. In addition, test patterns generated by different algorithms detects different faults. Therefore, it is necessary to select a proper set of test generation algorithms for a memory by taking possible faults into consideration. A typical N-class test generation for detecting cell stuck-at faults is the *modified algorithmic test sequence* (MATS), whose length is $4N$. This algorithm has three steps: (1) all memory cells in a RAM are written to logic 0, (2) each address is first read, with logic 0 being the expected value, and then written to logic 1, and (3) each address is read with an expected value of logic 1.

The test result of a memory can be plotted into its *fail bit map*, which shows the identified defective cell locations in the memory. From the shape or distribution of the defective cell locations, the possible cause of the defects may be found. For example, if the defective cell locations are not regular, the possible cause may be cell stuck-at faults. If all memory cells in a row or column fail, the possible cause may be address decoder faults.

2.4   Analog / Mixed-Signal Test

An analog or mixed-signal circuit, either in a standalone form or as part of a *system-on-a-chip* (SoC) device, is usually tested by explicitly checking its functions against specifications through measuring various performance parameters [1, 32]. Typical mixed-signal circuits include A/D converters and D/A converters. The static properties of an A/D or DA converter include linearity, gain error, offset error, monotonicity, miscode, *integral non-linearity* (INL), *differential non-linearity* (DNL). The dynamic properties of an A/D or DA converter include *signal-to-noise ratio* (SNR), *total harmonic distortion* (THD), *spurious-free dynamic range* (SFDR), and *effective number of bits* (ENOB). Obviously, such a parameter has a tolerance range instead of a single expected value. Therefore, it is necessary to determine whether a measured parameter falls within its design specification tolerance range. In the testing of an A/D converter, an *arbitrary waveform generator*

(AWG) is usually used to generate analog input signals and digital output signals are obtained by a tester. In the testing of a D/A converter, digital input signals are applied and analog output signals are obtained by a tester. Therefore, high-precision AWGs and digitizers are required. As a result, analog / mixed-signal test is usually expensive due to long test time and the need for complicated test equipment.

2.5 SoC Test

An SoC usually consists of embedded cores of various types and its test needs to target at all cores as well as the interconnects among them. However, both individual cores and interconnects in an SoC are usually difficult to access. This problem can be addressed with the IEEE 1500 Standard for Embedded Core Test, which is a scalable standard architecture for enabling test reuse and integration for embedded cores and associated circuitry [32, 33]. It uses a scalable wrapper architecture and access mechanism similar to boundary scan to facilitate test access to embedded cores and their interconnects. Note that IEEE 1500 is independent of the functionality of an SoC or its embedded cores. In addition, *built-in self-test* (BIST) also helps in easing the access requirement for embedded cores, especially logic and memory cores.

2.6 Board Test

Generally, various ICs need to be assembled onto a *printed circuit board* (PCB) in order to be used in an electronic device. This assembling procedure is complicated and prune to various defects, making it necessary to conduct *board test* [32, 33] The conventional method, namely *bed-of-nails*, of board test is to directly probe the solder points on the back of a board. This way, test stimuli can be applied to the input pins of component ICs and the test responses can be observed from their output pins. A modern PCB, however, usually consists of surface-mount ICs, whose pins cannot be accessed from the bottom of the board. This makes it impossible to apply the bed-of-nails method. This problem can be addressed with the boundary scan methodology, which has been documented as several IEEE standards. For example, the IEEE Standard 1149.1 for logic ICs inserts additional logic to form a boundary scan chain through all I/O buffers of logic ICs. This chain makes it possible to shift in test stimuli to internal pins and interconnections on a PCB. It also makes it possible to capture output responses at the input buffers on other ICs on the PCB and subsequently shift them out for observation. This way, access to all ICs and interconnections can be established without direct probe contact. The access to the boundary scan chain is provided by the *test access port* (TAP) through a four-wire serial bus interface and instructions applied through the interface. This boundary scan interface can provide access to the DFT features, such as BIST, of individual ICs on a PCB. In addition to IEEE Standard 1149.1, the IEEE Standard 1149.4 is available for mixed-signal ICs and the IEEE Standard 1149.6 is available for the I/O protocol of high-speed networks.

3 Volume Diagnosis for Yield Improvement

As mentioned earlier, diagnosis is used to identify the location and failure mechanism of the defect. To improve the yield, it is important to identify systematic defects which are caused by the same root cause. To recognize systematic defects, statistics methods are used on volume diagnosis results to identify whether there are common root causes. Yield can be recovered only after the common root causes are removed properly.

In high volume production test environment, there can be many failing dies and lots of failing data to collect and process. Besides ATE, the machines with sufficient CPU and memory to process these volumes of failing data need proper management to get enough throughput.

3.1 Failing Data Limitation in ATE and Its Impact to Diagnosis Accuracy and Resolution

In scan test, each pattern is independent. In other words, the test results are independent on the order of test patterns. The main advantage of this independency is in logic diagnosis. The failing data information is per pattern. The diagnosis operation can be done pattern by pattern. However, diagnosis suspect count per pattern can be quite large. Statistics method is used to find the common defect suspect from all failing patterns. The suspect count can be further reduced by using passing patterns.

3.2 How to Collect Failure Data for Diagnosis?

To achieve high speed and precise test results, ATE is quite expensive. The hardware to store test results is expensive as well. In general, ATE has limited failure data storage space. To do chain diagnosis, to achieve good diagnosis results, at least 100 failing patterns are needed. Each failing pattern has many failing cycles when scan chains are faulty. Some ATE does not have such big failing data storage. To do diagnosis, it has to retest repeatedly to collect enough failing data. For logic diagnosis, each failing pattern has small number of failing cycles. Typically, 1000 to 2000 failing cycles are sufficient to achieve good diagnosis resolution.

3.3   Failing Data Format

To achieve accurate diagnosis, it is important to know which observe data is failing or passing. However, due to limited ATE failure data storage, some information is lost. Since data collected from ATE only identify failing data, it can be mis-leading to assume non-failing observe data are passing. To avoid wrong diagnosis results, it is important for precisely label each observe data of each pattern as failing, passing or unknown. Base on ATE setting, the failing data truncation can be pin-based, cycle-based or even pattern based. Diagnosis accuracy depends on proper failing data truncation information.

3.4   Volume Diagnosis Server Farm Setup to Process Thousands of Failing Data in Production Flow

The resource to do a diagnosis job depends on the memory and the time it needs to generate the diagnosis report. For a fixed amount of volume diagnosis job, the throughput can be improved by either using smaller memory or smaller run time. In a typical server farm to process volume diagnosis, there are much more machines with smaller memory. To improve the throughput of volume diagnosis, it is more important to reduce the memory used for each diagnosis job than to speed up the run time.

3.5   Beside Failing Data Per Die, What Other Data to Collect for Statistics Yield Analysis

It is common that each die may go through same test under several operation corners such as different voltage, different clock frequencies, different temperatures. All these test environment data should be considered when analyzing volume diagnosis results to find common root causes. Also die locations in wafer, and lot information should be considered as well.

4   Fault Diagnosis of Customer Return

Despite best screening efforts during a product's manufacturing process prior to customer delivery, defects do escape and end up as product failures encountered by the customer. Generally speaking, customer expectations of quality vary depending on the market and product segment. When instances of failure exceed an acceptable level or consequences of failure are severe enough, the product is returned to the

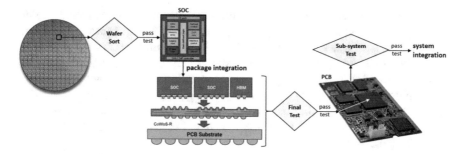

Fig. 12 Manufacturing test flow from wafer to PCB

manufacturer for root cause diagnosis. The goal is to prevent future occurrences of failure and to ensure minimal negative impact on the manufacturer's quality reputation. When the perception of quality takes a hit, it results in direct loss of business; and a great deal of effort and expense are needed to recover customer trust.

In looking across the full range of modern electronic products from hand-held mobile smart-phones to data center servers/routers/switches to artificial intelligence (AI)-based self-driving cars, one finds a complex assembly of electronic and mechanical hardware components with a stack of software layers above ultimately ending in interactions with the human user. The electronic components can be sophisticated system-on-chips (SoC) and multi-die integrated packages in their own right. When the user experiences a failure, it could be caused by defects in the software, in the hardware, or a combination of deleterious interactions among the components. Indeed, the likelihood of failure rises as an emergent property of increasing complexity [7].

Multiple testing steps are executed in the process of integrating and assembling components to create a final product. The purpose is to catch defects as early as possible since allowing escapes to the next step incurs additional cost. For example, consider the manufacturing test steps shown in Fig. 12 as a die component moves from wafer to package to printed circuit board (PCB). Dies on a wafer are tested and only those passing are diced and proceed to the packaging step. A defective die that escapes wafer test will be integrated with others in a multi-die package. If the integrated package fails final test, the defective die is successfully screened; but replacing it in the finished package may be too costly or impossible. Thus, the entire defective package has to be discarded rendering naught the costs of package integration and other companion good dies. If the defective die escapes package final test and causes failure in PCB sub-system test, the un-recuperated cost is even higher whether the PCB is re-worked or discarded. A well-accepted industry rule-of-thumb is that the cost of failure goes up 10X with each successive stage of integration and assembly.

To reduce the cost of failure, diagnosis should be performed at earlier steps in the test flow. Testing is never perfect and its inaccuracy has two aspects – allowing defects to escape and failing a good part. The latter aspect is commonly called overkill and it could be due to faulty test equipment or over-stringent test conditions. The yield at each test step, defined as the ratio of passing parts to tested parts, is carefully monitored to make sure it stays above an acceptable threshold. Falling below threshold is known as a yield excursion and must be resolved timely since profitability is negatively impacted. Low yield excursions could be due to overkill or a rise in defects at earlier manufacturing steps. Through failure diagnosis, the manufacturer is profit-driven to resolve low yield excursions as well as to enhance yield via defect reduction.

The current semiconductor and electronic product supply chain is a complex eco-system of vendor-customer relationships. For example, the semiconductor foundry is a supplier to the fabless design house who in turn is a supplier to the product system integrator who sells the system product to the user customer. The same entity can be both a customer and a vendor depending on whether one is facing upstream or downstream in the supply chain respectively. Consider a system integrator who sources components from multiple hardware suppliers. Incoming inspection through some form of system-level test (SLT) is performed on all received components. The system integrator judges each supplier on component quality while driving down costs in order to maximize profitability by selling system products to the user customers. Failure diagnosis of a customer return then becomes a shared responsibility up and down the supply chain.

Just as the cost of failure rises with each stage of assembly and integration, so goes the difficulty of failure diagnosis. For a complex system, diagnosing customer failure is notoriously laborious and time-consuming. A well-known case from the automotive space is the Toyota sudden acceleration issue that resulted in numerous fatalities and vehicle recalls. Investigations spanned several years identifying multiple causes including operator error, ill-fitting floor mat, sticky accelerator pedal, and possible design flaw in the electronic throttle control system [34]. For automotive electronic components where quality level is required to reach below 1 DPPM, root-cause analysis of customer returns may last more than a year and a significant portion of the cases still results in no-trouble-found (NTF).

In diagnosing a customer return, the first step is to reproduce the failure in a repeatable fashion. This may require replication of the customer's operating environment which is not always feasible. Once product defect, and not operator error, can be firmly established, then a series of experiments are carried out to isolate and narrow down possible causes. Defect isolation happens in both time and space. In the time dimension, some failures may take hours or longer to trigger. Long error detection latency (EDL) is a major challenge in complex system failure debug. Thus, reducing EDL is a topic of high research interest [20]. Usually both software and hardware have enhanced capabilities to do logging, take snapshots, and run self-checking and diagnostics to aid defect localization. Time-critical low-level software-hardware interactions in embedded systems is a frequent source of problems when the hardware itself is somewhat marginal. The problem can often be resolved by software changes to accommodate larger hardware variations.

When a system integrator is able to identify the hardware SoC component that is the prime suspect for causing system failure, it is extracted and sent to the SoC supplier for further diagnosis. Out of the system operation context, the supplier's challenge is to confirm that the SoC is indeed defective by running stand-alone tests on the isolated component. If no failure occurs when the original SoC production test patterns are applied, new patterns are added to target areas of the SoC that are suspected to be involved based on system failure symptoms. In the best case, the defect is exposed by high-coverage structural fault model-based test patterns in which scan diagnostic tools can help find the physical root cause. But the divide-and-conquer block-based approach of structural test may miss certain functional interactions in the SoC. As well in low-power designs with reduced operating voltage margins, subtle variation-related defects may only be detected by functional patterns or component-level SLT [23]. Catching defects with functional patterns raises the difficulty of diagnosis significantly. To aid functional failure analysis, embedded debug logic to gain deeper internal visibility [22] and sensors to obtain more granular measurements of internal health [17] have seen increasing adoption by SoC designers. In the worst case, the SoC supplier is unable to confirm that the component is the cause of system failure, so it ends up falling under the NTF category [6].

In summary, fault diagnosis of customer returns is still an unsolved and expensive challenge due to the inherent complexity involving many parties and factors. A customer failure is also a failure of the supply chain to deliver a quality product. Thus every effort should be made to screen defects at earlier steps, perform failure analysis, and improve upstream processes to minimize defects via yield learning. Though defects arise from random process variations, systematic aspects of the design and manufacturing process influence the probability of defect occurrence. Identifying systematic factors requires a minimum volume of failed samples to deploy volume diagnosis. The rarity of customer returns may not meet that minimum volume requirement. Indeed if there is a sufficient number of customer returns, it's a strong indication of serious inadequacies in quality control and yield learning in the supply chain.

5   Yield and Profitable Quality (*Changed from Initial ToC*)

When a product is manufactured in volume, some instances (so-called parts) may turn out to be defective and cause failures during deployment. Fig. 13 shows a simplified flow of design $\rightarrow$ fabrication $\rightarrow$ test $\rightarrow$ customer from which we shall derive the fundamental equation relating yield, quality and cost.

For a product, let N be the number of parts produced by manufacturing. Let's associate the descriptive labels *good* and *bad* with *non-defective* and *defective* parts respectively. Thus $N = G + B$ where G and B are respectively the number of good and bad parts. Intrinsic yield Y_0 of the manufacturing process is defined to be the ratio of good parts to total parts produced, i.e., $Y_0 = G/N$. In practice, Y_0 is not a

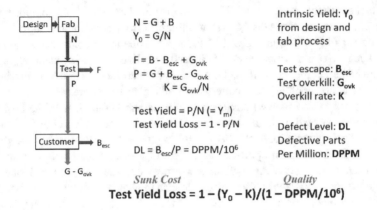

Fig. 13 Simplified flow of design→fabrication→test→customer process

known quantity though it is influenced by factors from the design and manufacturing process. Y_0 is estimated by testing each part to decide whether it is good (by passing the test) or bad (failing the test). Let P and F be the number of parts that passed or failed the test respectively where $N = P + F$. Then an estimate or measured value of Y_0 called test yield Y_m, is defined to be the ratio of P to N, i.e., $Y_m = P/N$.

If testing was perfect in separating good from bad, then Y_m is an exact measure of Y_0. But in reality, testing can make two kinds of mistakes: (1) let a bad part pass which is known as an *escape*; or (2) fail a good part which is called an *overkill*. Let B_{esc} be the number of escapes and G_{ovk} be the number of overkills. We obtain the following equations for F and P: $F = B - B_{esc} + G_{ovk}$ and $P = G + B_{esc} - G_{ovk}$. Let us also define the overkill ratio $K = G_{ovk}/N$. Only parts passing test are shipped to the customer. During product use, assume every bad product shipped results in customer failure. Then the measure of maximum failure rate as experienced by the customer is called *defect level DL* $= B_{esc}/P$ which is a direct reflection of product quality.

By algebraic manipulation of the definitions above, a relationship between Y_m and DL can be derived: $Y_m = (Y_0 - K)/(1 - DL)$. In the electronics industry, it is common to use the term *defective parts per million (DPPM)* instead of DL. DL is related to $DPPM$ by $DL = DPPM / 10^6$. For the discussion that follows, it is more convenient to shape key concepts by considering *test yield loss* which is $1 - Y_m$. Test yield loss represents the portion of manufactured parts that fail in the testing stage and thus discarded. It is detrimental to profitability because costs associated with the discarded parts are a direct loss. When test yield loss is significant, the associated cost dwarfs all other expenditures involved in test development and volume production test. The equation for test yield loss is:

$$\text{Test Yield Loss} = 1 - (Y_0 - K)/\left(1 - DPPM/10^6\right)$$

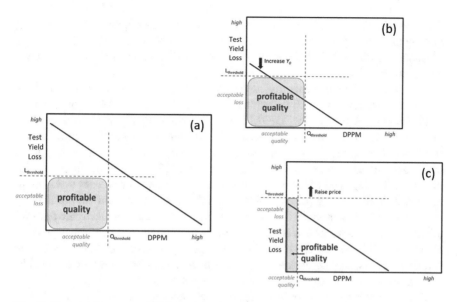

Fig. 14 Test yield loss versus DPPM in simplified form (**a**) unable to achieve profitable quality, (**b**) increase Y0 to achieve profitable quality, (**c**) raise price to achieve very low DPPM and profitable quality

SoC DPPM requirements vary depending on served markets. Some familiar examples are: ~500 DPPM for mobile smartphones, ~100 DPPM for data-center switches, and <1 DPPM for automotive. For all practical purposes, we can assume and work with DPPM values that are less than 10,000. Under that assumption, test yield loss can be approximated by a simplified equation:

$$\text{Test Yield Loss} \cong (1-Y_0 + K) - \text{DPPM}/10^6$$

In Fig. 14a, test yield loss is plotted as a function of DPPM which is simply that of a line with slope equal to -10^6. Suppose for a certain product, the customer is willing to accept DPPM values below a certain threshold indicated in the figure by the vertical dashed line at $Q_{threshold}$, and the manufacturer makes a profit with test yield loss below a threshold indicated in the figure by the horizontal dashed line at $L_{threshold}$. Satisfying both constraints defines a region called "profitable quality" indicated by the green square. As drawn in the figure, the line lies outside meaning it cannot meet both criteria. Satisfying $Q_{threshold}$ means no profit and unsustainable business, and satisfying $L_{threshold}$ results in unhappy and lost customers which also leads to an unsustainable business.

Can testing help? If overkill K is significant, reducing it can help shift the line lower, perhaps into the green region. But in most cases K is already at the lowest level. Improved defect screening by better defect coverage can certainly decrease DPPM, but the line itself does not move so test yield loss will simply rise, making profitability even further out of reach. The only viable option remaining as illustrated in Fig. 14b is to increase intrinsic yield Y_0 which shifts the line lower until a portion lies within profitable quality.

Very low DPPM is generally associated with products that must meet high reliability requirements such as automotive. Note that in Fig. 14b profitability is again not assured at very low DPPM levels. Defects at this level are in fact much more difficult and expensive to screen. The typical test methods employed such as outlier rejection and testing at stress voltage/temperature conditions generally cause overkill to increase thus raising K which shifts the line up in the wrong direction. Staying with older and more mature semiconductor process nodes helps to keep Y_0 high. Raising prices to accommodate overkill and maintain profitability as shown in Fig. 14c is another solution. It is indeed the case that electronic parts for automotive tend to adopt less advanced process technology and are more expensive. However with increased electrification and super-computing capabilities to enable self-driving cars, staying back a few process node generations is no longer tenable. Innovative approaches to raise Y_0 to achieve profitable quality are needed.

Strictly counting on manufacturing process improvements to raise Y_0 may be too confining. Opportunities should also be explored at the design level to achieve an overall cost-effective solution. One can adopt error tolerance schemes at the algorithmic, software, and architectural levels. Large and complex system designers have long assumed imperfect system components, and have devised schemes such as N-modular redundancy [3, 16] to survive component failures. Adaptive schemes such as dynamic voltage frequency scaling (DVFS) are well-established in the design of high-performance processors to adjust and maintain safe operating margins with minimum power consumption in the presence of manufacturing, environmental, and workload variability [13]. In silicon implementation, design-for-yield (DFY) methods dictate the design and usage of logic library cells and physical layout with the aim to minimize probability of defect occurrence in the presence of process variability [26]. Deciding which scheme to adopt to raise Y_0 is based on the results of failure analysis in Sect. 4 and systematic yield learning via volume diagnosis in Sect. 3.

6 Conclusions

This chapter has provided basic information that is related to fault diagnosis. First, it is important to understand the difference among the terms of "defect", "fault", "error", and "failure". Since fault diagnosis starts from chips that cannot pass test, it is helpful to understand the basics on logic test, including DFT, fault simulation, and test generation, as well as the basics on testing other types of integrated circuits, including memories, analog / mixed signal circuits, SoCs, and boards. Furthermore, basics on volume diagnosis for yield improvement, fault diagnosis of customer returns, as well as yield and its relationship to quality and profit are necessary to understand the various aspects of fault diagnosis to be described in subsequent chapters of this book.

References

1. Abramovici, M., Breuer, M., Friedman, A.: Digital Systems Testing and Testable Design. Wiley-IEEE Press, Piscataway (1990)
2. Armstrong, D.B.: A Deductive Method for Simulating Faults in Logic Circuits. IEEE Trans. Comput. **C-21**(5), 464–471 (1972)
3. Baharvand, F., Miremadi, S.G.: LEXACT: Low Energy N-Modular Redundancy Using Approximate Computing for Real-Time Multicore Processors. IEEE Trans. Emerg. Topics Comput. **8**(2), 431–441 (2020)
4. Bardell, P.H., McAnney, W.H.: Self-Testing of Multiple Logic Modules. In: Proceedings of IEEE International Test Conference, pp. 200–204 (1982, October)
5. Bushnell, M.L., Agrawal, V.D.: Essentials of Electronic Testing for Digital, Memory & Mixed-Signal VLSI Circuits. Springer, Boston (2000)
6. Conroy, Z., Richmond, G., Gu, X., Eklow, B.: A Practical Perspective on Reducing ASIC NTFs. In: Proceedings of IEEE International Test Conference (2005, November)
7. Dekker, S., Cilliers, P., Hofmeyr, J.-H.: The Complexity of Failure: Implications of Complexity Theory for Safety Investigations. Safety Sci. **49**(6), 939–945 (2011)
8. Eichelberger, E.B., Lindbloom, E., Waicukauski, J.A., Williams, T.W.: Structured Logic Testing. Prentice-Hall, Englewood Cliffs (1991)
9. Eldred, R.D.: Test Routines Based on Symbolic Logical Statements. J. ACM. **6**(1), 33–36 (1959)
10. Gadlage, M.J., Roach, A.H., Duncan, A.R., Williams, A.M., Bossev, D.P., Kay, M.J.: Soft Errors Induced by High-Energy Electrons. IEEE Trans. Device Mater. Reliab. **17**(1), 157–162 (2017)
11. Girard, P., Nicolici, N., Wen, X.: Power-Aware Testing and Test Strategies for Low Power Devices. Springer, New York (2009)
12. Hapke, F., Redemund, W., Glowatz, A., Rajski, J., Reese, M., Hustava, M., Keim, M., Schloeffel, J., Fast, A.: Cell-Aware Test. IEEE Trans. Comp-Aid. Des. Integr. Circuits Syst. **33**(9), 1396–1409 (2014)
13. Huang, B.-J., Fang, E.J.-W., Hsueh, S.S.-Y., Huang, R., Lin, A., Chiang, C.-H., Lin, Y.-H., Hsieh, W.-W., Chen, B., Zhuang, Y.-C., Wu, C.-Y., Chen, J.-M., Chen, Y.S., Wan, C.-T., Wang, E., Chiou, A., Kao, P., Tsai, Y., Chen, H.H., Hwang, S.-A.: An Octa-Core 2.8/2GHz Dual-Gear Sensor-Assisted High-Speed and Power-Efficient CPU in 7nm FinFET 5G Smartphone SoC. In: Digest of International Solid-State Circuits Conference, pp. 490–492 (2021, February)
14. Iyengar, V.S., Rosen, B., Spillinger, I.: Delay Test Generation 1. Concepts and Coverage Metrics. In: Proceedings of IEEE International Test Conference, pp. 857–866 (1988a, October)
15. Iyengar, V.S., Rosen, B., Spillinger, I.: Delay Test Generation 2. Algebra and Algorithms. In: Proceedings of IEEE International Test Conference, pp. 867–876 (1988b, October)
16. Kim, E.P., Shanbhag, N.R.: Soft N-Modular Redundancy. IEEE Trans. Comput. **61**(3), 323–336 (2012)
17. Landman, E., Brousard, N., Naishlos, T.: A Novel Approach to In-Field, In-Mission Reliability Monitoring Based on Deep Data. In: Proceedings of the International Reliability Physics Symposium, pp. 1–8 (2020, April)
18. Levendel, Y., Menon, P.: Transition Faults in Combinational Circuits: Input Transition Test Generation and Fault simulation. In: Proceedings of the International Fault Tolerant Computing Symposium, pp. 278–283 (1986, July)

19. Levi, M.W.: CMOS is Most Testable. In: Proceedings of the International Test Conference, pp. 217–220 (1981, October)
20. Lin, D., Hong, T., Li, Y., Eswaran, S., Kumar, S., Fallah, F., Hakim, N., Gardner, D.S., Mitra, S.: Effective Post-Silicon Validation of System-on-Chips Using Quick Error Detection. IEEE Trans. Compr-Aid. Des. Integr. Circuits Syst. 33(10), 1573–1590 (2014)
21. Malaiya, Y.K., Rajsuman, R.: Bridging Faults and IDDQ Testing. IEEE Computer Society Press, Los Alamitos (1992)
22. Mishra, P., Morad, R., Ziv, A., Ray, S.: Post-Silicon Validation in the SoC Era: A Tutorial Introduction. IEEE Des. Test. 34(3), 68–92 (2017)
23. Polian, I., Anders, J., Becker, S., Bernardi, P., Chakrabarty, K., ElHamawy, N., Sauer, M., Singh, A., Sonza Reorda, M., Wagner, S.: Exploring the Mysteries of System-Level Test. In: Proceedings of Asian Test Symposium, pp. 1–6 (2020, November)
24. Rajski, J., Tyszer, J., Kassab, M., Mukherjee, N., Thompson, R., Tsai, K.-H., Hertwig, A., Tamarapalli, N., Mrugalski, G., Eide, G., Qian, J.: Embedded Deterministic Test for Low Cost Manufacturing Test. In: Proceedings of the International Test Conference, pp. 301–310 (2002, October)
25. Reddy, S.M., Reddy, M.K., Agrawal, V.D.: Robust Tests for Stuck-Open Faults in CMOS Combinational Logic Circuits. In: Proceedings of the International Fault-Tolerant Computing Symposium, pp. 44–49 (1984, June)
26. Ripp, A., Buhler, M., Koehl, J., Bickford, J., Hibbeler, J., Schlichtmann, U., Sommer, R., Pronath, M., DFM/DFY Design for Manufacturability and Yield – Influence of Process Variations in Digital, Analog and Mixed-Signal Circuit Design. In: Proceedings of the Design Automation & Test in Europe Conference (2006, March)
27. Roth, J.P.: Diagnosis of Automata Failures: A Calculus and a Method. IBM J. Res. Dev. 10(4), 278–291 (1966)
28. Sayil, S.: Soft Error Mechanisms, Modeling and Mitigation. Springer, New York (2016)
29. Seshu, S., Freeman, D.N.: On Improved Diagnosis Program. IEEE Trans. Electron. Comput. EC-14(1), 76–79 (Feb. 1965)
30. Smith, G.: Model for Delay Faults Based upon Paths. In: Proceedings of the International Test Conference, pp. 342–349 (1985, October)
31. Ulrich, E.G., Baker, T.: Concurrent Simulation of Nearly Identical Digital Networks. IEEE Comput. 7(4), 39–44 (Apr. 1974)
32. Wang, L.-T., Wu, C.-W., Wen, X. (eds.): VLSI Test Principles and Architectures: Design for Testability. Morgan Kaufmann, San Francisco (2006)
33. Wang, L.-T., Stroud, C.E., Touba, N.A. (eds.): System-on-Chip Test Architectures: Nanometer Design for Testability. Morgan Kaufmann, San Francisco (2007)
34. Wikipedia, 2009-2011 Toyota vehicle recalls (n.d.). https://en.wikipedia.org/wiki/2009%E2%80%932011_Toyota_vehicle_recalls

Conventional Methods for Fault Diagnosis

Srikanth Venkat Raman

1 Introduction

Very few chips ever designed function or meet their performance goal the first time. Many fabricated chips may also fail to function because of defects, circuit/process sensitivity, infant mortality, or wearout throughout their product lifetime. When fabricated *integrated circuits* (ICs) of a new design do not function correctly or fail to meet their performance goal, the silicon debug process starts immediately to identify the root cause of the failure. For manufacturing defect-induced failures, the cause of failure may not be important as long as the failure rate is comparable to the process norm and acceptable. However, if the yield is low or the chip performance varies with process variation, the diagnosis process is utilized to identify the root cause so that corrective actions (with process and/or design changes) can be taken to improve yield. Failing to do so will reduce the profitability of the product or result in the product being unable to meet the volume demands of the market due to availability shortfall. When chips fail in the field and the field failure rate is high and above acceptable levels, customers will typically initiate action focused on the quality of the incoming product. Chips are usually sent back by customers and have to be analyzed for the root cause. Following this, corrective actions have to be put in place. If the cause of the problem is not understood and the customer is not convinced, it may create a *customer lines-down* situation or even *product recall* which can cripple the product and result in loss of market share relative to competitors.

In all of the above scenarios, speedy and accurate isolation and root-cause of the problem is needed. The process of isolation is also called as ***diagnosis***. The focus

S. Venkat Raman (✉)
Intel Corporation, Hillsboro, OR, USA
e-mail: srikanth.venkataraman@intel.com

© The Author(s), under exclusive license to Springer Nature Switzerland AG 2023
P. Girard et al. (eds.), *Machine Learning Support for Fault Diagnosis of System-on-Chip*, https://doi.org/10.1007/978-3-031-19639-3_2

of this chapter is on conventional *automated methods* for diagnosis utilizing model and simulation at different levels of abstraction including layout, circuit and logic levels utilizing changes that can be made to the design to make it more debug-able and diagnosable.

1.1 What Are Debug and Diagnosis?

The Webster's dictionary defines **diagnosis** as the investigation or analysis of the cause or nature of a condition, situation, or problem. In the context of *very-large-scale integrated* (VLSI) circuits and systems, diagnosis is a general term that applies to various phases of the design and manufacturing process that involves isolating failures, faults, or defects. The term **debug** refers to the process of isolating bugs or errors that causes a design to behave differently from its intended behavior under its specified operating conditions. Debug and diagnosis techniques try to address the following questions:

- What was wrong with this device?
- Which chip was bad on this board?
- Why did this system crash?
- Why did the simulation of the *arithmetic logic unit* (ALU) show that $2 + 2 = 5$?
- Why was this signal late?

The aim of debug and diagnosis is to locate the root cause of the device bugs and/or failures.

1.2 Where Is Diagnosis Used?

Diagnosis of electronic systems is performed at different stages of design and manufacturing of the constituent components of the system for several different objectives. However, the common aim of any *diagnosis* procedure is to locate the root cause of the device failures. Depending on whether the device is a VLSI chip, *multi-chip module* (MCM), board, or system, diagnosis is performed with different objectives. In the case of MCMs, boards, or systems, diagnosis is intended for identification followed by replacement of the faulty subcircuit (a chip on a board or a board in a system) or reconfiguration of the circuit around the failure. In the case of chips or integrated circuits, diagnosis is performed with a view to improving the manufacturing process.

Diagnosis followed by **failure analysis** is vital to IC design and manufacturing. Diagnosis identifies root cause defects for yield enhancement, finds design flaws that hinder circuit operation, and isolates reliability problems that could lead to early product failure. The primary focus of this chapter is on debug and diagnosis usage during the manufacturing and post-silicon phase of Digital ICs or VLSI chips.

Fig. 1 Debug and diagnosis applications across an IC life cycle

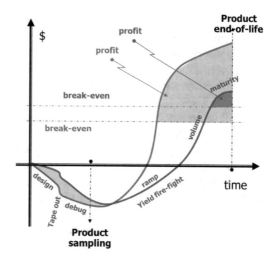

1.3 IC-Level Debug and Diagnosis

Figure 1 shows a typical life cycle of an integrated circuit or VLSI chip starting from the requirements up to the end of life. Three major phases of the life cycle are shown: (1) **design** along with validation and verification, (2) **ramp-up** to production with design revisions or steppings, and (3) **volume production**.

There are three major applications of diagnosis and debug techniques and technologies at the IC level to root-cause issues that are either design or manufacturing process related: (1) **design error diagnosis** during the design phase, (2) *silicon debug* during the ramp-up to production, and (3) **defect diagnosis** through production ramp and volume production. The focus of this chapter is on defect diagnosis (also called as fault diagnosis) and they are described in more detail in the following sections. Many of the techniques for fault diagnosis are also applicable for product silicon debug. However, additional techniques are needed for silicon debug which is outside the scope of this chapter. Only after successful debug and diagnosis to address both design and manufacturing issues, can a product be ramped successfully for volume production and be profitable as shown in Fig. 1.

1.4 Silicon Debug Versus Defect Diagnosis

Silicon debug is the first and necessary step in the first/initial silicon stage of a product life cycle for any complex VLSI chip. Silicon debug starts with the arrival of first silicon and continues till volume production. Once first silicon arrives, these early chips are sent to the design team for validation, test, and debugging of bugs not uncovered during the pre-silicon design process. Many bugs and errors, such as logic errors, timing errors, and physical design errors, may be introduced during the design process. These bugs and errors should have been uncovered during the design verification and validation processes that include simulation, timing verification, logic verification, and design rule checking. However, very often bugs and errors still escape the above checking and cause silicon not to behave as designed. These are often caused by limitations and flaws in the circuit models, simulations, or verification. Problems could range from logic or functional bugs to circuit sensitivities or marginalities, to timing or critical speed-path issues [1, 2]. Verification tests performed during the design phase may not cover all corner cases encountered in a real application leading to logic or functional bugs. Static timing analysis and dynamic timing simulations may not accurately and precisely predict every single critical path. This results in timing or speed paths being encountered in silicon that were not factored in design. As an interesting side note, one of the first examples of "debugging" was in 1945 when a computer failure was traced down to a moth that was caught in a relay between contacts [3].

In contrast, manufacturing defects are physical imperfections in the manufactured chip. The process of locating manufacturing defects is called **defect diagnosis**, **fault diagnosis**, or **fault isolation**. A key application of defect diagnosis is in supporting low yield analysis and yield enhancement activities.

One application of defect diagnosis to manufacturing yield enhancement is shown in Fig. 2. Initially the effort is mainly focused on memories, test chips, and in-line monitors. The regular and repetitive structure of a memory makes their diagnosis or fault isolation easier. Test chips and *static random-access memories* (SRAMs) are used to bring up a new process and the large, embedded memories in products are usually used to maintain or improve yield [4–6]. However, due to the differences in the possible defect types between logic and memories (layout topology, number of layers, and transistor density/count), memories may not capture all manufacturing process issues, leading to the use of logic diagnosis in **yield learning** in the intermediate and mature phases of the manufacturing process. A second application of defect diagnosis is in dealing with manufacturing excursions or low yield situations as shown in Fig. 2. Diagnosis is performed on low yield

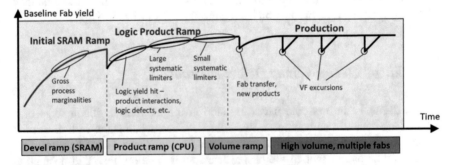

Fig. 2 Defect diagnosis applications to manufacturing yield

chips to isolate the root cause of failure (process abnormality leading to high defect density, process variation, design and process sensitivities, etc.) and follow up with appropriate corrective actions. Other applications of defect diagnosis include analysis of failures from the qualification of the product which includes reliability testing (both infant mortality and wearout) and failures from the field or customer returns.

2   Design for Debug and Diagnosis

Debug and diagnosis require a high degree of **observability**. The ability to observe erroneous events close to when they happen is important. In addition, **controllability** is needed to further validate what the cause of the problem is and/or to narrow down the circuits that manifest the symptoms.

Many *design for testability* (DFT) features (*e.g.*, scan) which enable both controllability and observability can be reused for debug and diagnosis. There are also specific DFD features (*e.g.*, clock controllability or reconfigurable logic) that are developed or tailored for debug.

The DFD features can be broadly bucketed into **logic DFD structures** which are added to the design to extract logic information and to manipulate the operation of a chip in a physically non-intrusive manner. There are several physical tools to extract logic and timing information from a chip in a minimally physical intrusive manner. Circuit or *focused ion beam* (FIB) editing is performed to either enable probing or make corrections to the circuit to verify root-cause fixes. DFD structures to enable probing and circuit editing are called **physical DFD structures** [7]. This chapter will only introduce the logic DFD structures needed to understand how logic diagnosis works. Physical DFD is out of scope and won't be discussed further. However, the interested reader is referred to the reference [7] for a broad introduction to the area.

Speed debug is an extension of **logic debug**. Throughout the description of these techniques, emphasis is placed on debugging for speed-related problems in addition to logic failures. Additionally, throughout this chapter, we summarize several successful industrial applications of diagnosis. For example, we will discussion applications where diagnosis helped reducing the search area for defect down to 3% of the original search area size. In another case, diagnosis was used to improve at-speed tests, increasing the speed by 300 MHz.

We do not cover memory related diagnosis techniques which are common for yield analysis purposes. However, references are provided should the reader want to study this area [4–6].

3   Logic Design for Debug and Diagnosis (DFD) Structures

Logic design for debug and diagnosis (DFD) *structures* are added to the design to extract logic information and to manipulate the operation of a chip in a physically non-intrusive manner. Key logic DFD features, including scan, observation-only scan, observation points and multiplexers, array dumps, clock control, partitioning and core isolation, as well as reconfigurable logic, are discussed in this section.

3.1   Scan

Scan is a widely used DFT feature. Scan is also very useful for debug and diagnosis as a DFD feature. While scan is typically utilized to enable *automatic test pattern generation* (ATPG), the increased observability of the design due to the insertion of scan also enables debug and diagnosis. For more information on scan and other DFT features, please refer to popular DFT textbooks [8–11].

For ATPG patterns generated automatically and applied using scan, automated analysis of the failures can be performed using diagnosis tools that provide fault candidates for further validation [12–14].

Scan diagnosis is an established technique identifying and localizing defects in digital semiconductor devices that fail manufacturing tests. We will cover this in some detail in Sect. 5.

Scan also helps in functional-pattern-based debug and diagnosis. There are two commonly used types of scan design: (1) **muxed-scan** or **clocked-scan** for flip-flop-based scan designs [11] and (2) *level-sensitive scan design* (LSSD) for latch-based scan designs [15]. In both types of scan design, a functional test can be executed up to a certain clock cycle and then stopped. Once the system flip-flops and latches are reconfigured as scan cells, the functional state of these system flip-flops and latches in the chip can be shifted out for analysis during the shift operation. This has to be coupled with *register-transfer level* (RTL) or logic simulation to compare against the expected state. For most scan designs (using muxed-scan, clocked-scan, or LSSD-like), unloading the scan cell contents is destructive in that the functional state is destroyed during shift. Hence, functional execution cannot be resumed after the data is unloaded. If a functional state at a further clock cycle needs to be investigated, then the functional pattern would have to be re-executed and then stopped at that later cycle. This can be time consuming if the time from reset to the failure point is long which is very common in system-level debug. An enhancement to this scheme is to wrap the scan chain around from scanout to scanin. If properly accounted for, the system flip-flops and latches will contain exactly the same contents as before it was shifted. Hence, functional execution can continue until the next clock cycle of interest [16]. This approach also requires that the scan shift operation be non-destructive with respect to all other storage elements (flip-flops and latches) in the circuit that are not scanned. Alternately, many designs add observation-only scan cells. This is described below. It is important to note that scan only provides a means to access internal state of the storage elements that are scanned and help determine incorrect state captured during the execution of a failing test. It does not tell exactly what caused the incorrect state. More debug is necessary from the point of observed failures to determine its root cause. These will be described in more detail in Sects. 4 and 5

3.2   Observation-Only Scan

Observation-only scan is a specialized version of scan that is typically used only for debug purposes [1, 3, 17, 18]. This is also called **scanout** or **shadow scan**. Figure 5 shows the schematic of a typical scanout cell. The *Sin* port is connected to the *Sout* port of the preceding stage to create a scanout chain, while the *Data* input port is connected to the signal of interest. When the *LOAD* signal is enabled (set to logic 1), the signal of interest is latched into a separate **scanout cell** (where a **snapshot** is taken just for a clock cycle of interest) and can be kept there until it is ready to be shifted out when enabling the *SHIFT* signal. The whole captured state from all the signals of interest connected to scanout cells can be shifted out through a special scanout chain for analysis. This type of observation-only scan is part of a separate scan chain that is typically clocked by the same system clock capturing the signals of interest at-speed. From a performance penalty standpoint, observation-only scan merely adds some small capacitance to the signal of interest and the functional signal does not have to pass through any additional multiplexers. Hence, this type of scan is more acceptable for functional debugging for high-performance circuits and systems.

Scanout cells are judiciously placed throughout the design to maximize observability of key states and signals. Potentially, they can be inserted at any signal line but that would increase the cost due to added area. A selective placement of this kind of scanout cells on key signals helps debug tremendously while keeping area overhead low. Due to its unique speed capturing nature, these cells can also be put on different segment of long signal lines to observe effects of signal degradation or noise. Very often, the area of the scanout cells can also be hidden (or further reduced) if they are placed under routing channels, where silicon is not utilized. Just like conventional scan design, the placement and routing of these cells can also be automated, so the scheme is perfectly compatible with both custom and *application-specific integrated circuit* (ASIC) design styles.

As indicated before, observation-only scan is primarily aimed at speed debug, because it is running concurrently with the system clock. At the specific clock cycle of interest, one would trigger the *LOAD* signal so the *Data* signal of interest would be captured into the scanout cell. On one hand, the captured state can be kept there until the test is completed and then shifted out for analysis. One the other hand, the captured state can also be shifted out simultaneously while the system is running. Repeated capturing and shifting is possible provided that the time to shift the whole

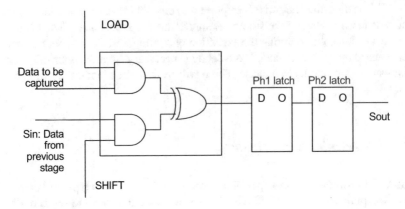

Fig. 3 A scanout cell

chain is shorter than the time it takes to the next capture point. Again, RTL or logic simulation is executed to allow the comparison of expected states.

Additionally, scanout systems typically have a **compressed signature** mode built into the cell whereas the content of an upstream cell is XORed with the content of the current scanout cell. This is accomplished by enabling with *LOAD* and *SHIFT* signals simultaneously as shown in Fig. 3. The final signature after the execution of a test can be shifted out of the scanout chain. The advantage of a signature *versus* a single snapshot is that it is more efficient to just check an accumulated signature to see if an error is captured during a time interval starting from one trigger event which starts recording to another trigger event which stops recording. This can help narrow down the cycles where errors occurred during the execution of a test. A signature compare operation can also be done with the chip running in a passing condition rather than running in a failing condition. An alternate observation-only cell design has been proposed in [19].

3.3   Observation Points with Multiplexers

Multiplexers (MUXes) are a common alternate way to observe on-chip signals. Once a set of important signals are identified for observation, multiplexers can be added so that individual signals can be mapped out to a test port (*e.g.*, through the *Test Data Out* [TDO] port defined in the IEEE 1149.1 boundary-scan standard [20]. However, the test control signals for the MUXes can add to the overhead especially when a large set of signals are observed. Also during debugging, the signal that one would like to observe may not be included for observation and would not be available during the debug process.

The authors in [21] proposed to have a layer of programmable logic that can be programmed to allow specific signals be observed. Figure 4 shows how this

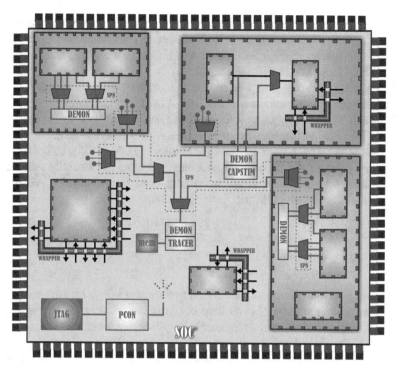

Fig. 4 Programmable observation of internal signals

can be done by using wrappers around individual blocks of circuits. The MUXes bring various signals of interest to each wrapper and also allow signals from various wrappers to be mapped to observation points. This alleviates the amount of overhead required while at the same time maximizing the likelihood for signals to be observed. However, the overall architecture of placing the programmable logic where it will provide a high degree of observability may still be expensive for certain designs.

3.4 Array Dump and Trace Logic Analyzer

In addition to logic states which can be observed using scan and scanout, other memory states such as those held in embedded arrays (*e.g.*, caches, register files, buffers/pointers, etc.) are also very important for debug and diagnosis. Since these structures are usually not scanned, scan dump cannot access the information stored in these arrays. So, an **array dump** mechanism is typically designed [18] in so that the contents stored in these arrays can be observed after a normal functional operation is stopped with clock control (Clock control will be covered in the next section). The information can be dumped on an external bus which can then be captured by the ATE (tester) or a logic analyzer. Alternately, the data may be accessed through some other test data pins (e.g., **test access port** [TAP] as defined in the IEEE 1149.1 boundary-scan standard [20]).

The inverse of array dump is to use the existing array(s) to store on-chip activities, such as internal bus traffic and specific architectural state changes. Multiplexers or even programmable interfaces can be designed to re-direct these types of information to be stored on specific on-chip arrays (*e.g.*, L2 cache on a microprocessor). This is called **trace logic analyzer** (TLA) in [22]. Of course, these arrays have to be resized so that it will not impair existing functionality. Alternately, we can design in dedicated arrays to capture these traces, but that will be too expensive just for debug purposes.

3.5   Clock Control

Clock control is a very important DFD feature for speed debug. Internal observation mechanisms like scan and scanout help obtain information in the spatial domain (where?), while clock control supplements it by enabling extraction of information in the time domain (when?).

Most modern VLSI chips with clock frequencies which run at hundreds of *megahertz* (MHz) to *gigahertz* (GHz) usually have an internally generated clock from a **phase-locked loop** (PLL). This generated clock is usually synchronized to an external system clock of much lower frequency. High-frequency oscillators are difficult to build and high-frequency clocks are difficult to route on a board. Thus, the solution of having an internal PLL multiply to a high-frequency internal clock is very common. Moreover, modern-day VLSI chips also typically contain multiple clock domains which drive specialized circuits or external I/O interfaces.

Starting, stopping, and restarting these internal clocks while keeping them synchronized to specific external and internal events is critical for debug [1]. Being able to start and stop the clock is an important first step in debug. While scan or scanout capture can extract internal logic state, the question of when to take this internal observation is answered by clock control. The stopping of the clock at a specific clock cycle and synchronized it to an internal or external event is the most basic of clock control features. If the clock stops at the wrong cycle, the internal observation results will be wrong resulting in the analysis being performed down the wrong path.

While it is possible to stop the clock after an external event, it is definitely more preferable to use internal events to control the clock since an external event may be too imprecise. This happens because the external event is synchronized to a much slower external clock and there may be an offset between the slow external clock and the much faster internal clock.

Specific **offset counters** are typically added to make the clock stopping points be much more flexible. These offset counters are often part of a much more comprehensive set of **debug registers**, designed specifically to capture information from or assert control to different parts of the chip, specifically for debug purposes [22]. These are also commonly referred to as **control registers** [18]. For example, one can specify that a scan or scanout capture be taken at 487 clock cycles after an exception event has occurred. The programming of these clock stop event(s) and offset are usually pre-shifted in through specific scan chains so that they are ready to execute when the debug tests are run.

Another useful function in addition to starting and stopping of the clocks is to issue a single clock pulse. This is also sometimes called **single stepping**. Single stepping allows observing the execution of the circuit in a controlled manner. A scanout or observation-only scan like capability which is non-destructive is needed to complement single stepping with internal observation of circuit state.

Besides starting and stopping of internal clocks, the capability of stretching a specific clock cycle is very useful in debug. When the clock is supplied externally, this is relatively simple. This can be accomplished by swapping in a different set of timing generators on a specific clock from the ATE. However, for internally generated clocks, circuit modification to the clock circuit is needed. Debugging performance problems (for example, "Why won't the chip run any faster than 500 MHz?") requires isolating the offending path or paths in the circuit. By sequentially stretching the clock from the failure point backward, the specific clock cycle where the failure was latched internally can be determined. This will provide the clues as to where the problem lies considering the pipelining nature of high-performance systems. Multiple failing paths may exist requiring multiple iterations of stretching clocks to find the root cause in an onion peeling fashion. Debug flows will be described in more detail in Sect. 4.

Stretching of individual clock phases can also be implemented to allow the pinpointing of the path to a particular clock phase. While for flip-flop-based designs, the specific phase may not be important but for latch-based designs, this is indispensable.

Figure 5 illustrates the clock controls for the Pentium 4 microprocessor [23]. An external system clock is fed into individual Core PLL and IO PLL, where the clock is multiplied. A skew detect circuit will make sure that the skew between the 2 PLLs can be monitored for deviation. Both clock systems can be adjusted for their duty cycles while the core clock allows for skew adjustment between its various distribution buffers since this clock is distributed all over the chip and is subjected to on-die process variation. Finally, this core clock can also be adjusted for its duty cycle by stretching either one phase or the other.

Another clock control feature that can help with debugging is the ability to introduce relative skew between different clock domains. As a chip can contain tens of millions to hundreds of millions of transistors (and include tens of thousands to hundreds of thousands of storage elements), the clock system has to be distributed from the PLL to all these storage elements with a very elaborate clock distribution system. The clock distribution system consisting of series of buffers must have

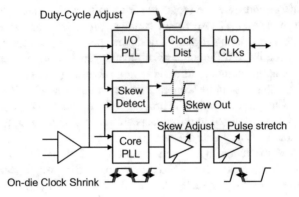

Fig. 5 Pentium® 4's on-die clock control

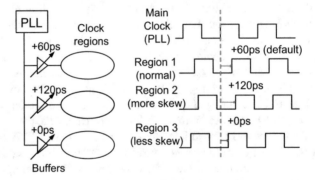

Fig. 6 Introducing intentional clock skew

the ability to deskew the various domains so that all the clocks arrive at their destination within a tight tolerance. Without this kind of clock distribution system, much performance could be lost due to the natural skews introduced by on-die process variations [23, 24].

Since the clock distribution system can be deskewed, intentional skews can also be introduced through additional adjustment above and beyond what is needed for deskewing [1, 3, 25, 26]. This intentional skew which is illustrated in Fig. 6 gives more sampling time for the storage elements using the delayed clocks. Of course, the storage elements using this delayed clock also launch their outputs later, so failures may be observed in their downstream logic as well. This feature provides additional information as to where the failing path may lie. If pulling in the clock edges of a clock region causes the test to pass, then the source or the driving storage elements of the failing path can be pinned to that clock region. Similarly, if pulling in the clock edges of a clock region causes the test to fail, then the destination or the receiving storage elements of the failing path can be pinned to that clock region.

3.6   Partitioning, Isolation, and De-featuring

Today's VLSI chips including **system-on-chips** (SoCs) and microprocessors are very complex devices. They typically consist of multiple heterogeneous sub-functional blocks operating together to deliver the final functionality that the chip is supposed to deliver. Debugging such a chip would be difficult if we cannot isolate or narrow down the problem at the macro level before taking internal observation. Partitioning is one mechanism to help find if the problem still exists, after some blocks are separated. This could be accomplished by logic partitioning or by re-writing software so as to confine the execution to a much smaller logic unit. For example, multiple execution units on the processor can be disabled so that the parallel execution of instructions can be restricted to a few or even down to one unit. The instructions can then be directed to individual units through subsequent enabling or disabling of specific units. Of course, this ability is limited to units that have some level of redundancy built-in. This is also called **de-featuring**.

3.7   Reconfigurable Logic

As mentioned in Sect. 3.3, the authors in [21] proposed placing a programmable logic (also called **reconfigurable fabric**) between blocks or interspersing them into the regular logic fabric to aid both debugging and making a fix. Patching can be done by re-programming the logic to replace or change existing logic functionality. This can be seen as half way to fully programmable logic. This will allow very fast time-to-market by allowing units to be shipped without having to wait for a new design revision and a new set of masks. Should a bug appear, a fix is just a software download.

4   The Diagnosis Flow and Process

Figure 7 shows a generic diagnosis flow. The diagnosis process involves starting with the test results which capture all the observed failures and mapping the test results to defects or errors, which are then used as a starting point for repair (replacement or redesign) or finding the root cause for process improvement depending on the goals of the diagnosis process. Depending on the objectives of the diagnosis flow and the type of systems under diagnosis, the defects or errors may include defective components (for example, a faulty IC on a board, a faulty cell in a **random-access memory** [RAM], a faulty column of cells in a *field programmable gate array* [FPGA], or a faulty board in a system), defective interconnections (shorts or opens), logic implementation errors, timing errors, or IC manufacturing defects.

Fig. 7 A generic diagnosis
flow

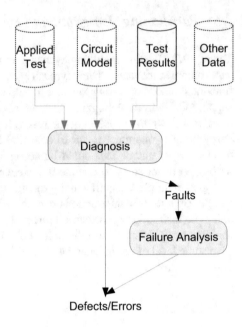

The diagnosis process would typically use information about the circuit model and applied tests along with other data that may available. Fault models [11, 27] are also typically used as a means to arrive at the final defects or errors for consideration.

Defects are fabrication errors caused by problems in the manufacturing process or human errors. They may also be physical failures which are caused by wearout and/or environmental factors. Manufacturing defects are unpredictable in both location and effect, and the processes that cause them are continuous over a wide range of variables, *e.g.*, a short between two adjacent wires may occur anywhere along their length with a range of possible resistances, capacitances, etc. Ideally, the test process would take every defect into account, develop a test for each, and apply these tests during manufacturing test. However, because the space of all possible defects is continuous and unpredictable, there is no way to apply a finite number of tests that are guaranteed to detect everything in that space. The complexity of defect behavior makes a strict defect-based test approach impossible; some simplification is necessary.

Defects themselves could be approximated. This approach is usually taken both during testing and during diagnosis [8, 11]. The infinite defect space is approximated by a finite set of faults. A fault is a deterministic, discrete, change in circuit behavior. It is important to stress that the fault is an approximation of defective behavior, not a true representation. By this definition, a fault can never been "found" during diagnosis. Fault models are used as a means to arrive at the final defects or errors for consideration. Figure 8 illustrates this process. Faults are often thought of as localized within a circuit (*e.g.*, a particular gate is broken), but they may also be thought of as transformations that change the Boolean function

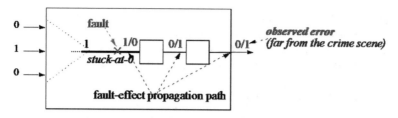

Fig. 8 A fault with its propagated fault effect leading to an observed error

implemented by a circuit. Many fault models are timing-independent, while a few include timing behavior explicitly. The most commonly used fault models nowadays include *stuck-at faults*, *bridging faults*, *delay faults* (including *path-delay faults* and *transition faults*), and *functional faults*. The choice of fault models depends on its intended use (*e.g.*, test generation, manufacturing quality prediction, defect diagnosis, characterization for defect tolerance, etc.).

Fault models are an integral part of the fault diagnosis process and thus help find the root cause of the defective device under consideration. In its most basic form, a fault model is used to predict the behavior of faulty circuits, compare these predictions to the actual observed behavior of defective chips, and identify the predicted behavior that most closely match the observations. An analogy to this is a detective story. A fault can be likened to a criminal or suspect that exists in the circuit model. The fault may be **permanent, intermittent** (with alibis most of the time), or **transient** (hit-and-run) depending on the nature of the fault model. An *error* (or *fault effect*) is created by applying stimulus to *activate* the fault (provoke the criminal) at the fault site. The fault effect may *propagate* through the circuit and be detected. A *detection* implies that the fault effect has propagated to an observation point and can be observed as an error or failure. Figure 8 illustrates fault effect propagation. The goal of the process is to enable further analysis by identifying promising locations for further study.

4.1 Diagnosis Techniques and Strategies

The diagnosis process can employ a variety of different methodologies. Some of the common types of diagnosis are enumerated:

- **One-step or non-adaptive**: The test sequence is fixed and does not change during the diagnosis process.
- **Multi-step or adaptive**: Tests applied in a subsequent step of diagnosis depend on results of the previous steps.
- **Static**: Diagnostic information is pre-computed for all possible faults before testing.

- **Dynamic**: Some diagnostic information is computed during diagnosis based on the actual *device under test* (DUT) response.
- **Cause-effect**: Computes faulty responses for a set of faults.
- **Effect-cause**: Analyzes the actual DUT response to determine compatible faults.
- **One-step**: Without replacement.
- **Multi-step**: Alternate retest and replacement steps.

In line with the detective story, the common elements of most diagnosis methods include:

1. Being prepared
2. Assume, Suspect, and Exclude
3. Tracking them down

Being prepared involves storing information that may potentially be useful for analysis later. In criminal investigation this may involve fingerprinting the entire population, communities or just those individuals with prior criminal record. This is the strategy employed by cause-effect fault dictionaries [8, 11]. However, the questions of the feasibility of fingerprinting the entire population or determining which communities to fingerprint are tedious at best. The challenges with this criminal investigation strategy also have an analogy in diagnosis. Which faults or faults models (*e.g.*, stuck faults, bridging faults, delay faults, etc.) should be considered? Is it feasible to store responses of all faults?

How to process all the pre-stored information during analysis? Questions in criminal investigation may include finding matches for fingerprints found at the crime site, what happens when the criminal has not been previously fingerprinted, or what happens when several criminals committed the crime and their fingerprints are mixed?

Analogous situations occur in diagnosis: Using a single-stuck-at fault dictionary when in reality the defect is a short that behaves close to a bridging fault, or using a single-stuck-at fault dictionary, when the defect behaves like multiple stuck-at faults. Dealing with these situations requires diagnostic models and algorithms that can deal with partial matches. Several diagnostic models and algorithms have been developed and successfully employed [12–14]. More diagnosis approaches can also be found in [11].

The *"assume, suspect, and exclude"* strategy used in criminal investigation applies equally well to the diagnosis problem. Make some initial assumptions about the criminals (faults in the case of diagnosis). Do they act isolated (single faults) or in gangs (multiple faults)? Are they always bad (permanent faults) or only sometimes (intermittent faults)? Where are they coming from (stuck-at faults, bridging faults, functional faults, etc.)? The next step involves rounding up the plausible suspects and then reducing the set of suspects. In the words of Sherlock Holmes: *"One should always look for a possible alternative and provide against it. It is the first rule of criminal investigation."* In criminal investigation this involves checking existing data for alibis. In diagnosis, this may require post-test fault simulation on the data collected from the on-chip logic DFD and clock control features discussed earlier. One could also perform experiments to provide alibis (adaptive diagnosis). Quoting Sherlock Holmes again: *"When you have eliminated the impossible, whatever remains, however improbable, must be the truth."* This step is performed using the on-chip DFD features and collecting failing information on a tester. This leads to the final step. When there are no more suspects, change the initial assumptions and restart. In diagnosis, this may involve starting over with a new fault model.

The final strategy involves tracking down the suspects or backtracing in diagnosis. "There is no branch of detective science which is so important and so much neglected as the art of tracing footsteps." Tracking down or backtracing involves starting from the available acts (observed errors in the device), trace back through the city (tracing through a circuit) and in time (tracing through previous test vectors), following paths that lead to the criminal(s) (actual fault(s)). This step uses the physical DFD structures on-chip and the physical tools described earlier to retrieve information from inside the chip.

5 Automated Logic Diagnosis Using Scan

In this section we explore the world of automated scan diagnosis of digital semiconductors devices. After shortly outlining the basic technology behind diagnosis, the other parts of this section describe the work to set up and use scan diagnosis.

5.1 How Diagnosis Works

Scan diagnosis is an established technique identifying and localizing defects in digital semiconductor devices that fail manufacturing test. Scan diagnosis leverages some of the same technology that is used in automatic test pattern generation (ATPG). Most scan diagnosis tools can only diagnose failures in ATPG or BIST patterns that leverage internal scan chains or scan-based DFT such as logic BIST. A typical diagnosis system requires a gate level representation of the design, ATPG test patterns, and fail data from the tester. In addition to scan ATPG tests, functional tests may be used for screening manufacturing defects. A snapshot of observable scan cells or scan-dump can be captured at a given cycle of a functional test, and analysis of faults around the window of failing cycles can be used effectively to diagnosis silicon defects. The inputs required for a typical diagnosis system as shown in Fig. 9.

Each failing pin/cycle on the tester corresponds to a failure captured in a scan cell for a particular scan pattern. The first step in the diagnosis process is to trace the design backwards from the failing scan cell then to identify which candidate (also called as suspects) could cause these failures. This is illustrated in Fig. 10. A candidate or suspect is typically a particular fault type (such as stuck-at, bridge, or open) in a particular location (net). An example suspect would be "stuck-at-0 fault

Fig. 9 Inputs to a typical diagnosis system

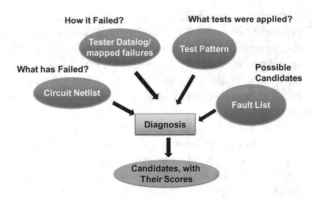

Fig. 10 Failure region for suspect simulations

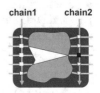

Fig. 11 Scoring suspects against tester observed failures

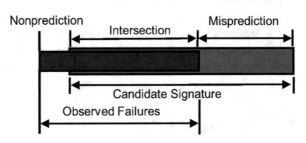

on the A input of the OR gate /design/module/inst023." The suspect lists are then compared across multiple failures (multiple scan patterns, multiple scan cells), and this process helps narrow down the suspect list. Simulation is also performed to see if any of the suspects would cause any additional failures other than those observed on the tester.

Suspects are then ranked and scored based on how well the suspect explains the failures. For instance, suspect /design/module/inst023/A stuck-at-0 might explain 8 out of the 10 failures on the testers and would cause 3 additional patterns to fail. This suspect would receive a lower score than a suspect that explains all the 10 failures and would not cause any additional failures. Since the suspects are based on fault models and not the actual defects, it is common to observe less-than-perfect diagnosis results. Figure 11 illustrates the process of scoring the suspects by comparing their simulated responses against observed tester failures.

The two most common metrics in measuring diagnosis results are **accuracy** and **resolution**. *Accuracy* is a measurement of how well the diagnosis tool has identified the true defective net(s) and the true defect type. For a given list of candidates, accuracy is a binary decision: Is the true candidate on the list, yes or no. *Resolution* measures the area of the defective location captured by a suspect, that is, the bounding box of a defect. For logic-based diagnosis tools, the resolution equals the combined area of all reported nets. Logic-based diagnosis tools do not know where two nets potentially bridge or where an open on a net might be located; thus, the whole net must be searched to find the defect. Layout-aware diagnosis with its access to the polygon level easily finds and reports, for example, the bridge location in x/y/layer terms, or the few polygons on a net, where an open defect might be.

5.2   A Typical Diagnosis Flow

A typical diagnosis flow requires three sets of inputs: A gate level description (i.e. model) of the design, the test stimuli applied on the tester and the failures recorded by the tester (a.k.a. datalogs) when these stimuli are applied.

Diagnosis flow starts by processing the datalogs. The datalogs record failures in terms of pin names and test cycle numbers which are dependent on the specific tester that is being used. By analyzing the DFT architecture of the design, these failures are mapped to scan operation ID numbers that have failed, and particular scan cells or primary output signals in the design where the faulty behavior was observed. In the end, a list of failures in terms of test stimuli and design signals, which are independent of the tester architecture, is obtained.

Diagnosis execution starts by consuming the list of failures produced by datalog processing step. The tool identifies the cone of logic that supports the observed failures and reasons that the defect location must be contained in that cone. It enumerates candidate locations, simulates each candidate using the test stimuli and ranks them by comparing the simulated behavior to actual faulty behavior. In the end, a list of candidate locations is reported by decreasing likelihood of containing the actual defect.

5.3   Making Diagnosis Work in a Full Work Flow

In this section we described the various requirements and other considerations that typically need to be taken into account to set up a full working scan diagnosis system along with capturing fail information necessary for diagnosis directly from production manufacturing test flow. Scan test content is applied on the tester along with the other manufacturing test content which includes cache tests, functional tests, IO tests, and parametric tests. Failures can be captured in-line during production tests by setting up a fail flow which is exercised on failing parts (dice at wafer sort or packaged parts). However, in order to minimize overall test time budgets, the increase in test-time due to fail data collection needs to be limited relative to the total test-time The is because the time to collect failures on the test

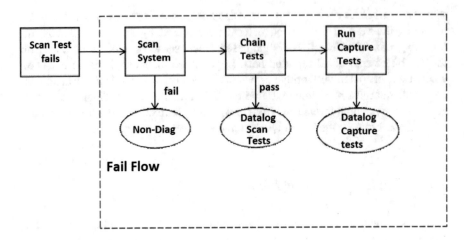

Fig. 12 Failure Bucketing of Scan Tests

equipment on a failing test is usually much longer than the time to execute a good test. The fail data collection test time is also proportional to the amount of failing information collected.

A test flow to collect fail data is illustrated in Fig. 12. The scan test content includes a scan-system test to determine if scan-system is functional. After the scan-system test, scan chain shift tests and scan ATPG capture tests are applied. If the scan system fails, then the scan chains themselves can't be operated. So in the event of a failure with scan system, the diagnosis process exits without collecting scan chain or scan ATPG failures. If the scan chain passes, then scan ATPG failures are captured. If not, this implies the scan chains themselves are non-functional, and failures from scan chain failures are captured.

Usually just the production tests can be used for fail data collection. In addition, some other tests are usually added with negligible impact to tester memory. For example, a simple walking 0 and 1 pattern (a sequence of 1100 repeated) is usually used to test a scan chain. Since this pattern writes both 0's and 1's to each scan cell in the chain, and additionally also creates a 0->1, 0->0, 1->0 and 1->1 transitions on each cell. For additionally diagnosability, in addition to the traditional (1100) repeat scan chain pattern which detects stuck at and all transition fault types, some additional chain tests can be added. If a transition defect cause delays longer than one bit, then shift out value of this traditional chain test pattern will be a constant 0 or constant 1, which could be incorrectly interpreted as stuck-at-0 or stuck-at-1. For example, if a defect caused a 2-cycle delay on a 0->1 transition, then the traditional 1100 repeating pattern will result in a repeating 0000 pattern or an 0 output response. To avoid this pitfall, (0) repeat and (1) repeat chain patterns are also included. These two patterns will not cause any failures for transition type defects. Also, a slow changing pattern – for example 16 zeros followed by 16 ones repeat pattern – could be added. This pattern can detect large delay defects.

Since failing data collection requires additional test time, a trade-off needs to be done to ensure adequate data collection while minimizing costs. To maintain diagnosis quality and resolution, we need to collect adequate failure information from failing scan test. However, too many failures collected can result in wasted test time with minimal additional benefit.

The data logging the first 100 failing scan operations are usually adequate for diagnosis of logic failures. An additional strategy to reduce fail data volume and the test time impact was to use sampling. In the initial phase of the yield learning when volume of material being processed was low, all failing dies on all wafers in a lot were analyzed. Later in the middle phase with volume picking up, fewer wafers are sampled and data logged.

6 Improving Fault Diagnosis for Accuracy, Precision and Resolution

In this section, we cover the key improvements to the basic scan logic diagnosis technology. These improvements enable the failure analysis engineer as well as the yield analysis engineer to use diagnosis as a powerful software instrument supplementing their portfolio of tools.

Traditionally, scan diagnosis tools relied on the stuck-at fault model and could diagnose down to a logic net in the design [14]. Although this approach is useful for localization, it has some limitations. Stuck-at patterns typically detect a vast variety of defect types, including bridges and opens, but this fault model is not always sufficient for effective diagnosis, since an actual defect may not behave as a stuck-at fault. This typically results in a large number of candidates, typically referred to as suspects. Even with a single suspect net, that net could cover a large area of the die.

Over the last few years, these limitations have been addressed through technologies such as

- Interconnect-open diagnosis [28]
- Composite fault models [13]
- Layout-aware diagnosis [29, 30]
- Cell-internal diagnosis [31],
- Scan chain diagnosis [12],
- At-speed diagnosis [32], and
- Iterative diagnosis [33].

With these advances software tools are now capable of diagnosing defects down to a logic net segment and physical polygon, and they can differentiate between defects internal to the cells vs. defects in the interconnect. They can also accurately classify a wide range of defect types. In addition, iterative diagnosis can be used to increase the initial diagnosis resolution. Diagnosis can now be used in conjunction with on-chip compression [34, 35] and logic built-in-self-test (BIST) [36] as well. These methods will be covered in more detail in the next section after covering the basics of fault diagnosis in this section along with work flows to set up diagnosis.

These advances have significantly improved the value of diagnosis for failure analysis. Current diagnosis tools produce results directly applicable to the failure analysis task at hand. They represent the results for example in terms of physical location, including the x/y/layer coordinates of a possible defect location. They also consume production test patterns even for designs using embedded pattern compression techniques, no longer requiring special 'diagnosis only' test patterns bypassing the compression circuitry. In the remainder of this article, we will review some advances in more detail and discuss their merits.

Beyond the immediate needs of the failure analysis engineer, diagnosis advances have also paved the way for using this technology in other areas like yield loss analysis and subsequent yield improvements or test quality improvements.

6.1   The Case for Opens

On the logic level, the diagnosis algorithms can figure out that an observed defective behavior of the die does not match a stuck-at behavior. The question is then, what kind of (static) defect could explain the observations? The only other two explanations are an unintended connection of a net to a neighboring net or an unintended disconnection of a net, i.e., a bridge or an open. The actual, physical reason for a short or the bridge is not relevant for the diagnosis tool. For example, if two nets have a (low-resistive) electrical connections of whatever physical cause, a diagnosis tool will classify this defect as a bridge.

For the case of opens, diagnosis usually determines the defective net very precisely. However, it usually cannot tell the open from a bridge based only on the observed failing bits, since both have similar observable behavior at the logic level.

For opens, the tool will report back to the user nothing better than the entire net – and somewhere on it is the open. In some cases, this result might already be sufficient for the failure analysis engineer to proceed for example with probing. In many cases however, a better resolution is desired. In the case of opens, this means reporting to the user, which segments of the net could have the open, and which segments can be assumed defect free. Figure 13 illustrates this issue. While diagnosis may simulate faults – for example, stuck-at at signal locations or ports of cells – real defective locations can have a many-to-one mapping from the defect space to the fault space.

Intel presented a method [28] to isolate interconnects opens by extending the stuck-at fault model to create a net fault model. It was shown that interconnect opens to could be isolated to single nets with information on portions of the interconnect that contain the defect as illustrated in Fig. 14. Instead of diagnosing down to dozens of stuck-at suspects, the diagnosis algorithm can isolate down to a single physical net.

Fig. 13 Defects versus fault simulated during diagnosis

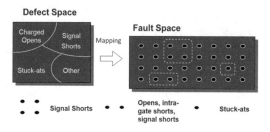

Fig. 14 Diagnosing defects down to physical nets

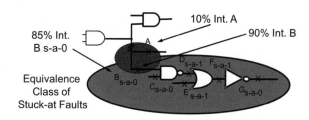

AMD reports in [29] case studies on bridges and opens, where so-called *layout-ware diagnosis* is used to improve the accuracy and resolution. In layout-aware diagnosis the common, logical analysis engine is paired with an integrated layout-based engine, which has access to exact routing of each net, each placement of cells, and all other layout components. Through this tight integration, diagnosis can correlate logically determined results against physical (i.e., layout) possibilities. Through this, the tool can quickly eliminate impossible candidates. Further, the layout-aware diagnosis can find additional suspects derived from layout, which are beyond the reach of traditional, logic-only diagnosis.

Finally, layout-aware diagnosis also gains access to the exact topology of a net, i.e., it knows through which layers it routs, where single and double vias are located, and where the net branches out. Using the net topology, an open segment of a net can be determined, instead of declaring an open somewhere on the net. For nets with branches, these segments are usually significantly smaller than the total net. Figure 15 illustrates the difference of being layout aware during diagnosis versus just considering logical models for diagnosis.

In one of AMD's case studies a suspective net can be divided into 14 segments based on its topology. This net runs through three metal layers and spans a maximum of 36 μm with a total search area for the suspect of 1192 μm². Layout-aware diagnosis of the device leaves only a single net segment that could contain the open defect. Also, all bridge candidates from an earlier non-layout-aware diagnosis run were eliminated. Overall, layout-aware diagnosis reduced the suspect area down to 11% (130 μm²), with a maximum span of 12 μm.

Fig. 15 Layout-aware
diagnosis versus logical only
diagnosis

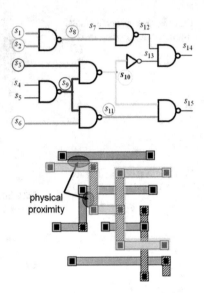

6.2   The Case for Bridges

In the same study AMD [29] reports layout-aware results for a bridge-type of defect. In this case, the non-layout-aware diagnosis reports a possible open or dominant bridge as the defect behavior. Dominant bridge means that the reported net is the victim net of a bridge defect. Victim nets are the ones over which the faulty values are propagated to an observation point, which is a primary output of the die or a scan flip-flop, whereas the aggressor net never carries any faulty values to any observation point. Hence, the aggressor net is not visible for a non-layout-aware diagnosis tool. The reported victim net could potentially bridge to any of the neighboring nets at any location. The non-layout-aware diagnosis tool cannot provide more detail. Furthermore, it is possible that an open-type defect could be the cause for the observed defective behavior of the device.

For the failure analysis process, this means that not only does the engineer need to look out for two very different types of defects, an open or a bridge, but also knows only one side of the potential bridge. The engineer still has to look for the other net and the defect somewhere along the victim net, potentially forming bridges to any of the neighboring net polygons. In other words, although the diagnosis report contains only one dominant bridge suspect, the number of potential defect locations the failure analysis engineer has to inspect is bounded only by the number of neighboring polygons.

While the logic report is ambiguous between a dominant bridge defect and an open defect, the layout-aware diagnosis report in AMD's case study shows only a bridge type defect. All open candidates were dismissed. Further, layout-aware diagnosis identifies all of the *relevant* aggressor nets. In this particular case, two neighboring nets are found, and all other impossible aggressor-victim net combinations are eliminated.

Overall, instead of searching a suspect area of 1012 μm^2 (the whole net), failure analysis can focus the activities to only the 26 μm^2, due to the tight bounding boxes computed by layout-aware diagnosis. This is a reduction down to only 2.6% of the original area. This area is divided only between two metal layers, instead of four, in which failure analysis is looking for a bridge-type defect. The total span of the suspect area is reduced from 39 μm to 16 μm.

6.3   Diagnosing At-Speed Failures

In order to gain the high level of outgoing product quality, it became common practice to execute additional type of tests. Building on top of the stuck-at fault, other fault models are used. In devices manufactured at 130 nm and below, it is common to use a structural at-speed test, typically leveraging the transition fault model (some use path-delay). The title of this section reflects that not all at-speed failures identified during testing, and subsequently diagnosed, are actual, physical defects. The Freescale test case summarized below is a good example of companies using the convenience and performance of at-speed diagnosis for design optimization, timing analysis, timing debug, and other related tasks.

In [32] Freescale reports the successful application of diagnosis of at-speed test patterns. Interestingly, this paper reports that the root cause for the failure of the dice on the ATE was not a physical defect, but an incomplete description to the ATPG tool of false and multi-cycle paths. Freescale claims that eliminating these paths based on the diagnosis result accumulated in a 300 MHz increase in speed. For some observed failures, the root cause was a design problem which needed design changes.

A slightly different aspect of at-speed diagnosis is reported by Infineon in [37]. In this paper, special 'diagnosis friendly' test patterns are used to improve the efficiency of physical fault isolation of timing failures using a time-resolved emissions system. This paper presents a case study, where the identified defect was most likely a resistive via. Analysis of the topology of the net in question further reduced the number of possible defective vias by 40%. Failure analysis eventually identified and verified a resistive via within the identified area.

6.4   Cell Internal Defect or Interconnect Defect?

With shrinking technology nodes, a significant number of manufacturing defects are inside the library cells (standard cells) themselves. This is in part caused by the increasing use of custom cells designed to deal with higher process variations. The ability to differentiate between defects in the interconnect ("back-end defects") and

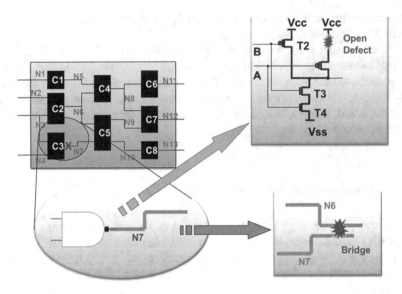

Fig. 16 Differentiating between cell-internal and interconnect defects

internal to the cells ("front-end defects") is therefore crucial to ensure that failure analysis can be performed in a timely fashion [38].

Intel [31] presented techniques to distinguish between cell-internal and interconnect failures, and methods to isolate to within a cell using transistor-level information and switch-level simulation. Figure 16 shows that while conventional diagnosis would reason down to a NAND gate and a net (N7), cell-internal diagnosis would go further to reason to locations within the NAND gate. Cell-internal diagnosis requires characterizing defective behaviors within cells using cell-level simulations at a spice, fast-spice or switch-level to create dictionaries of faulty-behaviors within cells for all cells in a cell library. This is then used during diagnosis as illustrated in Fig. 17.

According to TSMC [38], the process improvement comes from two sources. (a) Knowing the defect is in the cell, any de-layering can go straight to the metal layer where the cell interconnects are routed, without the need to investigate any higher metal layer, and (b) having failure analysis investigate fewer layers reduces the overall cost of the examination. Further, being able to distinguish cell-internal defects from interconnect defects also provides useful cell statistics for yield analysis and subsequent improvements of the cell library. Additional references on this topic are [39, 40].

In [38], TSMC outlines at first the methodology how defect locations can be determined to be inside or outside of cells, then presents results based on emulation of defective cell behavior. In the end of the report however, the methodology is put to the test on a 90 nm technology, AMD graphics chip. From a set of diagnosis reports, 7 were selected that had a cell as the one of the highest suspects reported.

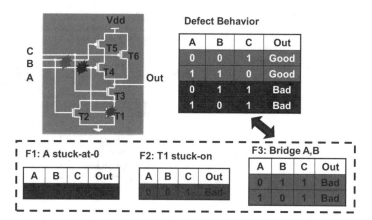

Fig. 17 Reasoning down to cell-internal defects

For these dice failure analysis was performed. In all 7 cases, a defect was found in the indicated instance of the cell. TSMC concluded additionally that using the cell-internal diagnosis classification, the overall failure analysis process was faster and less costly, because diagnosis eliminated the need to investigate any layer higher than metal 1.

6.5   Defects in Scan Chains

Thus far the discussion was about finding the defect in the functional logic. In this section, we focus on defects in the scan chains themselves. It is well known that the cause of a significant portion of dice failing logic test relates to 'issues' in the scan chains themselves. Literature reports numbers in the 10% to 30% range [41, 42]. Not all of these issues are actual defects in the physical sense. Some issues are simply setup or hold-time violations, i.e., the scan chains are operated incorrectly. Nonetheless, after testing on an ATE, this is not obvious. Hence, a diagnosis tool for scan chain failures must consider and report all possibilities.

The method of scan chain diagnosis was already discussed in the 1990s, with an alternative approach to scan chain diagnosis reported by Intel in [12]. Additionally, it is possible to diagnose errors from scan enables and multiple failing scan chains [41, 42]. Scan chain diagnosis is a very active field of academic and industrial research. The reader is encouraged to use the literature references as a starting point.

In principle, scan chain integrity testing comprises of shifting in a repetitive sequence of 0s and 1s, typically '0011' and observing the shift-out values at the output of the scan chain. From these observed chain test values, one can conclude defects in the scan chain. To name a few possibilities, if the scan out values are '0000', most likely a stuck-at 0 fault is somewhere on the scan chain. If the scan out

Fig. 18 Reasoning down to suspect scan cells within failing chains

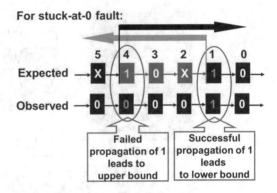

values are '0001', a possibility is a 'slow to rise', whereas a '0111' could indicate a 'fast to rise' problem.

There are many different conclusions scan chain diagnosis tools compute nowadays. Reporting the cause is only half of the answer of a scan chain diagnosis tool. The other half of the answer is the location of the problem. Scan chain diagnosis usually reports the location by the ordinance number of the scan cells, counted from the scan output of the chain (the cell closest to the scan out usually has the ordinance number zero). In this sense, a scan chain diagnosis report might say 'slow to rise problem between cells 25 and 21'.

It is the goal of these diagnosis tools to compute the range of possible locations as tight as possible. But they do not always succeed. In the papers listed above, many algorithmic methods are proposed how to tighten the range, some papers present case studies underlining the success of their method. Figure 18 illustrates the basic idea of reasoning down within a scan chain the scan cells that may be suspects.

Overall, diagnosis tools are in good shape when it comes to scan chain failures. One recent key improvement is the inclusion of scan logic tests in addition to the scan chain integrity test. At first it seems odd to execute scan test patterns, knowing that some scan chains are not operational. However, it has been demonstrated that the additional information gained from this is sufficient to considerably improve the quality of scan chain diagnosis. Some methods, like [43, 44], are actually able to diagnose and distinguish a scan logic and a scan chain failure at the same time.

6.6   Further Improving Resolution by Reducing Suspects

Up to now, we assumed that diagnosis uses the production test pattern set. There is good reason to start with this pattern set, since it detected the defect in the first place. Yet, the purpose of a production test pattern set is to detect as many faults as possible with the smallest number of tests. In other words, each test pattern potentially detects many different faults simultaneously. As long as there are several failing patterns, a diagnosis tool can determine the intersection of these sets of possible faults. Usually this reduces the number of reported suspects to a meaningful quantity. In some cases, however, there are just too many suspects reported to be meaningful for any failure analysis work.

LSI describes in [33] the concept of *iterative-diagnosis*. In this method, additional test patterns are generated, with the goal of reducing the number of suspects reported by diagnosis, before any (destructive) failure analysis work should be commenced. These new patterns together with the previous ones are brought to the ATE and applied to the device-under-test. The new test response is diagnosed again, and if the number of suspects is still too large, the process repeats. It is an iterative process, which stabilizes after a few cycles.

The analysis performed in [33] concludes that in 41% of the cases this method is successful in reducing the number of suspects. Actual failure analysis case studies reported in this paper underline, according to LSI, the usefulness of iterative diagnosis for both, scan and chain defects.

It should be noted that the principles LSI uses in [33] are applicable to all kinds of diagnosis including at-speed and layout aware. Especially for layout-aware diagnosis, which by itself usually reduces the suspect count, the benefit of iterative diagnosis is in addition to the benefit the user gets from using layout in diagnosis.

7 The Next Step: *Machine Learning in Diagnosis*

The conventional objective of identifying the failure locations during fault diagnosis has been augmented with various physically-aware techniques that are intended to improve both diagnostic resolution and accuracy as described in the previous section. Despite these advances, it is often the case however that resolution, i.e., the number of locations or candidates reported by diagnosis, exceeds the number of actual failing locations. Imperfect resolution greatly hinders any follow-on, information-extraction analyses (e.g., physical failure analysis, volume diagnosis, etc.) due to the resulting ambiguity. Additionally, various uncertainties in the fabrication process bring several challenges, resulting in diagnosis with undesirable outcomes or low efficiency, including, for example, diagnosis failure, bad resolution, and extremely long runtime. It would therefore be very beneficial to have a comprehensive preview of diagnostic outcomes beforehand, which allows fail logs to be prioritized in a more reasonable way for smarter allocation of diagnosis resources. Machine learning methods have been extensively studied and evaluated to address these challenges [45–47].

CMU researchers [46, 47] developed, a unsupervised learning methodology that uses ordinarily available tester and simulation data is described that significantly improves resolution with virtually no negative impact on accuracy. Simulation experiments using a variety of fault types (SSL, MSL, bridges, opens and cell-level input-pattern faults) reveal that the number of failed ICs that have perfect resolution can be more than doubled, and overall resolution is improved by 22%. Application to silicon data also demonstrates significant improvement in resolution (38% overall

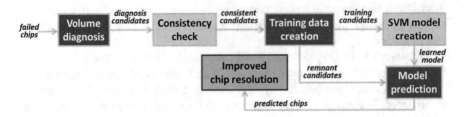

Fig. 19 Overall flow for the PADRE method using ML

and the number of chips with ideal resolution is nearly tripled) and verification using PFA demonstrates that accuracy is maintained. The main steps in the method called physical aware diagnosis resolution enhancement (PADRE) is shown in Fig. 19.

More recently CMU researchers have tackled other diagnosis challenges including failing diagnosis, bad resolution and long runtimes. It would therefore be very beneficial to have a comprehensive preview of diagnostic outcomes beforehand, which allows fail logs to be prioritized in a more reasonable way for smarter allocation of diagnosis resources. In [45], they propose a learning-based previewer, which is able to predict five aspects of diagnostic outcomes for a failing IC, including diagnosis success, defect count, failure type, resolution, and runtime magnitude. The previewer consists of three classification models and one regression model, where Random Forest classification and regression are used. Experiments on a 28 nm test chip and a high-volume 90 nm part demonstrate that the predictors can provide accurate prediction results, and in a virtual application scenario the overall previewer can bring up to $9\times$ speed-up for the test chip and $6\times$ for the high-volume part.

8 Summary and Future Challenges

Silicon debug and diagnosis is a complex field that employs a wide set of technologies (architecture, logic, circuit and layout design, simulation at different levels, instrumentation, optics, metrology, and even chemistries of various sorts). Logical deduction and the process of elimination are of paramount importance and central to debug and diagnosis.

As technology scales and designs grow in complexity, the debug and diagnosis processes are constantly challenged. While much progress has made, many challenges lie ahead.

Tools to enable automated diagnosis continue to evolve while tools for debug are still in a relatively infant state. Tools that identify sensitive areas of the design, add DFD features in an automated manner, generate tests for validation, debug and diagnosis, and automate the debug process need to advance.

Technologies presented in this chapter will need to continue to advance to keep up with CMOS scaling. Chip and manufacturing complexity will continue to grow every year along with pressures on design resources, ever faster time-to-market and time-to-profitability. Effective and efficient debug and diagnosis are certainly critical to the success of products. Better debug and diagnosis capabilities and faster fixes to problems is a constant imperative.

With the background on conventional modeling and simulation-based methods for fault diagnosis in this chapter, the readers should have a steppingstone to understand further how conventional methods for fault diagnosis can be improved using machine learning based techniques in subsequent chapters.

References

1. Josephson, D., Poehlman, S., Govan, V.: Debug Methodology for the McKinley Processor. In: IEEE International Test Conference, pp. 451–460 (2001, October)
2. van Rootselaar, G.J., Vermeulen, B.: Silicon Debug: Scan Chains Alone Are Not Enough. In: IEEE International Test Conference, pp. 892–902 (1999, September)
3. Gizopoulos, D. (ed.): Advances in Electronic Testing: Challenges and Methodologies Series: Frontiers in Electronic Testing. Springer, Boston (2006)
4. Gangartikar, P., Presson, R., Rosner, L.: Test/Characterization Procedures for High Density Silicon RAMs. In: IEEE International Solid-State Circuits Conference, pp. 62–63 (1982, May)
5. Hammond, J., Sery, G.: Knowledge-Based Electrical Monitor Approach Using Very Large Array Yield Structures to Delineate Defects During Process Development and Production Yield Improvement. In: Proceedings of International Workshop on Defect and Fault Tolerance in VLSI, pp. 67–80 (1991, November)
6. Segal, J., Jee, A., Lepejian, D., Chu, B.: Using Electrical Bitmap Results from Embedded Memory to Enhance Yield. In: IEEE Design and Test of Computers, pp. 28–39 (2001, May–June)
7. Livengood, R.H., Medeiros, D.: Design for (Physical) Debug for Silicon Microsurgery and Probing of Flip-Chip Packaged Integrated Circuits. In: IEEE International Test Conference, pp. 877–882 (1999, September)
8. Abramovici, M., Breuer, M.A., Friedman, A.D.: Digital Systems Testing and Testable Design. IEEE Press, Revised Printing, Piscataway (1994)
9. Bushnell, M.L., Agrawal, V.D.: Essentials of Electronic Testing for Digital, Memory & Mixed-Signal VLSI Circuits. Springer, Boston (2000)
10. Jha, N., Gupta, S.: Testing of Digital Systems. Cambridge University Press, London (2003)
11. Wang, L.-T., Wu, C.-W., Wen, X. (eds.): VLSI Test Principles and Architectures: Design for Testability. Morgan Kaufmann, San Francisco (2006)
12. Guo, R., Venkataraman, S.: An algorithmic technique for diagnosis of faulty scan chains. IEEE Trans. Comput-Aided Des. **25**(9), 1861–1868 (2006)
13. Venkataraman, S., Drummonds, S.B.: Poirot: Applications of a logic fault diagnosis tool. IEEE Des. Test Comput. **18**(1), 19–30 (2001)
14. Waicukauski, J.A., Lindbloom, E.: Failure diagnosis of structured VLSI. IEEE Des. Test Comput. **6**(4), 49–60 (1989)
15. Eichelberger, E.B., Williams, T.W.: A Logic Design Structure for LSI Testability. In: IEEE Design Automation Conference, pp. 462–468 (1977, June)
16. Hao, H., Avra, R.: Structured Design for Debug - The SuperSPARC-II Methodology and Implementation. In: IEEE International Test Conference, pp. 175–183 (1995, October)

17. Carbine, A.: Scan Mechanism for Monitoring the State of Internal Signals of a VLSI Microprocessor Chip. U.S. Patent No. 5,253,255 (1993, October 12)
18. Carbine, A., Feltham, D.: Pentium Pro Processor Design for Test and Debug. In: IEEE International Test Conference, pp. 294–303 (1997, November)
19. Sogomonyan, E.S., Morosov, A., Gossel, M., Singh, A., Rzeha, J.: Early Error Detection in Systems-on-Chip for Fault-Tolerance and At-Speed debugging. In: IEEE VLSI Test Symposium, pp. 184–189 (2001, April)
20. IEEE Std. 1149.1-2001: IEEE Standard Test Access Port and Boundary Scan Architecture. IEEE Press, New York (2001)
21. Abramovici, M., Bradley, P., Dwarakanath, K., Levin, P., Memmi, G., Miller, D.: A Reconfigurable Design-for-Debug Infrastructure for SoCs. In: ACM/IEEE Design Automation Conference, pp. 2–12 (2006, July)
22. D. C. Pham, T. Aipperspach, D. Boerstler, M. Bolliger, R. Chaudhry, D. Cox, P. Harvey, P. M. Harvey, H. P. Hofstee, C. Johns, J. Kahle, A. Kameyama, J. Keaty, Y. Masubuchi, M. Pham, J. Pille, S. Posluszny, M. Riley, D. L. Stasiak, M. Suzuoki, O. Takahashi, J. Warnock, S. Weitzel, D. Wendel, and K. Yazawa, "Overview of the architecture, circuit design, and physical implementation of a first-generation cell processor," IEEE J. Solid-State Circuits, Vol. 41, No. 1, pp. 179–196, Jan. 2006.
23. Kurd, N.A., Barkarullah, J.S., Dizon, R.O., Fletcher, T.D., Madland, P.D.: A Multigigahertz clocking scheme for the Pentium 4 microprocessor. IEEE J Solid State Circuits. **36**(11), 1647–1653 (2001)
24. Fetzer, E.S.: Using adaptive circuits to mitigate process variations in a microprocessor design. IEEE Des. Test Comput. **23**(6), 476–483 (2006)
25. Josephson, D., Gottlieb, B.: The Crazy Mixed up World of Silicon Debug [IC validation]. In: IEEE Custom Integrated Circuits Conference, pp. 665–670 (2004, October)
26. Mahoney, P., Fetzer, E., Doyle, B., Naffziger, S.: Clock Distribution on a Dual-Core, Multi-Threaded Itanium Family Processor. In: Digest of Papers, IEEE International Solid-State Circuits Conference, pp. 292–599, (2005, February)
27. Aitken, R.C.: Finding Defects with Fault Models. In: IEEE International Test Conference, pp. 498–505 (1995, October)
28. Venkataraman, S., Drummonds, S.B.: A Technique for Logic Fault Diagnosis of Interconnect Open Defects. In: IEEE VTS, pp. 313–318 (2000)
29. Chang, Y.-J., et al.: Experiences with Layout-Aware Diagnosis. In: Electronic Device Failure Analysis (2010, May)
30. Mekkoth, J., et al.: Yield Learning with Layout-aware Advanced Scan Diagnosis. In: International Symposium of Testing and Failure Analysis (ISTFA) (2006)
31. Amyeen, M.E., Nayak, D., Venkataraman, S.: Improving Precision Using Mixed-level Fault Diagnosis, IEEE International Test Conference (2006), Paper: 22.3
32. Tendolkar, N., et al.: Improving Transition Fault Test Pattern Quality through At-Speed Diagnosis. In: ITC (2006)
33. Gearhardt, K., Schuermyer, C., Guo, R.: Improving Fault Isolation using Iterative Diagnosis. In: 34th International Symposium for Failure Analysis, pp. 390–391 (2008, November 2–6)
34. Cheng, W.-T., Tsai, K.-H., Huang, Y., Tamarapalli, N., Rajski, J.: Compactor Independent Direct Diagnosis. In: Proceedings of Asian Test Symposium, pp. 15–17 (2004)
35. Stanojevic, Z., Guo, R., Mitra, S., Venkataraman, S.: Enabling Yield Analysis with X-compact. In: IEEE International Test Conference, pp. 734–742 (2005)
36. Cheng, W.-T., Sharma, M., Rinderknecht, T., Liyang, L., Hill, C.: Signature based diagnosis for logic BIST. In: IEEE International Test Conference (2007, October 21–26)
37. Burmer, C., Guo, R., Cheng, W.-T., Lin, X., Benware, B.: Timing Failure Debug using Debug-Friendly Scan Patterns and TRE. In: 34th International Symposium for Failure Analysis, pp. 383–389 (2008, November 2–6)
38. Sharma, M., Cheng, W.-T., Tai, T.-P., Cheng, Y.S., Hsu, W., Chen, L., Reddy, S.M., Mann, A.: Faster Defect Localization in Nanometer Technology Based on Defective Cell Diagnosis. In: IEEE International Test Conference (2007, October 21–26)

39. Fan, X., Moore, W., Hora, C., Gronthoud, G.: A Novel Stuck-at Based Method for Transistor Stuck-Open Fault Diagnosis. In: IEEE International Test Conference, pp. 253–262 (2005)
40. Fan, X., Moore, W., Hora, C., Konijnenburg, M., Gronthoud, G.: A Gate-Level Method for Transistor-Level Bridging Fault Diagnosis. In: Proceedings of the IEEE VLSI Test Symposium (2006)
41. Basturkmen, N.Z., Guo, R., Venkataraman, S.: Diagnosis of Multiple Scan Chain Failures. In: Proceedings of European Test Symposium (2008)
42. Lee, K.L., Basturkmen, N.Z., Venkataraman, S.: Diagnosis of Scan Clock Failures. In: Proceedings of VLSI Test Symposium, pp. 67–72 (2008)
43. Tang, X., Guo, R., Cheng, W.-T., Reddy, S.M., Huang, Y.: On Improving Diagnostic Test Generation for Scan Chain Failures. In: Asian Test Symposium (2009)
44. Wang, F., Hu, Y., Li, H., Li, X., Jing, Y., Huang, Y.: Diagnostic Pattern Generation for Compound Defects. In: ITC (2008), paper 14.1
45. Huang, Q., Fang, C., Mittal, S., Blanton, R.D.: Towards Smarter Diagnosis: A Learning-based Diagnostic Outcome Previewer. ACM Trans. Des. Automat. Electron. Syst. - Special Issue on Machine Learning **25**(5), 1–20 (2020)
46. Xue, Y., Poku, O., Li, X., Blanton, R.D.: PADRE: Physically-Aware Diagnostic Resolution Enhancement. In: IEEE International Test Conference, pp. 1–10 (2013)
47. Xue, Y., Li, X., Blanton, R.D.: Improving diagnostic resolution of failing ICs through learning. IEEE Trans. Comput-Aided Des. Integr. Circuits Syst. **37**(6), 1288–1297 (2018)

Machine Learning and Its Applications in Test

Yu Huang, Qiang Xu, and Sadaf Khan

1 Introduction

Machine Learning is a subfield of artificial intelligence. It is based on the concept of acquiring skills and synthesizing knowledge from past experiences (historical data) and then making decisions without being explicitly programmed or with human intervention. Machine learning solutions can extract the knowledge out of raw data using an algorithm and can automatically improve themselves through experiences. Whenever a machine learning model is exposed to new data, it can produce results based on learned information from old data.

There are two main breakthroughs that led to the emergence of machine learning in recent past decades. The first one is the realization made by Arthur Samuel (1959) that rather than instructing computers everything about how to perform tasks, it might be possible to teach them to learn for themselves. The second is the availability of data. In this digital world, we have a huge amount of information available with better storage capacities and better computational power. The real challenge is to extract meaningful information from all the data. These breakthroughs brought a surge in development in the machine learning area, where engineers focused on designing algorithms for computers and machines that make them learn, think, and extract knowledge from data as human beings do. The recent success of machine learning has paved its way to the EDA field where many optimization problems are tried to solve using machine learning. This chapter will

Y. Huang (✉)
Huawei Technologies Co. Ltd., Shenzhen, China
e-mail: huangyu61@hisilicon.com

Q. Xu · S. Khan
The Chinese University of Hong Kong, N.T. Hong Kong S.A.R, Ma Liu Shui, Hong Kong
e-mail: qxu@cse.cuhk.edu.hk; skhan@cse.cuhk.edu.hk

© The Author(s), under exclusive license to Springer Nature Switzerland AG 2023
P. Girard et al. (eds.), *Machine Learning Support for Fault Diagnosis of System-on-Chip*, https://doi.org/10.1007/978-3-031-19639-3_3

specifically focus on machine learning solutions that are tried to improve and solve VLSI testing challenges.

There are hundreds of existing machine learning algorithms, and many new algorithms are being developed every year. Each ML algorithm consists of three main components:

- *Model*: It represents a real problem in the form of a machine learning algorithm.
- *Learning*: This is the second phase in which a model tries to learn from training data.
- *Inference*: This is the last stage in which a trained model is used to extract meaningful information from test (unseen) data and provide us with the results.

2   Basic Concepts of Machine Learning

Machine Learning methods mainly fall into three categories based on their learning process. There are three basic learning approaches: *supervised learning, unsupervised learning,* and *reinforcement learning.* The selection of the learning approach depends upon the type of data and problem.

2.1   Supervised Learning

In this approach, each data point is paired as input and corresponding output label. Input is independent variables and output is dependent variables. Using this kind of data, a machine learning algorithm learns the process of mapping input to its corresponding output during training. The training process continues until the algorithm achieves a desired accuracy and training loss is minimized. Some examples of supervised machine learning algorithms are *linear regression*, *random forest,* and *decision trees*.

2.2   Unsupervised Learning

In an unsupervised learning approach, the input dataset doesn't contain any label or target point. The algorithm learns the desired information by deducing the structure present in the input data e.g., *k-means clustering* algorithm. These algorithms are used to identify hidden patterns, similarities, and differences in its input data.

2.3 Reinforcement Learning

This kind of learning approach is different from supervised and unsupervised learning. Here the idea is based on taking intelligent moves by an agent in order to maximize the cumulative reward under a particular environment. It is used in various machines and softwares to find the best possible solution in a given situation. It learns the possible moves by exploration (unexplored knowledge) and exploitation (existing knowledge) approaches.

3 Popular Algorithms for Machine Learning

This section describes the popular and common machine learning algorithms used in practice.

3.1 Linear Regression

This is a supervised machine learning algorithm. It learns a relationship between two variables by fitting them in a linear equation (straight line) from training data. One variable is considered as an independent variable (input) and the other is considered as a dependent variable (output). In simple words, it gives us the value of dependent variables using an independent variable. It predicts the values in continuous form. Equation (1) shows the linear equation.

$$y = mx + c \tag{1}$$

Here x denotes the independent variable (input) and y denotes the dependent variable (output). m is the slope of the line and c is y-intercept. m and c are the parameters that linear regression tries to find out to fit the x and y in a straight line. So, the core of this model is to find parameters that represent a line. Equation (2) shows the linear equation in terms of machine learning.

$$h(X) = W_0 + W_1 \cdot X \tag{2}$$

W_0 and W_1 are the weights. The linear regression finds the weights W_0 and W_1 that make the best-fitting line for input data at the lowest cost.

For example, to train a housing price prediction model based on size of house using linear regression, data consists of m training examples. Each training example contains an input (size of house) output (price of house) pair (x, y). Table 1 shows sample data consisting of four records.

Table 1 Example sample training data for housing price prediction model

No. of training examples (m)	Input: Size of house in feet^2 (x)	Output: Price ($) of house (y)
1	125	460
2	275	600
3	100	325
4	375	800

Fig. 1 Plot of a sample hypothetical training data for housing price prediction model

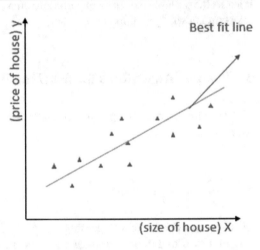

The plot of sample data is shown in Fig. 1. As Y is the linear function of X, Eq. (3) shows its function in terms of X.

$$Y = h(X) = W_0 + W_1 \cdot X \tag{3}$$

So, during training, linear regression tries to learn W_0 *and* W_1 that for each training example x can predicts the output h(x) close to original y i.e., minimize$_{W_0, W_1}$ (h(x) – y).

For this purpose, it uses a cost function that computes the average of square differences (errors) between predicted values and original values for all training examples as shown in Eq. (4). If the square difference is computed for single training example, then it is called *loss function*.

$$\text{Cost Function} = J\ (W_0, W_1) = \frac{1}{m} \sum_{i=1}^{m} \left(h\left(x^i\right) - y^i \right)^2 \tag{4}$$

Now, the objective is to learn W_0, W_1 that minimizes the cost function i.e., minimize$_{W_0, W_1}$ ($J\ (W_0, W_1)$). Let's assume $W_0 = 0$. Figure 2 shows the intuition of cost function for different values of W_1 .The most appropriate results (h(x) = y and $J(W_0, W_1)$ is minimum) are achieved when $W_1 = 1$.

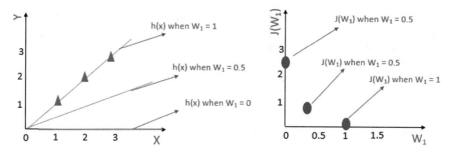

Fig. 2 Plot of h(x) and $J(W_1)$ for different values of W_0

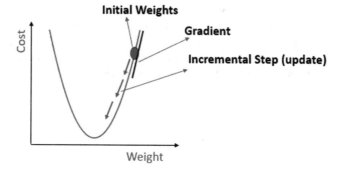

Fig. 3 Weight update process using Gradient Descent

To learn the desired W_0, *and* W_1 parameters that makes the cost minimum, it uses an algorithm *Gradient Descent* (GD). GD uses the following steps to learn the weights.

- Start with some W_0 *and* W_1 (usually with 0)
- Keep Changing W_0, *and* W_1 to reduce $J(W_0, W_1)$ using Eq. (5).

$$W_j := W_j - \alpha \frac{\partial}{\partial W_j} J(W_0, W_1) \quad for \ j = 0 \ and \ 1 \tag{5}$$

- Repeat it until W_0, *and* W_1 are found that minimizes $J(W_0, W_1)$

Gradient Descent computes the first order derivate of cost function that tells the slop at current W_0, and W_1. Then, it updates the W_0, *and* W_1 by α times the value of gradient in the opposite direction of gradient (slope). α is a learning rate that tells how big a step should be taken to update the W_0, and W_1. This process is repeated several times until W_0, *and* W_1 are found that give the minimum value of cost function. Figure 3 shows an example of an update process.

The slopes tell the direction to update and learning rate α tells how much we should move in that direction. If α is too high then weights may keep bouncing, without reaching the minima whereas if it is too low then update process will be quite slow and training process may take long time.

Hence, the choice of α is critical and it is provided as an external configuration to the model. The parameters that are provided externally to the model in machine learning are *hyperparameters*. Different values of α are usually tried as an external configuration one by one and the α that yields the best result is selected as the final value.

So, during training, it runs multiple iterations until the loss converges over all training examples (minimum cost is achieved) and it finds the best values of weights. And that weights are used to predict the output for unseen data during inference.

The independent variable could be more than 1. This form of the model is called *multi linear regression*. For example, linear regression used for the prediction of house prices based on location (area, surroundings) and size. Equation (6) shows a linear regression problem with four independent variables

$$Y = W_0\, x_0 + W_1 \cdot x_1 + + W_2 \cdot x_2 + + W_3 \cdot x_3 \tag{6}$$

x_0, x_1, x_2, x_3 all denote the independent variables.

3.2 Logistic Regression

This is also another supervised machine learning algorithm. It is used to predict the values in binary form (yes/no or true/false) based on the prior observed data (training data). This is an important algorithm in the machine learning domain for problems involving binary classification of data. The main difference between linear regression and logistic regression is that the former is used to predict values in continuous form, which means there are infinite numbers of possible outcomes, whereas the latter is used to predict constant (categorical) outputs.

It also makes predictions by analyzing the relationship between dependent variables and one or more independent variables. For example, this algorithm can be used to predict whether a student will be admitted to the college or not, based on the student's grades in high school, score in the college entrance test, and participation in extracurricular activities.

In logistic regression, we pass the output of the linear equation to a sigmoid function as shown in Eqs. (7) and (8). The sigmoid function maps the output of a linear equation (real number) in the range of 0 to 1.

$$y = mx + c \tag{7}$$

Fig. 4 Plot of the sigmoid function

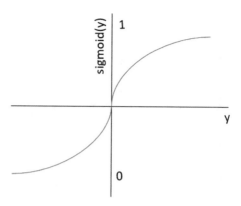

$$y' = sigmoid\,(y) \tag{8}$$

Where,

$$sigmoid(y) = \frac{1}{\left(1 + e^{-y}\right)} \tag{9}$$

Hence the Eq. (8) becomes

$$y' = sigmoid(y) = \frac{1}{1 + e^{-(mx+c)}} \tag{10}$$

The equation shows that the sigmoid function outputs values in the range of 0 to 1. if y is quite large then sigmoid (y) will be close to 1 and if y is small then sigmoid(y) will be close to 0. For the values around 0, sigmoid(y) will give a value 0.5. The graph of the sigmoid function is shown in Fig. 4. Hence, by setting a decision boundary on the output values of the sigmoid function, the results are classified into two categories.

For example, to create a model that predicts whether an input image is of a cat or not, a threshold is set on the output of the sigmoid function. For example, the threshold value is set to 0.5. The input images with the predicted values above the threshold are classified as 1 (actual prediction, cat). The input images with the predicted values below the threshold values are classified as 0 (false prediction, not cat).

3.3 Support Vector Machines (SVM)

It's another supervised machine learning algorithm for classification tasks. It works by finding an optimal hyperplane that separates the two different classes based on

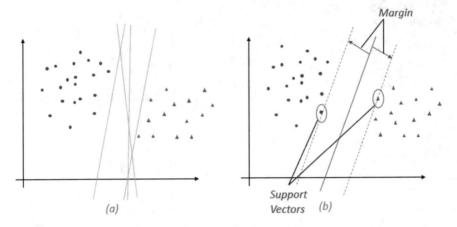

Fig. 5 (**a**) some possible hyperplanes (**b**) the optimal hyperplane using SVM

the feature vectors of each example in training data. In a 2-d space, the hyperplane is just a straight line.

More than one hyperplane can separate the different classes, as shown in Fig. 5a. The goal of SVM is to find the *optimal plane* that classifies all objects with high confidence. The optimal space is the one which maximizes the margin (distance) between two classes as shown in Fig. 5b. The margin (distance) between classes is measured by taking the difference between the points of both classes that are nearest to the separating planes. These points are called support vectors.

The separating hyperplane in 2-d is represented by the following equation $wx + b = y$. If $y > =1$ then the vector x belongs to class 1 and if $y < = -1$ then the vector x belongs to class 2. Suppose support vectors of both classes are $x1$ and $x2$. The linear lines passing through these points are represented by Eqs. (11) and (12). +1 indicates one class and −1 indicates the other class.

$$wx1 + b = 1 \tag{11}$$

$$wx2 + b = -1 \tag{12}$$

To calculate the margin between classes, the difference between Eqs. (11) and (12) is taken.

$$w \ (x1 - x2) = 2 \tag{13}$$

$$\frac{w}{|w|} \ (x1 - x2) = \frac{2}{|w|} \tag{14}$$

So, SVM tries to learn weights that maximize the value $2/|w|$.

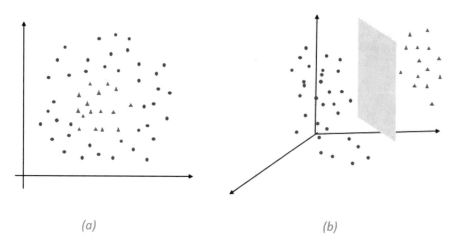

(a)   (b)

Fig. 6 (**a**) non-linear data in 2-d (**b**) separation of non-linear data using SVM in 3-d

If the data is not linearly separable as shown in Fig. 6a then SVM performs a transformation that maps the lower-dimensional data to a higher dimension where the data is linearly separable. The decision boundary in higher dimensions will be a hyperplane as shown in Fig. 6b. The transformation is performed using a kernel function. There are different kinds of kernel functions used in SVM. One common type of kernel function is polynomial kernel as described in Eq. (15). A polynomial kernel checks the pairwise similarity between original data points by taking the pairwise dot product between points. The general equation of polynomial function is given below.

$$K\ (x,\ y) = <x.y>^d \tag{15}$$

Where x and y are input vectors from training data and $<>$ denotes the dot product. $d > = 1$ shows the degree of polynomial.

3.4 Support Vector Regression (SVR

SVR is a regression model that predicts the output in continuous form. It is based on the concept of SVM with slight modifications. It tries to find a hyperplane that contains the maximum data points (best fits the data).

In Fig. 7 the decision boundaries are drawn around the data. SVM tries to find a plane that consists of maximum points within the decision boundaries.

The equation of hyperplane is given by $Y = wx + b$.

Fig. 7 SVR based hyperplane and corresponding decision boundaries

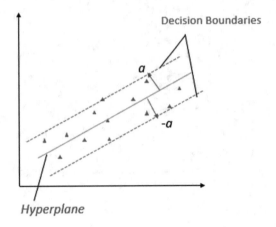

Let's say decision boundaries are at a distance a from the hyperplane. Then the equation of the decision boundary above the hyperplane is given by $wx + b = +a$ and the equation of the hyperplane below the decision boundary is given by $wx + b = -a$. So, the goal of SVR is to satisfy the following equation $-a < Y-wx + b < +a$. This constraint is satisfied by minimizing $1/|w|^2$.

3.5 Artificial Neural Networks (ANNs)

Artificial Neural Networks (ANNs) come under Deep Learning (subfield in machine learning) and are designed to mimic the human brain. The goal of deep learning is to design software programs that work or respond like human minds. ANNs are the basic unit of computation in deep learning. As its name suggests, this concept is inspired by neurons in human brains. ANNs are developed to replicate the functionality of human brains. It consists of multiple single processing nodes called 'neurons', that are divided into a stack of different layers.

The first layer is called the input layer and the last layer is called the output layer. There can be one or more than one intermediate hidden layer. All these neurons are interconnected and feed-forward which means data can be moved in one direction only. One neuron in some intermediate layer might be connected with several neurons in its previous and next layer. It receives information from the previous layer neurons and passes it to the forward layer. Each link between neurons contains a weight. The input signal on each link is multiplied and added by their respective weights. Neural networks learn (or are trained) the appropriate weights by processing training examples. Figure 8 shows a fully connected artificial neural network with 2 hidden layers.

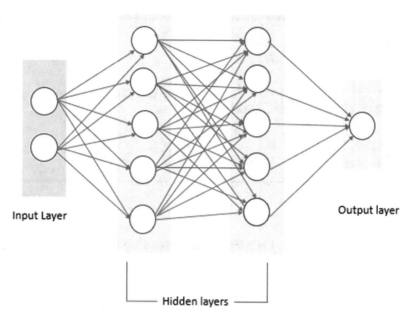

Fig. 8 A fully connected neural network

3.6 Recurrent Neural Networks (RNNs)

RNN is a type of artificial neural network used to process time-dependent and sequential data. It is commonly used in machine translation, natural language processing, and speech recognition applications. It works on the principle of memorizing the output of the previous step and feeding it as the input of the current step. Hence, the output at current step t depends upon input at t and input at t-1. This is desirable in RNN because while processing sequential data, the output at a step t depends upon the output at step t-1. e.g., during machine translation, the generation of new words in the sentence depends upon the context of previous words while in a feed-forward neural network, the decisions are solely based on the current input only and it doesn't memorize the past data.

Figure 9 shows the architecture of an RNN in which X is input, Y is output, and A, B, C are the learnable parameters in neural networks.

There are four types of Recurrent Neural Networks:

- *One to One RNN*: This type of neural network is known as the Vanilla Neural Network. It is used for machine learning problems, containing a single input and a single output.
- *One to Many RNN*: This type of neural network is used to solve problems having a single input and multiple outputs, e.g., image caption.

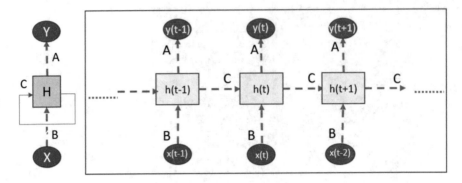

Fig. 9 Basic RNN Architecture

- *Many to One RNN*: This RNN takes a sequence of inputs and generates a single output e.g., sentiment analysis in which a given sentence is classified as a positive or negative sentiment.
- *Many to Many RNN*: This RNN takes a sequence of inputs and generates a sequence of outputs. Machine translation is one of the examples.

RNNs suffer from the problem of vanishing gradients. The information is carried to the next steps through gradients. When gradients become too small, the parameter updates become insignificant. This makes the learning of long data sequences difficult. It causes the loss of information through time. To solve this problem, Long Short-Term Memory Networks (LSTM) are developed.

3.6.1 Long Short-Term Memory Networks (LSTM)

LSTMs are capable of remembering information over a long period of time. Like RNNs, the LSTMs also have chain-like structures. In RNNs, the repeating block consists of a very simple structure such as tanh, but it is quite complex in LSTM as shown in Fig. 10.

LSTMs work in a 3-step process.

- *Step 1*: The first sigmoid function decides how much information should be removed (that is unnecessary) by computing the sigmoid function on previous state h(t-1) and current input x(t).
- *Step 2*: This step decides how much data contributes to the current state. It consists of two parts. In the first part, the sigmoid function decides which information to pass by giving the value in the 0 to 1 range and the second part consists of tanh which decides the importance of all information in the range of −1 to 1.
- *Step 3*: This step decides the output from this module. It contains a sigmoid layer, which decides what parts of the cell state make it to the output. Then, the result

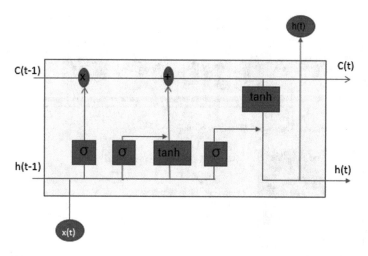

Fig. 10 LSTM Cell Architecture

from previous two steps is passed through a tanh that maps it between −1 and 1, and multiply it by the output of the sigmoid gate.

Besides LSTMs, there are also some invariants of RNN such as GRU [1] or Bidirectional-RNN [2].

3.7 Convolutional Neural Networks (ConvNets / CNNs)

Convolutional Neural Network is also another popular deep learning model mainly used for image processing, classification, and segmentation, etc. The size of the image to process can be very large. e.g., a colored 1000 × 1000 image would have dimension 1000 × 1000 × 3. It brings challenges in training the model in terms of memory requirement, computational power, and sufficient training data. CNN works by reducing the size of an image into a form that is easier to process, without losing important information (required features extraction in an image for good prediction).

This is achieved by a convolutional operation that is the basic unit of CNN. CNN consists of multiple convolutional layers and each layer performs a convolutional operation on the input image using a filter (also called a kernel) to extract the necessary features and to reduce the image size. This is called *feature learning*. The final learned features are then passed to FC layers for underlying task prediction. Filters in the convolutional layers can be of any size.

Conventionally, 3 × 3 and 5 × 5 filter sizes are used. Filter moves across the whole image from the top left to the bottom right and performs a matrix multiplication operation between its value and the portion P of the image. This

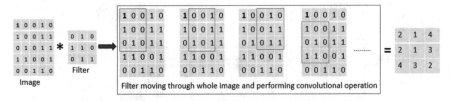

Fig. 11 Convolutional Operation on a 5 × 5 image

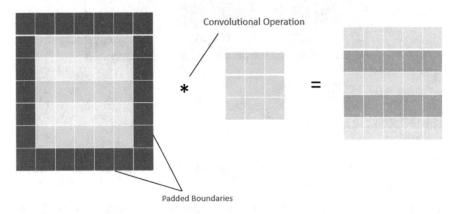

Fig. 12 Convolutional Operation on padded image

operation produces a new convolved value for that portion P of the image (convolved feature) as shown in Fig. 11. Initially, filters contain random values and keep on updating themselves during training until they are set to the required value.

It is noted that on the boundary rows and columns, one time convolution operation is performed but on intermediate rows and columns, convolutions are performed multiple times. Hence, to apply multiple convolutions on boundaries values as well, padding is performed. It adds an extra row and columns having values 0 to all sides of images as shown in Fig. 12. Convolution operations on padded images preserve the original image size.

Besides the convolutional operations, ConvNet also uses a pooling operation to reduce the image size and to extract meaningful features. Max pooling is a commonly used pooling operation in which maximum value is extracted only from a selected image portion. Figure 13 shows an example of max pooling.

3.8   Graph Neural Networks (GNNs)

Graphs are common and useful data structures that consist of a set of vertices, also called nodes, and a set of edges i.e., connections between nodes. This data structure appears frequently in science and engineering as many applications are

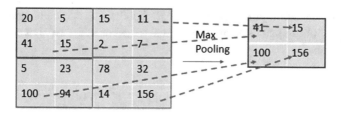

Fig. 13 Max Pooling Operation

represented by graphs e.g., social networks and recommender systems. A graph is an irregular structured data i.e., size and shape of the graph might vary in the dataset. All previous deep learning models were designed to process the regular structure data (having fixed shapes and sizes). So, they were not applicable to graphs. Hence, Graph Neural Networks (GNNs) are developed to solve problems represented in the form of graphs.

Every node and edge contains some properties that are represented by a feature vector associated with them. For example, when we represent molecules as graphs, atoms inside molecules are represented by nodes, and bonds between atoms are represented by edges. Node feature vectors contain information like atom type, size or number of electron, and edge feature vectors contain information like bond type and strength. Besides the node features, graphs also contain structural information embedded in themselves.

In machine learning, GNNs are used for node-level prediction tasks, edge-level prediction tasks also called the link predictions (predict whether there is a connection between two particular nodes or not), and also for graph-level prediction tasks. The fundamental of GNN is to learn a suitable representation of graph data called representation learning. So, a GNN learns a function f that takes the graph as input and maps every node n of graph g into d-dimensional continuous vector space. This d- dimensional vector is called the embedding vector that is the new representation of nodes in the graph. This embedding contains the local structure information as well as the node features information of the graph. These embeddings can be used in many downstream machine learning tasks.

Embeddings are generated through message passing layers in GNN. For every node in the graph, the message passing layers gather the information of neighbor nodes, combine them together with the current node and update the state of the current node. This process is done simultaneously for each node.

Let's formally define this process mathematically. Let G = (V, E) is a graph in which V is the set of nodes and E is the set of edges. N(v) is the set of all neighbor nodes of a node v that are directly connected with the node v. x_v is the feature vector associated with node v. GNN learns the function f that produces the embedding vector h_v for node v which contains the information of node v itself and the information of its neighborhood as shown in Eq. (16).

$$h_v = f\{x_v, h_N(v)\} \tag{16}$$

To learn the information of the neighborhood for a particular node v, GNN aggregates embeddings of its multi-hop local neighbor nodes $N(v)$ as shown in Eq. (17).

$$h_{N(v)}{}^k = Aggregate \left\{ h_u{}^k : u \in N(v) \right\} \tag{17}$$

k in Eq. (17) indicates the current message passing layer. Total number of messages passing layers is a hyper parameter and it indicates the depth of the model that means it tells us how deep we should go in the graph to aggregate neighbor's information. At each layer we have embeddings. For layer-0, embedding of node v is its input feature vector x_v. Layer-k embeddings get information from nodes that are k hops away. There are different ways to perform the aggregate action. The basic method is to take the average of information from neighbors as shown in Eq. (18).

$$h_{N(v)} = \Sigma_{u \in N(v)} \left(h_u / |N(v)| \right) \tag{18}$$

The next step is to combine this information with the node embedding of the current node to update the embedding for the current node. This is done by an update operation as shown in Eq. (19).

$$h_v{}^{k+1} = Update \left\{ h_v{}^k, h_{N(v)}{}^k \right\} \tag{19}$$

Figure 14 shows an embedding generation process for a node A with depth K = 2. Once, the embeddings are learned, they can be used to solve the underlying task e.g., for graph classification, the embeddings can be passed to a FC layer(s) that performs the classification.

3.9   Reinforcement Learning (RL)

It is a subset of machine learning that is far different from supervised and unsupervised learning. It targets more complex problems and is more dynamic in nature than classical machine learning methods. These are the set of algorithms that are smart enough to take decision based on their local environment (surroundings) without any supervision.

A RL model consists of an *agent,* responsible for taking actions under a particular environment to maximize an overall reward. In a particular state s, it takes an action a, then the environment gives him the feedback in form of a new state (change in environment due to action) and a reward corresponding to that action. It learns the set of correct actions by trial-and-error process over time. Figure 15 describes the execution flow of reinforcement learning. It tries different set of actions in an environment and evaluates the total reward gained. After many trials, it learns which

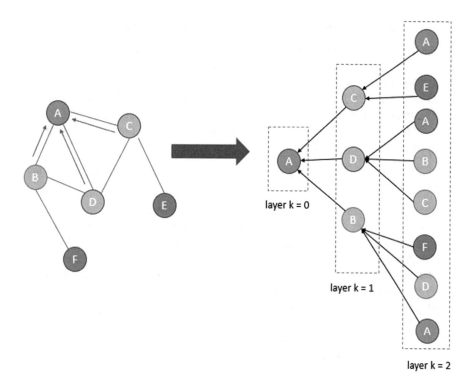

Fig. 14 Embedding generation for a node A with depth K = 2

Fig. 15 Execution flow of reinforcement learning

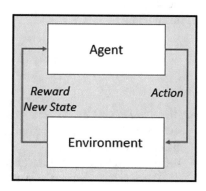

kind of action yields a higher reward, and then it establishes a pattern of behavior under that environment. RL is a multiple-decision process as it forms a decision-making chain (set of actions) required to finish a specific job. It has applications in various domains such as robotics and games. For example, playing a chess game. In this game, chessboard is the environment, and the agent is a computer program playing the game, and it wants to learn the set of moves that results in winning the game. AlphaGo Zero is another game that also uses RL to learn the game of GO.

So, the main components of RL algorithms are the followings:

- *Agent*: It is the main component that is trained to make a series of correct decisions in an environment to complete a task.
- *Environment*: It is the surroundings with which the agent interacts.
- *Set of States* (*S*): It is defined as the set of all possible states in an environment. The state s ϵ S indicates the current situation of the environment.
- *Set of Actions* (*A*): It is the collection of all possible actions that an agent can perform in an environment.

RL algorithms are developed based on the Markov decision process (MDP) that contains the following components:

- *State Transition P(s,a,s′)*: It is the probability of change in the current state *s* of the environment due to an action *a* taken by an agent. It is shown by P(s,a,s′) where *s* is the current state, *a* is the action taken by agent and s' is the new state resulted due to the action a.
- *Reward R(s,a,s′)*: It is the evaluation of an action and indicated by R(s,a,s′). Whenever an agent in a state *s* performs an action *a* that results in a new state s', it gets a reward in return. It can be positive or negative depending upon the kind of action taken by the agent.
- *Policy Function* π (*a*| *s*): It is the main strategy learned and used by the agent to decide which action to take based on the current state. It is the mapping from state s $\in$ S to action a $\in$ A.
- *Utility*: The utility is the total sum of all rewards (discounted) that an agent receives from the environment while following a policy Π. In RL, we focus on maximizing it.

$$\text{Utility } U = r_0 + r_1 Y^2 + r_2 Y^3 + \ldots \tag{20}$$

- *Discount Factor (Y)*: A discount factor between [0, 1] is used to adjust the future reward so that the overall reward is a finite value and future rewards worth less than immediate rewards.
- *Value function $v_\pi(s)$*: It is the expected utility received by following a policy Π from state s. It is used for the evaluation of a policy. A value function is calculated for each state *s* and it determines how good it is for the agent to be in a particular state by following a policy π which is denoted by $v_\pi(s)$.

$$\text{Value } v = E\,[U] \tag{21}$$

- *Q-value(action-value)*: It is just like a value function except it takes an additional parameter, the present action *a*. $Q\pi(s, a)$ refers to the expected utility of taking an action *a* from state *s*, then following policy π (Fig. 16).

Fig. 16 State transitions
using value and Q functions
while following policy

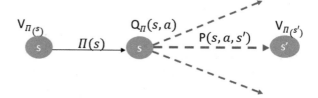

The intuition of Q value is when we are in a state s and wants to take an action a,
where there are multiple possible state, we can end up with. It is calculated using
Eq. (22).

$$Q_\Pi(s, a) = \Sigma_{s'} P(s, a, s') [R(s, a, s') + Y V_\Pi(s')] \qquad (22)$$

This Q function depends upon the current reward plus the discounted policy value
from new state, where P (s, a, s′) is the transition probability over all possible states
s′ and R(s, a, s′) is the immediate reward gained due to action *a* and state transition
from s to s′. $Y V_\Pi(s')$ is the discounted value of policy from new state s′. From Q
value, we can formulate the value function as follows:

$$V_\Pi(s) = \begin{cases} 0 & s = end\ state \\ Q_\Pi s(a) & s \neq end\ state \end{cases} \qquad (23)$$

To solve a MDP based RL problem, there are two methods: value-iteration and
policy-iteration. Both methods assumes that the agent knows about the state
transitions and reward probabilities before interacting with the environment.

3.9.1 Policy Iteration

In policy iteration, we start by choosing an arbitrary policy. Then, we iteratively
evaluate the policy at each state and improves it until convergence. It is an iterative
algorithm that consists of the following steps:

- First, it initializes all $V_\Pi{}^0(s)$ to zero for all states s.
- During iteration t = 1, ...m, for each state s, it calculates the state value based
  on transition probabilities, immediate reward and state value learned so far (till
  previous iteration) using Eq. (24).

$$V_\Pi{}^t(s) = \Sigma_{s'} P(s, a, s') [R(s, a, s') + Y V_\pi{}^{t-1}(s')] \qquad (24)$$

It iteratively repeats the process until values do not change much.

3.9.2   Value Iteration

In RL, an agent tries to learn a policy that yields the best action to take at each state and hence maximizes the total reward. To learn this, we need to determine the optimal value function that gives maximum expected utility. The optimal policy is achieved by following the set of actions that yields the higher reward (maximize the Q value).

$$Q_{opt}(s, a) = \Sigma_{s'} P(s, a, s') [R(s, a, s') + Y V_{opt}(s')] \qquad ((25)$$

Where

$$V_{opt}(s) = \begin{cases} 0 & s = end\ state \\ \max_{a \in Actions(s)} Q_{\Pi} s(a) & s \neq end\ state \end{cases} \qquad (26)$$

This leads to optimal policy V_{opt} (s) (that has the maximum value) that is equal to

$$V_{opt}(s) = \arg\max_{a \in Actions(s)} Q_{opt}(s, a) \qquad (27)$$

This algorithm is similar to policy evaluation algorithm. During each iteration i, for each state s, it learns V_{opt} (s) by following the action that yields the maximum utility.

In Policy iterations and Value iterations algorithms, the agent exactly knows the transition and reward functions i.e., agent knows how his actions are going to affect the environment.

There are some other RL algorithms that are used in situation when agent doesn't know how the environment works (transition probabilities & reward probabilities) such as Model based Monte Carlo.

3.9.3   Model Based Monte Carlo

In this model, an agent learns from the history of interactions with the environment. It has a data that consists of series of states, actions, and corresponding rewards e.g., $s_0; a_1, r_1, s_1; a_2, n\ r_2, \ldots a_n, r_n, s_n$. From this data, it builds the transition and reward functions.

$$P(s, a, s') = \frac{\#of\ times\ (s, a, s')\ occurs}{\#of\ times\ (s, a)} \qquad (28)$$

Transition function is created by calculating the probability of how many times a state transition happens out of all action performed, as shown in Eq. (28).

$$R\left(s, a, s'\right) = r \ in \ \left(s, a, r, s'\right) \tag{29}$$

Reward function is created by just observing the reward whenever a state transition occurs, as shown in Eq. (29).

Once the transition and reward functions are obtained, it can use policy and value iteration algorithms to get the optimal policies. But the possible issue in this approach is that data may not contain the information of all kinds of state transitions. In this case, it will not give the optimal solution.

3.9.4 Model Free Monte Carlo

This is a model free process which does not try to model the transition and reward functions from data. Instead, it directly tries to estimate Q_π because it gives the possible utility (sum of discount rewards). So, given a bunch of sequences of actions, states, and rewards in the data, it computes the $Q_\pi\ (s,\ a)$ by taking the average of expected utilities achieved at state s in each record.

Besides Model Free Monte Carlo, there are some more algorithms such as SARSA, and Q-Learning.

4 Applications in Test

Many prior published papers in test area are using machine learning technologies. In the following sections we will give a brief review of such papers.

In Li et al. [3], parameter optimization (e.g., the PRPG length, MISR length, and scan-chain length) in a scan-compression architecture was studied. These parameters are very important to be specified properly, otherwise the benefits of the advanced scan-compression architectures cannot be fully utilized. As a key component for test application, the PRPG has to be carefully designed for test-cost reduction. To address this problem, the authors proposed an approach based on a support-vector regression (SVR) model to select the optimum PRPG length for a given scan-compression architecture. Based on previous results on existing designs, a training library is constructed and a trained SVR model is generated. Once a new design is given, the proposed SVR-based PRPG selection model can automatically output the optimum PRPG length, and the corresponding PRPG can then be generated and inserted into the new design.

Figure 17 illustrates the overall flow of the proposed approach. The approach consists of two phases: (i) Model Construction and (ii) Core Generation (to determine scan compression architecture for each core). In the model-construction phase, experimental data on previous industry designs are collected and used to construct and train the prediction model. In the core-generation phase, the PRPG length is specified and the corresponding TCSA is generated for the new design. Previous

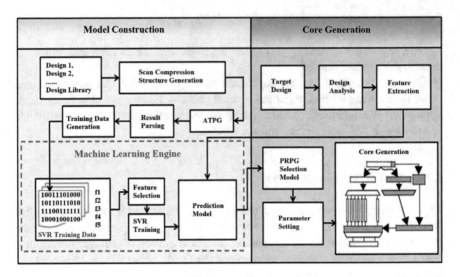

Fig. 17 Overview of the proposed approach in Li et al. [3]

designs are used to construct the design library in Fig. 17. The scan compression architecture is then generated for each design with different PRPG lengths. After ATPG is run on the cores with generated scan compression architectures, test reports are parsed and the collected features are used to build the training library. The features are selected by a correlation-based feature selection technique to reduce computation time and increase prediction accuracy.

After applying machine-learning techniques, the prediction model is finally constructed. When a new design is given, the selected features of the new design are first extracted. The prediction model is used to predict the test cost based on those selected feature values. Finally, an optimum PRPG length can be selected based on prediction results and the CODEC is generated. Large industry designs have been used to demonstrate that the proposed PRPG-selection method is efficient and can effectively reduce test cost. However, the generalizability of the proposed method is not illustrated because the authors only have used big designs of similar GPUs.

In Zorian et al. [4], a DFT recommendation system for ATPG Quality of Results (QoR) is proposed. As the number of test cycles is a significant factor in maintaining good ATPG QoR, the goal of the proposed DFT recommendation system is to optimize test cycles required by a given design for a targeted test coverage to give the most-optimal performance. For full-scan compression designs, two key configuration parameters, the number of scan chain in/out ports (SIO) and the maximum scan chain length (MaxLen), usually influence the QoR. While the optimal configuration can be obtained by trial-and-error, this process is iterative, laborious, and inefficient. The authors developed a supervised machine learning (ML) recommendation system that predicts the optimal DFT configuration for scan-compression configuration, which produces the minimum number of test cycles at the targeted test coverage for the given design.

Figure 18 presents the workflow of the proposed method. It consists of three phases:

- *The data collection phase*: Initially, the input Verilog RTL is synthesized using DFTMAX. Subsequently, Synopsys TestMAX Advisor is used to extract the design's features, such as number of flip-flops, number of clock domains, number of faults, number of primitive gates, static and random test coverage. In parallel to the Advisor process, the test-case generator creates a set of test cases, where each test case (TC) has different SIO and MaxLen setting. TestMAX™ DFT for generating the DFTMAX codec have been used for the DFT insertion. DFTMAX codec is created for a range of SIO. For each SIO, it selected a range of MaxLen resulting in hundreds of testcases generated for each design. Finally, all features and test cycles for all test cases will be stored in Synopsys's internal data analytics library: Common Logging Library (CLL).
- *The ML model generation phase*: An important factor on the model performance is the hyperparameter optimization process, which is performed before starting the model training process. In the proposed method, a tree-based boosting regression model was used. Some of the important hyperparameters for this model include training rate, maximum tree depth, maximum number of features and minimum samples split. The grid-search method was used for hyperparameters optimization, which explores exhaustively all possible combinations of the selected parameters. Model training step is performed on the training designs using all test cases to generate the ML model. The last step is the model evaluation process, which uses the testing designs set to evaluate the generated models and generate the features importance list that will be used in the next iteration. For evaluating the model's performance, it is desirable to see how well it was able to predict the configuration that would yield the lowest number of test cycles. The model's accuracy does not necessarily need to be calculated based on its ability to predict the exact test cycle count for a given testcase. Instead, if the SIO/MaxLen combination that had the lowest test cycles is the SIO/MaxLen combination that the model predicted would have the lowest test cycles, is was considered as a hit.
- *The optimal configuration inference phase*: The inference process is performed on the current design using the generated ML models from the previous phase. The input to this phase, in addition to current design, is the range of the available SIO. In order to extract the required features for the ML models, TestMAX Advisor is processed on the synthesized Verilog of the current design without any DFT. The output is the features vector, which is the input of the ML Model. The output of the inference phase is a list of ranked scan compression configurations based on the increasing predicted test cycles. The configuration with the lowest test cycles is ranked at the top.

The training set had 5544 testcases spanning 24 different designs, with design sizes ranging from 100k–400k sequential cells. The testing set consisted of 1155

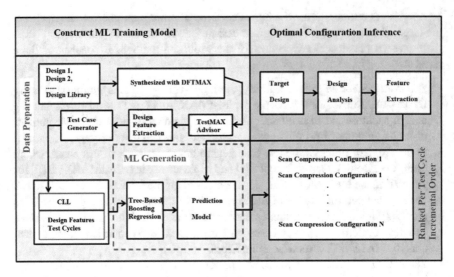

Fig. 18 The workflow of the proposed method in Zorian et al. [4]

unique testcases over variations of 5 designs, with design sizes ranging from 100k–400k sequential cells. The prediction accuracy is over 85%. The robustness of the proposed approach needs to be validated by more designs of different types and different sizes.

In Wu et al. [5], it used deep learning to determine a set of optimal parameters such as the number of scan chains, scan channels, power budget, etc. such that it can reach the highest test coverage with a minimum amount of test data volume whilst satisfying various other constraints. Based on the prediction results, the optimal test architecture is selected, yielding a more efficient approach compared to the traditional trial-and-error methods.

Figure 19 shows the proposed compression configuration selection flow in [5]. The first phase is to "Construct DL Training Model". During the training of a 7-layer deep Neural Network (DNN), a set of input training data with known outputs (labels) is fed to the DNN and a set of predictions are obtained. In this work, two DNNs were built, one with test coverage as its single output, and the other with data volume as its single output. Both DNNs used the same set of design features as inputs. The designs features include #Scan chains, #EDT channels, #Total gates, #Primary Inputs and #Primary Outputs, #Testable faults, #Scan cells. More importantly, it used SCOAP score distribution to measure testability of faults in a circuit. After building two DNNs for Test Coverage (TC) and Data Volume (DV), it applied many test cases (with various compression configurations) with selected design features. DNNs will calculate the outputs, i.e., predicted TC and predicted DV, through forward propagation using the weights determined in the training phase. This prediction is several orders of magnitude faster than running ATPG. After that, it goes into the next step "Optimal Configuration Inference &

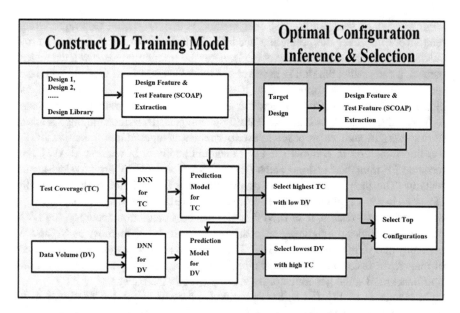

Fig. 19 Compression Configuration Selection Flow in Wu et al. [5]

Selection". Among a set of predicted TC and DV, it will select the ones with acceptable TC and lowest DV. Here, it prefers to avoid the cases with low TC even if their DV is the lowest. Finally, it runs ATPG for these selected configurations to get actual TC and DV and make a final selection of the configuration to be used. The prediction accuracy was measured by comparing this final selection with the one obtained using Compression Predictor.

Based on the DNN model trained by 7 circuits, the experimental results show that, for TC prediction, the average error is 0.7% for a new circuit without any training data from the new circuit. Whereas for DV prediction, the average error is 21.25% for a new circuit without any training data from the new circuit. The average error can be reduced to 15.22% when it added 15 configurations to the training data. It means the generalizability of the proposed method still needs to be enhanced for DV prediction for new unseen circuits.

In Liu et al. [6], the authors proposed to use three machine learning methods to predict circuit timing considering PSN effects and speedup the identification of risky test patterns. Given a set of automatic test pattern generator (ATPG) patterns (in .pat format), PSN-aware timing analysis is performed by either a commercial tool (for example, RedHawk), or a specialized tool such as IR-drop-aware timing analyzer (for example, IDEA). Then, machine learning is used to train a model. In this work, the authors tried three machine learning methods: (1) neural network (NN), (2) support vector regression (SVR), and (3) least-square boosting (LSBoost). It was found that NN is good at capturing the non-linear nature of timing prediction. SVR is a kernel-based method with sparse solution that saves memory and handles large number of data points. LSBoost method is a multi-model technique which recursively improves the prediction accuracy indicated by root mean square error (RMSE) and correlation coefficient (CC). After the training phase, given a new test pattern, it can predict the timing based on the model learned from the training phase.

Though machine learning is fast, it is difficult to predict circuit timing if the circuit size (dimension) and the number of test patterns (data points) are too large to be handled by modern machine learning techniques. To reduce the huge amount of raw data, the authors proposed to use four features: input/output transition (IOT), flip-flop transition in window (FFTW), switching activity in window (SAW), and terminal FF transition of long paths (PATH). SAW and FFTW are physical-aware features. The idea is based on the observation that IR-drop hot spots are usually clustered in a small physical region. With the feature dimension reduction, based on the experimental results, it showed that NN has best prediction accuracy and SVR has the least under-prediction. LSBoost uses only a third of memory as NN. SAW uses only one sixth of memory as IOT. The proposed method is more than six orders of magnitude faster than traditional circuit simulators. However, since it only used four benchmarks, the generalizability is not proved in the paper.

In Dhotre et al. [7], the authors proposed a flow to target the identification of worst-case test power-related issues by means of prediction. They proposed the use of a Machine Learning (ML)-based prediction mechanism to characterize the test power behavior of all tests, thereby avoiding the detailed resource-consuming analysis of all test patterns.

In this flow, first, a few pre-selected tests are accurately simulated as in the regular flow. These tests are referred to as training test vectors (T_t). The training test vectors as well as the corresponding analysis results, i.e. the simulation data, are then used to train an ML model f. Then, this trained model is used to predict the power analysis result of the set of prediction target test vectors (T_p) without an explicit simulation of T_p. This methodology allows us to predict the overall power profile of a test. The application is done in two different ways:

- The overall (global) power consumption of a test t is targeted and predicted to identify critical tests.
- Since a low global power consumption does not guarantee the absence of hot spots, the power consumption is related to the layout of the chip.

In this way, local hot spots can also be predicted. The proposed methodology has been implemented using various supervised learning algorithms such as Linear Least-Square (LLS) regression, Ridge Regression, Nearest Neighbors Regression, Neural Network Regression. The experimental results have been evaluated. The prediction time is very small compared to the actual accurate analysis time. At the same time, the predicted values are highly reliable and show only a small variance (<5%) to the actual values obtained by the accurate simulation. This enables the processing of all tests and their corresponding power profile prediction in order to find potentially power-unsafe tests. The proposed approach therefore increases the overall possibility to cover critical tests during the sign-off stage. Nevertheless, the experiments were only performed on IWLS benchmark circuits. Its application on large industry designs has not been investigated yet.

In Ma et al. [8] the authors proposed a Graph Neural Network (GNN) based solution for observation point insertion problem. The addition of the observation points increases the readability of faults in circuit under test (CUT). They represented the netlist as directed graph and formulated the observation point insertion as a binary classification problem i.e., for each node the problem is whether to add an observation point or not. Each node in the input graph contains a 4-d feature vector [LL, CC1, CC0, CO] where LL is the logic level of the node in the netlist and CC1, CC0 and CO are the SCOAP measurements. Every node has a binary supervision. 0 means easy to observe nodes and 1 means difficult to observe nodes.

To classify a node in the graph, the first step is to generate its embedding that contains node information as well as neighborhood information. Hence, an aggregator and encoder are used. Equation (30) shows the aggregator used in this work. A depth D is defined as a hyper parameter to mark the radius of the neighbors. They proposed to use weighted sum of embeddings as an aggregator with the different weights for predecessors and successors. Encoder consists of a nonlinear transformation to combine the current node information with the aggregated neighborhood information.

$$g_d^v = e_{d-1}^v + w_{pr} * \Sigma_{u \epsilon PR(v)} \, e_{d-1}^u + w_{su} * \Sigma_{u \epsilon SR(v)} \, e_{d-1}^u \qquad (30)$$

Once the embedding of each node is achieved, it is passed to a classifier for final prediction. In a typical circuit, the negative labels (easy to observe nodes) are more than the positive labels (difficult to observe nodes) in a typical circuit. It creates an imbalanced classification problem. So, they imposed larger weights on positive nodes such that the penalty of misclassifying them would be significant and used a multistage classifier that consists of four FC layers. This classifier filters out the negative nodes more confidently in each stage. To handle the training time on large circuits, they used a parallel training scheme with multiple GPUs in which each GPU processes one graph and outputs from all GPUs are combined to calculate the loss and then do backpropagation to update the model.

After training GCN predicts the possible hard to observe nodes. But each node could have a different impact on the circuit. So, to choose the node with the largest impact they proposed an impact evaluation method of each node i.e., reduction in the positive prediction of its fan-in cone. After selecting the node with the largest impact, GCN is again deployed to get the next observation point candidate list. This process is repeated until the maximum number of observed points are inserted or desired FC is achieved. Training dataset contains 4 circuits in which each circuit consists of 1.3–1.4 million nodes.

The inference runtime of this method on a circuit containing 1 million nodes is quite fast i.e., 1.5 s. However, as the supervision in this work comes from commercial DFT tools, this method has achieved the same fault coverage (FC) as of commercial tools but with 11% less observation points and 6% less pattern counts. Figure 20 shows the architecture proposed in [8].

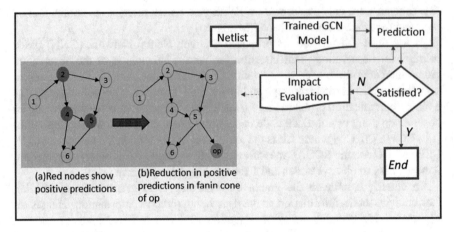

Fig. 20 Architecture proposed in Ma et al. [8]

The study in Sun et al. [9] showed the effectiveness of applying artificial neural networks (ANNs) in *test point insertion* process in order to predict the impact a candidate test point would have on the circuit in terms of increase in fault coverage (FC) under the pseudo random test.

ANN always takes input in the consistent form i.e., all sample input points should have the same dimensions and constant order but logic circuits vary in sizes and designs. To apply ANN to a logic circuit, they convert all gates in the circuit in a constant format of 2 input and 2 fanouts. The input of ANN is the subcircuit around a test point (TP) location. The size of the subcircuit is defined by two hyper params N and M. N indicates the number of logic levels before TP location and M indicates number of levels after TP logic level.

The input feature vector consists of CC, CO and gate type of each gate in the subcircuit. CC and CO are COP testability measures. First the CC value of the candidate test point is placed in the feature vector, followed by the CC values of the gates feeding the desired TP location, then followed by the CC values of the gates inputs feeding these gates and so on until the limit N is reached. The process is repeated starting from TP location and moving forward in the circuit till M levels. After that the CO values are placed in the feature vector in the same order and in the last gate type is placed using the same order.

Subcircuits around all test point locations are fed into the ANN and it outputs the increase in FC by activating the corresponding TP in the circuit. The TP that increases the maximum FC is selected and the process is repeated until one of the following conditions is met.

- The desired FC is reached
- The desired number of TPs are inserted
- Computational cost is reached
- No FC increases by remaining TPs.

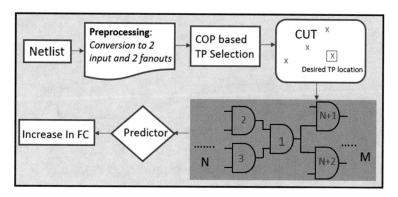

Fig. 21 Architecture proposed in Sun et al. [9]

To generate training data, the authors marked arbitrary locations in the dataset as candidate test points locations, extracts subcircuits around those locations and calculates COP values for all gates in subcircuits. The supervision of ANN is the increase in the FC after activating the candidate TP in the circuit. This supervision is achieved using fault simulation in which 2^i patterns are simulated where i indicates the number of inputs of subcircuits.

To avoid complexity in the solution they train three separate and identical ANN models for control-0, control-1 and observation points. ISCAS'85 and ITC'99 benchmarks are used in this study for training and inference. The TPs selected by this solution achieved the superior FC as compared to COP based TPs and the TP evaluation time is also less than COP based TP selection method. Figure 21 shows the workflow proposed in [9].

The solution in [10] used the same design of ANN as proposed in [9] to predict the signal probabilities (logic of gate being 1) of all gates for VLSI circuits with many reconvergence fanouts. For primary inputs, circuits are assigned a fixed probability value of 0.5 and for the remaining nodes, the subcircuit around desired controllability location (the gate whose probability is to be determined) is given to ANN as input. The subcircuit is extracted starting from desired location to N levels backward. Feature vector of each gate in the circuit contains the following information (Fig. 22):

- Gate type
- Probabilities value (either calculated by ANN or fixed for primary inputs)
- Boolean Value

Boolean value in the feature vector shows if the gate output is repeated in the sub-circuit more than once. This adds a bias for fanouts in the model. For supervision, every node contains actual probabilities values, that is achieved by simulating 10,000 random patterns on circuit. They trained their model on ISCAS'85 bench-marks. The trained model is evaluated on all ISCAS'85 and ITC'99 circuits. TPs

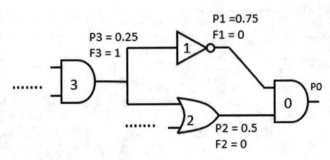

Input Feature Vector: G0, P1,F1,G1,P2,F2,G2, ...
Output: PO

Fig. 22 Sample sub-circuit with corresponding input and output feature according to method proposed in Millican et al. [11]

selected by using the predicted controllability from this model give higher FC as compared to FC by TPs selected using COP.

In [11] the author further applied the ANN design proposed in the [9] to evaluate the TPs in LBIST to increase delay fault coverage. The quality of LBIST usually suffered from *Random Pattern Resistant (RPR) f*aults (i.e., the faults that are only detectable from a small set of vectors). An example of RPR fault is stuck-at 0 output of 32-bit AND gate. To detect this fault, all inputs of AND gate must be one but the probability of this to happen is very low i.e., $(0.5)^{32}$.

To obtain the training data, pairs of random patterns are applied on a sub-circuit without TP and when TP is activated to check the increase in delay fault coverage. To control the training data generation time each vector is simulated with the probability describe in Eq. (31).

$$p\,(vector\;simulated) = p_v = \prod\nolimits_{\forall i \in I} \left(1 - \left(1 - C_v'\right)^V\right) \tag{31}$$

So, each pair of patterns (v1, v2) is applied $p_{v1}.p_{v2}$ times. Again, in this work, the authors trained three separate ANNs for Control-1, Control-0 and Observe Point. During the inference, the ANN predicts which TP is more likely to increase the maximum FC. Hence, ANN can replace the actual fault simulation required to select the best TP among all candidate TP.

Sometimes obtaining large training data (many circuits) is difficult when owners of data are unwilling to share due to privacy concerns. To overcome this issue and to get large training data, the generate random DAGs and convert them in to the circuits. A given DAG is converted into circuit by (1) making all vertices with no input *PIs* (2) making all vertices without output *POs*. (3) making all other vertices with randomly selected gate type (if input of vertex is 1 then it is randomly converted into buffer or inverter).

This author trained their model on both post-synthesis logic netlists of the ISCAS'85 and the ITC'99 benchmarks and random circuits. The authors randomly

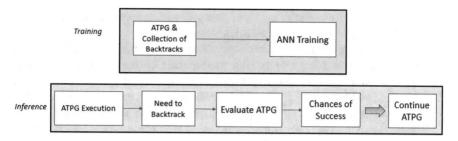

Fig. 23 Pipeline of solution proposed in Roy et al. [14]

divided these circuits into train and test sets. The proposed model achieved the higher transition delayed fault (TDF) and stuck-at-fault (SAF) coverage as compared to conventional solutions [12, 13]. ANN trained on random circuits gives the higher TDF and less SAF as compared to benchmark trained ANN. but still its performance was better than heuristic based TPI [12, 13].

Roy et al. [14] explored the flexibility of ANN to replace the conventional heuristics to decide the backtracking direction in Automatic Pattern Generation Algorithm (ATPG). Figure 23 shows the pipeline of solution proposed in [14]. ATPG algorithms are used to generate a set of test vectors that when applied to logic circuits can identify faults. Finding the set of vectors that capture all possible faults is an NP-hard problem. ATPG algorithms use heuristics to reduce the test generation time. This work replaces the conventional backtracking heuristics in ATPG with artificial neural networks (ANNs), and observes its impact on ATPG CPU time.

The ANN used in this work consists of an input layer (feature vector), a single hidden layer and an output layer. For a particular feature line, the input feature vectors consist of (1) Gate Type as one hot encoding (2) COP based CC(controllability of logic being 1) (3) COP based CO(observability). (4) LL of circuit line being traced (Normalized value with maximum depth).

The idea is, in ATPG while doing backtracking on a particular signal line, first evaluate ANN, it predicts the probability of backtracking on this given line will result in a successful test. When backtracking through a gate with multiple inputs, the ANN is evaluated at each circuit input, and the input with higher predicted value is selected.

Training data is generated by recording the history of successful (test found) and unsuccessful backtracking (*PI* assignments are undone) during PODEM based ATPG execution on benchmark circuits c6288, b05, and c3540 for 100 hard to detect faults. Only hard to detect faults are used for training data generation because easy to detect faults may not require large backtracking. These hard to detect faults are selected using COP based detection probabilities ($CC \times CO$ for stuck-at 1 and $(1 - CC) \times CO$ for stuck-at 0). Faults with low probabilities are selected.

After training ANN, authors compared the performance of *(1) heuristic based PODEM 1 untrained ANN (2) ANN trained with different combinations of input features (3) ANN trained with all input features* with respect to total number of backtrackings. Setting 3 outperforms all configurations.

In the second experiments, authors compared the performance of their model for same 100 hard to detect faults with distance based and COP based heuristics in terms of number of backtracks, # of backtraces and CPU time on circuits c6288, b04, c432, b08, b03, and b01 from ISCAS'85 and ITC'99 benchmarks. In this experiment, their model outperformed all circuits except b04 (slight decrease in performance).

The third experiment is performed on all stuck-at faults in 21 evaluation circuits and 3 training circuits from the ISCAS'85 and ITC'9 benchmarks. The result shows that for most circuits, the proposed model reduced the number of backtracks but a smaller number of backtracks didn't consistently result in fewer backtraces. Also, CPU execution time is less but this is not as much as expected, which means the evaluation time of ANN is high.

Roy et al. [15] further improves the solution proposed in [14]. The authors proposed to use *Principle Component Analysis* (PCA) as a preprocessing step to reduce the dimensionality of input features and generate new dataset with maximum variance (minimum information loss). In this work, 8 input features are used for all signal lines in the three training circuits, c6288, c3540 and b05. Feature vector consists of fanout (0 if the signal line has single destination and 1 if signal line has multiple destination), Gate type (normalized value), COP (CC, CO), SCOAP (SC0, SC1, SO) and Distance that shows the shortest path between signal line and PIs.

PCA is performed on this data to get the eigenvector for each dimension. As n dimensional input data will produce at maximum n eigenvectors and n corresponding eigenvalues. These vectors form a $n*n$ matrix T. Mean adjusted feature data is multiplied with T to produce a new *n-dimensional* vector of principal components (*PC*) for the corresponding line. Similar transformation is applied to all lines of the circuits. To reduce the dimensions of the new dataset, a subset of p data elements is selected out of n. In this work, they used *Pearson Correlation Coefficient* (PCC) to get the value for p. *PCC* found that out of 8 input features SC1, SC0 and Distance are strongly correlated. So, two out of these three dimensions can be dropped. Hence the final feature vector consists of a 6-dimensional vector. This new dataset is used to train ANN proposed in previous work [14]. Similarly, during inference, these features are computed for each line, and the values are transformed into PCs, in a similar way as was done for training circuits.

The authors compared the performance of previous ANN [14] and new ANN solution on ISCAS'85 and ITC'99 benchmark circuits and found that for most of the circuits, new ANN performs well than the previous one in terms of # of backtracking, # of backtraces and CPU runtime of ANN.

Pradhan et al. [16] provided a mechanism to predict the sensitivity of X (unknown) logic values on fault coverage of circuit. Digital circuits are likely to suffer from various issues such as uncertain timings, inadequate sensor feedbacks, limited controllability of past states. These kinds of issues induce an unknown logic value X at various nodes in the circuit that impacts the controllability and observability of the circuit and results in loss of fault coverage because the X values on circuit nodes can prevent the fault sensitization and propagation. The authors

Fig. 24 A sample circuit with P1, P2 and P3 cones

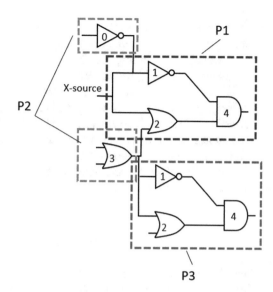

predicted X value sensitivity on fault coverage using support vector regression solely based on few structural features of circuits. This analysis is helpful to a chip designer while deciding for fixing X value.

This work focuses on a single stuck-at-fault model and defines sensitivity of X on fault coverage in the form of the *detectability loss* (DT) i.e., percentage of undetected fault in the presence of a certain X value. They used SVR for prediction of DT due to each X value. The authors partitioned the circuit into three disjoint sets as shown in Fig. 24.

- Partition P1 consists of X-cone (fanout cone of X source) in which X source can prevent the sensitization and propagation of faults in P1.
- Partition P2 is the sub-circuit that can propagate a signal to the X-cone that means X source can also prevent the propagation of faults in P2 to output ports.
- Partition P3 is the rest of the circuit that means X source has no impact on the faults in P3.

From these partitions it is clear that X nodes have impact only on the behavior of P1 and P2 cones. By analyzing the different circuit structures from ISCAS'89 and ITC'99 benchmark, the authors derived the following set of the input features that captures the behavior of P1 and P2 cones as much as possible.

- *Number of Nodes in P1 (nP_{1nodes})*: This shows the size of P1. This is normalized by all number of nodes in the circuit.
- *Number (n_{D1}) of Nodes in X-Depth-1 (D1)*: X-Depth means the shortest path to a node from X-source and X-Depth-1 (D1) means the nodes that are 1 depth away from X-source. So, n_{D1} in D1 is the number of fanouts of X source. This value is normalized with respect to total number of inputs.

- *Average Level (A_{D1}) of Nodes in X-Depth-1 (D1)*: This value is computed by taking the average of depth of all nodes in X-Depth-1. This feature is normalized with respect to the average depth of the output ports in the X-cone.
- *Maximum Level (M_{D1}) of the Nodes in X-Depth-1 (D1)*: This shows the maximum depth of nodes in X-Depth-1. The maximum level among the output ports belonging to X-cone of the concerned input is used as normalization factor.
- *Number of Outputs Ports (nP_{1op}) in P1*: This value is normalized with respect to the number of output ports in the circuit.
- *Normalized Sum of Levels (NP_{1op}) of the Output Ports in X-cone*: The output port at different levels contributes differently to DT loss. This normalized sum is defined by Eq. (32).

$$NP_{1op} = \frac{\Sigma_{u\epsilon \, P_{1op}} \, level(v)}{\Sigma_{u\epsilon \, M} \, level(v)} \tag{32}$$

Where, M is the subset of number of output nodes in circuit such that $|M| = nP_{1op}$. While picking this subset, output nodes on higher levels are selected.

- *Average Ratio of Depth-to-X-Depth (C_oP_2)*: For a node v, this shows the ratio of original depth of v (shortest path to v from PI) to the X-Depth of v (shortest path to v from a X-source). This value is normalized by total number of nodes in the circuit.
- *Influence of P2 on the X-Cone (I_{P2})*: This feature indicates the number of gate inputs in P1 that are coming from P2 and normalized with the total number of gate inputs in P1.
- *Maximum X-Depth (MP_1) Among the Nodes in XCone*: The impact of an X-source is likely to decrease with the increase in X-depth of a node. This value is not normalized because it is independent of circuit size.
- *Categorical Feature*: Some circuits netlists contains an iterative structure. The iterative structure effect the DT-Loss due to the presence of an X-source differently. So, a binary feature is added to indicate such structures. The X-sources extracted from the iterative structure are assigned value 0 and the rest of X-sources are assigned value 1.
- *X-Depth-Level Ratio (DLRO) of Output Ports in X-Cone*: It is computed using Eq. (33).

$$DLRO = \frac{\left(\Sigma_{u\epsilon \, P_{1op}} X - depth \, (v) \right)}{level(v)\Big)/nP_{1op}} \tag{33}$$

- *Number of Input Ports Influencing X-Cone*: It is expressed as the total number of input ports that are feeding X-cone and normalized with total number of input ports in circuit.

Inverters, buffers, and the logic gates, "XOR" and "XNOR are X-insensitive. So, these vertices do not contribute as node to any input feature calculation. As circuits vary in sizes, all features are normalized by the total number of nodes in the circuit so that all input features become comparable across all circuits in the dataset. This work uses the Kahn's [17] algorithm and breadth-first search (BFS) for circuit traversal to generate input feature set. As it requires the single circuit traversal, the complexity for feature generation is $O(|E| + |V|)$. Once all the features of each X-node are obtained then a SVR model is trained that predicts the DT-Loss of each X-input in a circuit. The input consists of N data points (N number of X sources obtained from all training circuits) and corresponding DT-Loss (obtained by TertraMAX) for supervision.

This work used the different circuits from ISCAS'85 and ITC'99 as train and test set. Once the model is trained, the predicted DT-Loss is graded (from higher to lower order). To analyze the grading, the author proposed to use Kendall's tau distance method. The result showed that this proposed method fairly graded the X-sensitivity in terms of DT-Loss but in very less time as compared to actual ATPG and Fault simulator tools.

Chen et al. [18] reports an experience of a novel test selection approach for verification of a commercial processor. The objective of this work is to analyze the applicability and effectiveness of novel test detection in a practical environment. Full chip functional verification relies on extensive simulation. This verification involves test generation, checking and coverage collection. To generate the test that gives the desired coverage is a challenging and costly task. For processor verification, *constrained random test generators* are used in which users provide constraints and biases in the form of test templates and directives to the test generator. The idea of novel test detection is to learn a model that can select the novel tests (that provides the huge coverage as compared to the rest of the tests) from a large pool of tests generated by a constrained random test generator. Then only selected novel tests are applied for verification that reduces the simulation cost.

The authors implemented the approach proposed in [19] in a company's in-house simulation environment for verification of a dual-thread low-power 64-bit power architecture-based processor and focused on toggle coverage of the Complex Fixed Point unit (*CFX*) and the Load Store Unit (*LSU*).

This approach proposed in [19] is based on a SVM model to check the similarity between simulated test and unsimulated test. The function used to check the similarity is given by Eq. (34).

$$M(T) = \sum_{1 \le i \le m} \alpha_i K (T, t_i) - \mathrm{p} \tag{34}$$

Where T is the un-simulated test. t_i is the simulated vector and α_i shows the importance of test t_i and p is the constant that shows the boundary of the measured outlier value for a test T. If $M(t)$ is negative that means the test T is different from t to t_m. For a particular test T, a large negative value shows the test is more novel.

$K(T; t_i)$ is the kernel function that actually measures the similarity between T and t_i. In this work, tests are assembly programs. Hence, a kernel function K is used to measure similarity between pairs of assembly programs. The work in [19] proposed a graph-based kernel function. That works by transforming one test graph into another using several operations (insertion, deletion, and substitution of vertices and edges) and calculates the cost of this transformation because each operation has a definite cost that is defined in the cost table. This cost shows the graph edit distance (GED) of the two graphs. The larger distance indicates the programs are dissimilar. The cost of each operation can be design dependent or unit dependent. It takes days or even weeks to understand the behavior of each instruction with respect to desired coverage matrix and then costs are assigned to each operation. Experiment showed that this approach reduces the simulation time for CFX coverage by up to 80%. But the creation of a cost table is a major drawback of this approach.

To avoid the drawback of graph-based kernel approach, the author proposed a following coverage-based kernel function describe in Eq. (35).

$$K_c\left(t_i, t_j\right) = \frac{\mid S_{ti} \cap S_{tj} \mid}{\mid S_{ti} \cup S_{tj} \mid} \tag{35}$$

Where t_i and t_j are tests and S_{ti} and S_{tj} are subsets of covered items by t_i and t_j, respectively. During the inference, the coverage S_{tj} by un-simulated test is not available. So, the authors build an instruction coverage database that contains the coverage information against each instruction. For a given un-simulated test T consisting of a sequence of instructions, for each instruction I they retrieve the coverage from the database based on the instruction instance that is closest to the instruction I. The closeness is calculated using Hamming-distance. The union is performed on coverage by all instructions in a test T to compute the test T coverage. But this kind of union does not consider coverage contributed by multiple instructions collectively. To overcome the limitation of single-instruction database, the authors build a new database in which test program instances consist of three instructions. Then, they use the coverage information stored in this 3-instruction database to estimate the coverage of test programs with a longer length.

Novel test detection is an iterative process as shown in Fig. 25. During each iteration, the novelty for test selection is determined with respect to coverage on undetected areas. To achieve this, the authors applied weight $w_i = 1$ on each item in coverage set S. Every time an item is covered, w_i is adjusted to w_i/a where a is a hyperparameter. Such a weight adjustment scheme reduces the importance of a covered item gradually. This approach resulted in up to 96% potential saving in simulation time.

From the results, the authors observed a special test that caused a high jump in the coverage. To understand this special test and generate more special tests, the authors proposed a feature-based diagnosis scheme in which features such as instruction types, operand values and the changes of these values in a program are derived for each test. They analyze the features and extract the rules that state the unique property of the special test. The authors manually inspected the extracted

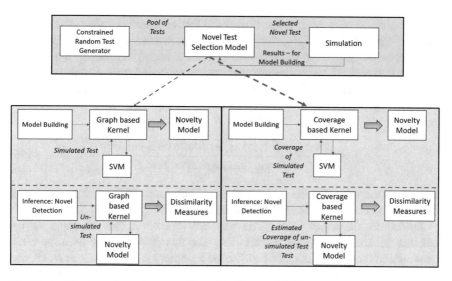

Fig. 25 Architecture of experiments performed in Chen et al. [18]

rules and used them to modify the test template that is then used to produce more tests similar to the special test.

Chen et al. [20] proposed a multi goal reinforcement learning (RL)based solution for test program generation to test a MIPS 32 processor (PUT) in a software based self-test environment. Test program consists of a set of instruction sequences that brings the processor to a fault sensitizing state. This work focused on a transition delay faults (TDF). They used Constraint ATPG (CTAPG) to collect the fault sensitizing states (FFS) S_f for each fault $f \in F$. After collecting the FFS for each fault f, they trained an RL model to predict which instruction sequences can set registers to a desired values specified by FSSs. They only considered the registers that are relevant to reach FSSs and carrying control signals. So, in total 46 registers were found to be relevant.

Before training the RL model, the first step is to encode the processor's behavior (possible FFS). The PUT state s is represented as $s = v1v2v3\ldots$, where v_i indicates the logic value of the i-th relevant register. 2-bit one hot encoding is used to represent the logic value. As this work involves 46 processors, the total length of state encoding is 96. They also encoded the fault sensitizing goal space g_f, for each $f \in F$ into G. Each fault sensitizing goal contains the goals to set each relevant register to the desired value. For example, a particular g_f is represented with encoded S_f for each register bit, and a goal to control the i-th register (goal state) is represented as $00\ldots00vi$, $1hot00\ldots00$. Here, only i-th register is set to v_i. It gives high reward regardless of other registers.

The encoded process is then followed by an RL learning algorithm that consists of a generator R* (agent). The agent produces the instruction sequence for each fault-sensitizing goal g_f. It maps current PUT state s_t and learning goal g to the future rewards R(st, i, g) for each candidate i in the instruction set I of PUT.

To train the generator R∗, the authors used a neural network. NN takes the encoded current PUT states concatenated with the encoded current goal states as its input and produces the estimated future total rewards for each instruction candidate $i \in I$. For reward, they used the formula describe in Eq. (36).

$$r_{gf} = \frac{(2 * \#of\ registeres\ set\ to\ desired\ values)}{\#of\ relavent\ registers\ for\ f} - 1 \qquad (36)$$

R∗ will get reward of 1 if the generated instructions makes PUT reach S_f, i.e., set all the registers relevant to sensitize f to the desired values specified by S_f .

Sometimes the set of instructions may fail to reach a FSS for some fault f, but may be able to sensitize another fault. For this purpose, they used the multi goal reinforcement learning that first choose a primary fault f and S_f to reach, and interacts with the environment accordingly to get experiences then samples another set of goals $G1 \subset G$ and replay the original experiences to get additional experiences from the perspectives of the other goal $g \in G1$.

For all the faults that haven't been sensitized during the learning process, the authors included its corresponding sensitizing goal g_f for replay. Also, they keep track of how many times relevant registers are set to desired values. For those, who haven't been successfully set for adequately enough times, are included for replay. The trained estimator R* is used to generate fault sensitizing sequences then do fault simulation to check the undetected faults. For the fault propagation, the authors use the strategies from [21, 22]. If the required fault coverage is not satisfied, the authors deploy fault constraint extractor (FCE) to extract the constraints from the collected traces for the CATPG of the next iteration.

The experiments are performed on full-forwarding, single-issue MIPS32 processor that has 25, 847 cells. The hyperparams adjusted for NN based RL model are batch size, and 0.3 discount factor is used to ensure the predicted reward is a finite number. The fault coverage achieved by this work is 94.94% without any area overhead but it is 2.57% less than the full scan-based approach. This work has shown some preliminary results on MIPS32, but results are not better than existing solutions and the feasibility of this solution is yet to be explored for modern processors (Fig. 26).

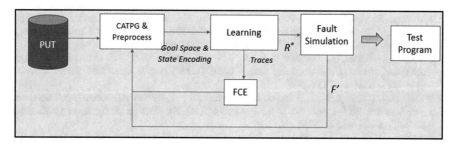

Fig. 26 Architecture of solution propose in Chen et al. [20]

5 Conclusion

In this chapter, we first introduced the basic concepts of machine Learning, followed by some popular algorithms for machine learning. Next, we reviewed some published papers in recent years that use machine learning to solve real problem in test. With machine learning some DFT and ATPG related problems can be solved in a more efficient way.

References

1. Cho, K., van Merrienboer, B., Gulcehre, C., Bahdanau, D., Bougares, F., Schwenk, H., Bengio, Y.: Learning Phrase Representations using RNN Encoder-Decoder for Statistical Machine Translation. arXiv:1406.1078 (2014)
2. Schuster, M., Paliwal, K.K.: Bidirectional recurrent neural networks. IEEE Trans. Signal Process. **45**(11), 2673–2681 (1997)
3. Li, Z., et al.: Test-Cost Optimization in a Scan-Compression Architecture Using Support-Vector Regression. In: VLSI Test Symposium (VTS) (2017)
4. Zorian, A., et al.: Machine Learning-Based DFT Recommendation System for ATPG QOR. In: International Test Conference (ITC) (2019)
5. Wu, C.-H., et al.: Deep Learning Based Test Compression Analyzer. In: Asian Test Symposium (ATS) (2019)
6. Liu, Y., et al.: PSN-aware circuit test timing prediction using machine learning. IET Comput. Digit. Tech. **11**(2), 60–67 (2017)
7. Dhotre, H., et al.: Machine Learning-based Prediction of Test Power. In: European Test Symposium (ETS) (2019)
8. Ma, Y., et al.: High Performance Graph Convolutional Networks with Applications in Testability Analysis. In: 2019 56th ACM/IEEE Design Automation Conference (DAC), pp. 1–6 (2019)
9. Sun, Y., Millican, S.: Test Point Insertion Using Artificial Neural Networks. In: 2019 IEEE Computer Society Annual Symposium on VLSI (ISVLSI), pp. 253–258 (2019). https://doi.org/10.1109/ISVLSI.2019.00054
10. Immanuel, J., Millican, S.K.: Calculating Signal Controllability using Neural Networks: Improvements to Testability Analysis and Test Point Insertion. In: 2020 IEEE 29th North Atlantic Test Workshop (NATW), pp. 1–6 (2020) https://doi.org/10.1109/NATW49237.2020.9153082

11. Millican, S., Sun, Y., Roy, S., Agrawal, V.: Applying Neural Networks to Delay Fault Testing: Test Point Insertion and Random Circuit Training. In: 2019 IEEE 28th Asian Test Symposium (ATS), pp. 13–135 (2019). https://doi.org/10.1109/ATS47505.2019.000-7
12. Ghosh, S., Bhunia, S., Raychowdhury, A., Roy, K.: A novel delay fault testing methodology using low-overhead built-in delay sensor. IEEE Trans. Comput-Aided Des. Integr. Circuits Syst. **25**(12), 2934–2943 (2006)
13. Tsai, H.-C., Cheng, K.-T., Lin, C.-J., Bhawmik, S.: A Hybrid Algorithm for Test Point Selection for Scan-based BIST. In: Proceedings of 34th Design Automation Conference (DAC), pp. 478–483 (1997)
14. Roy, S., Millican, S.K., Agrawal, V.D.: Machine Intelligence for Efficient Test Pattern Generation. In: 2020 IEEE International Test Conference (ITC), pp. 1–5 (2020). https://doi.org/10.1109/ITC44778.2020.9325250
15. Roy, S., Millican, S.K., Agrawal, V.D.: Principal Component Analysis in Machine Intelligence-Based Test Generation. In: 2021 IEEE Microelectronics Design & Test Symposium (MDTS), pp. 1–6 (2021). https://doi.org/10.1109/MDTS52103.2021.9476085
16. Pradhan, M., Bhattacharya, B.B., Chakrabarty, K., Bhattacharya, B.B.: Predicting \$\{X\}\$ - Sensitivity of circuit-inputs on test-coverage: a machine-learning approach. IEEE Trans. Comput-Aided Des. Integr. Circuits Syst. **38**(12), 2343–2356 (2019). https://doi.org/10.1109/TCAD.2018.2878169
17. Kahn, A.B.: Topological sorting of large networks. *Commun. ACM.* **5**(11), 558–562 (1962)
18. Chen, W., Sumikawa, N., Wang, L., Bhadra, J., Feng, X., Abadir, M.S.: Novel Test Detection to Improve Simulation Efficiency — A Commercial Experiment. In: 2012 IEEE/ACM International Conference on Computer-Aided Design (ICCAD), pp. 101–108 (2012)
19. Chang, P.-H., et al.: Online selection of effective functional test programs based on novelty detection. In: Computer-Aided Design (ICCAD), 2010 IEEE/ACM International Conference on, pp. 762–769 (2010, November)
20. Chen, C.-Y., Huang, J.-L.: Reinforcement-Learning-Based Test Program Generation for Software-Based Self-Test. In: 2019 IEEE 28th Asian Test Symposium (ATS), pp. 73–735 (2019). https://doi.org/10.1109/ATS47505.2019.00013
21. Zhang, Y., Li, H., Li, X.: Automatic test program generation using executing-trace-based constraint extraction for embedded processors. IEEE Trans. VLSI Syst. **21**(7), 1220–1233 (2013)
22. Singh, V., Inoue, M., Saluja, K.K., Fujiwara, H.: Instruction-based self-testing of delay faults in pipelined processors. IEEE Trans. VLSI Syst. **14**(11), 1203–1215 (2006)

Machine Learning Support for Logic Diagnosis and Defect Classification

Hans-Joachim Wunderlich

1 Introduction

Logic diagnosis involves both fault detection and fault localization, while defect classification addresses identifying the root causes of a malfunction, for instance resistive bridges and opens. Fault localization and defect classification are highly relevant for improving a design, its layout, the manufacturing process, and finally, the yield. Complex circuits in recent technologies are subject to high variations of their physical and geometrical parameters, which in turn lead to variations in their electrical behavior and timing profile. The move towards the most recent FinFET technologies has also been motivated by the excessive increase of variability in planar CMOS but cannot stop this trend completely [15]. Fault detection, fault localization, and defect classification are particularly complicated in the presence of variations, since they introduce indeterminism and uncertainty. For test and diagnosis, it is challenging to distinguish additional delays due to variations from small delay faults caused by defects, which may affect reliability [23].

Variability is usually faced by robust design techniques for avoiding or masking errors, but the resulting robust designs complicate fault detection and diagnosis further. These techniques mitigate the effects of some transient errors caused by indeterminism. In addition to external causes like particle strikes, transient errors have more and more intrinsic causes like metastability or synchronization errors between different clock regions and voltage islands. Robust designs tolerate certain marginalities which are benign, and test should not compromise yield through possibly false alarms. Hence, rare and random erroneous test results have to be distinguished from intermittent and permanent faults. Intermittent faults are

H.-J. Wunderlich (✉)
University of Stuttgart, Stuttgart, Germany

© The Author(s), under exclusive license to Springer Nature Switzerland AG 2023
P. Girard et al. (eds.), *Machine Learning Support for Fault Diagnosis of System-on-Chip*, https://doi.org/10.1007/978-3-031-19639-3_4

especially hard to be diagnosed as they appear to be random while they have systematic but unknown activation conditions.

As detailed diagnosis and physical failure analysis come with considerable efforts and costs, as much information as possible should already be extracted from the production test results. Machine learning helps during this phase to monitor and evaluate the production test results before initiating more expensive diagnostic steps.

There is no single machine learning scheme which fits best for all of these various challenges. In the next section of this chapter, we discuss attempts to distinguish maleficent defects from benign variations. The following section presents a neural network-based fault classifier. The final section uses Bayesian networks for classifying intermittent, transient, and permanent faults.

2 Variations Versus Defects

A defect is defined at the physical level and characterized by additional, missing or wrong material at a certain location, and a fault is a model of such a defect in the structural view, while an error describes a wrong output in the behavioral view. Additional material, for instance, could lead to a bridging fault if the material were metal, or it could lead to an open fault, if gate oxide as an insulator were placed in a via. Similarly, missing metal could result in an open fault, and a missing insulator could be modeled as a bridging fault. Obviously, the amount of missing or additional material has some impact as well, and we observe *resistive* opens and *resistive* bridges [33, 41, 49] which mostly result in a wrong timing and in delay faults. Wrong material, the combination of missing and additional material, is often mapped to a parametric fault as well, for instance in the case of a wrong doping profile.

Also, in the correct, fault-free case, we observe a wide spread of the behavioral parameters of components like threshold voltage, sub-threshold swing, resistance, leakage or delay [13, 60–62]. The significant increase in variability was one of the reasons to move to new CMOS technologies like FinFET apart from the need of denser scaling, yet the new technologies can only decelerate the increase but do not stop it. We still face the challenge to differentiate between the effects caused by defects and the delays due to variations.

2.1 Marginalities and Reliability

Marginalities due to resistive defects tend to increase during the operation of a circuit. For example, gate-oxide defects can result in small delays, which increase after some time and lead to gate-oxide breakdown [34]. This process differs from the usual aging mechanisms like Bias Temperature Instability (BTI), Hot-Carrier Injection (HCI), or Electro-Migration (EM), which shift parameters in transistors and interconnects and degrade the circuit performance over its entire lifetime [37], since it evolves much faster and will result in so-called *Early Life Failures (ELF)*. Already the Stanford ELF-project reported that a large portion of marginal hardware is manifested as *Small Delay Faults (SDF)* [34] which may not be covered by a standard delay test. If these marginalities were not detected before shipping, additional costs for recall and repair have to be expected. Moreover, diagnosis and possible actions for reliability improvements would be delayed, and the delayed improvements of the process and design will cause additional costs.

Testing for small delay faults can be considered as a cost-effective means to rule out or at least reduce the need for an expensive burn-in process during manufacturing test and has gained increased attention [34, 57], especially for FinFET technologies [40]. A small delay fault may not be detectable by a standard delay test if it can be propagated only along short paths, and it will be undetectable even by advanced timing-aware Automatic Test Pattern Generation (ATPG) algorithms. Faster-than-at-Speed Test (FAST) targets these so-called Hidden Delay Faults (HDF) by overclocking the circuit, typically using multiple frequencies up to three times higher than the nominal frequency [1, 36, 66]. Silicon experiments demonstrate the effectiveness of this strategy [14, 58].

FAST significantly increases the reliability of shipped products, but to achieve a quick yield ramp-up of a manufacturing process, it must be combined with proper diagnosis techniques. Here, the goal is to assess the severity of the reliability problems and distinguish defective chips from chips that are slow due to parameter variations in order to avoid unnecessary yield loss. In previous works, this problem was addressed for SDFs due to resistive opens and gate-oxide defects [54, 66]. The proposed solutions exploit the observation that slow chips show a different behavior than defective chips for varying supply voltages.

Identifying the root cause of a slow gate under variations is essential for optimizing both yield and reliability. In FinFET, metal-gate-granularity (MGG), and line-edge-roughness (LER) which consists of fin-edge-roughness (FER) and gate-edge-roughness (GER) are the dominant sources of variations [60], whereas random-dopant-fluctuation (RDF) and oxide-thickness-variation (OTV) have less impact. MGG has some impact on both the sub-threshold swing and the threshold voltage, while LER mainly causes variations of the threshold voltage. Both lead to changes in the timing behavior but are usually not considered as reliability threats in contrast to resistive defects, even if both introduce a similar additional delay.

Resistive opens are usually due to incorrect contacts, insufficient overlaps or even cracks in some wires. They often exhibit some increase in the current density and tend to enlarge over time. *Resistive bridges* may increase electromigration, and the resistance may decrease further. They may cause leakage and hence thermal heating which affects the reliability not only of the respective component but of an entire region on the chip. To increase reliability, circuits with some resistive defects have to be sorted out, but circuits just slow due to variations have to be kept in order not to sacrifice yield.

In the next subsection, we analyze the behavior of a cell with and without a resistive open under variations and show that the differences are large enough for an accurate classification based on simulation data. If such a cell is deeply embedded in a combinational circuit, masking effects will apply, and more complex machine learning schemes are needed to achieve still an amazingly accurate classification of the circuit. These findings are based on the material published in [42–44]. All the results are based on simulation data, and a sufficiently precise simulation model is required to transform them into volume test data in production.

2.2 Variation-Aware Defect Characterization at Cell Level

Let us illustrate the problem to be solved in this subsection with the help of the NAND-gate from FinFET Free PDK 15 nm [39] in Fig. 1 which contains a resistive open and transistors with variations. In Fig. 2, we pick three specific cases from the results of a Monte Carlo HSPICE simulation. The lower black curve shows the delays of the NAND gate under varying supply voltages V_{dd}, the delays decrease with increasing V_{dd}. The applied random parameters are within the specifications and lead to a fast instance of this gate. If the parameters are changed but still within the specification, we receive a slow NAND gate whose behavior is reflected in the dashed green upper curve. If we insert the open of Fig. 1 with an appropriate resistance into the fast NAND gate, the result is the dotted red curve. For each operation voltage V_{dd}, this curve is within the specified performance range of the slow and the fast NAND instances, and testing at a single voltage value cannot detect this defect. However, the shape of the red curve differs from the two fault-free cases sufficiently to allow a machine learning-based classification.

This effect can be analyzed further just with textbook information. According to [15], the delay τ depending on the supply voltage V_{dd} of a CMOS transistor can roughly be expressed by

$$\tau(V_{dd}) = Const \frac{V_{dd}}{(V_{dd} - V_t)^\alpha} \tag{1}$$

Fig. 1 NAND-gate with a resistive open and transistors under variations

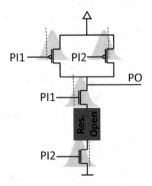

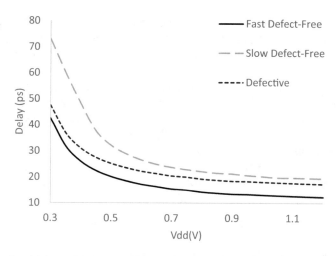

Fig. 2 Simulated delays $\tau(V_{dd})$ vs. V_{dd} for three instances of a NAND gate

Here, the constant *Const* depends on the capacitive load at the gate output, the gate oxide thickness and permittivity, the channel length and width, and the carrier's mobility. Sakurai's index α can be taken to 1 in current technologies, and V_t is the transistor threshold voltage. For FinFETs, threshold voltage fluctuation and sub-threshold swing are the major effects of parameter variations [16, 61, 62]. Since circuit timing is mainly affected by the threshold variations, we investigate the effects by assuming $V_t' = V_t + \delta$. Then the new delay will be

$$\tau'(V_{dd}) = Const \frac{V_{dd}}{(V_{dd} - V_t')^\alpha} = Const \frac{V_{dd} - \delta + \delta}{(V_{dd} - \delta - V_t)^\alpha}$$

$$= \tau(V_{dd} - \delta) + Const \frac{\delta}{(V_{dd} - \delta - V_t)^\alpha} \qquad (2)$$

The term $\tau(V_{dd} - \delta)$ corresponds to a shift of the black curve to the right in Fig. 2, while adding $Const \frac{\delta}{(V_{dd} - \delta - V_t)}$ shifts the curve upwards scaled by $\frac{\delta}{(V_{dd} - \delta - V_t)}$ and finally results in the dashed green line.

A resistive open, however, will mainly affect the effective resistance R_{eff} in the usual RC-mode $\tau = C_l * R_{eff}$, and for $R_{eff}' = R_{eff} + \delta$ we get the new delay $\tau' = C_l * (R_{eff} + \delta)$. Hence, a resistive defect of size δ will shift the black curve of Fig. 2 upwards by $C_l * \delta$ resulting in the dotted red curve. As already mentioned, it is not possible to detect such a small resistive open just by a single measurement at only a single voltage value, yet measurements at multiple voltages may allow to reconstruct the curve $\tau(V_{dd})$ and to classify a cell as defect or defect-free.

The qualitative discussion above only applies to resistive opens, and the analysis of resistive bridges will be more complicated. A resistive bridge introduces a new component into a cell which connects at least two existing components and can also affect the behavior of neighboring cells. Such defects disturb a larger region of the circuit and may also lead to significant changes of the delay curve $\tau(V_{dd})$ to be classified.

In order to distinguish an admissible shape of $\tau(V_{dd})$ from defective ones the cell library has to be augmented by the description of these curves for some variation corner cases. The description of the different shapes of $\tau(V_{dd})$ can be stored either in the form of tables as described in [42], or by the coefficients of polynomials approximating the functions $\tau(V_{dd})$ [55].

In the next section, we explain how test results of combinational circuits can be classified by machine learning algorithms to distinguish between admissible and defect indicating delay curves of embedded cells.

2.3   Circuit Level Defect Characterization

In general, a cell under test is not directly accessible, and the function $\tau(V_{dd})$ will not be observable. A slow cell may be embedded into a combinational circuit, in which all the other cells show variations as well. This leads to indeterministic behavior and complicates classification. As seen above, rather small changes of the delays of a defective cell have to be propagated under varying voltages and are subject to the usual masking effects which are *logical masking*, *electrical masking* and *timing or latch-window masking*.

To detect a delay, a propagation path has to be sensitized, and appropriate test patterns have to be applied to avoid logical masking which denotes blocking such a propagation path either by a controlled gate or due to reconvergencies. CMOS is a robust, self-restoring technology that filters short pulses and reshapes the slopes of transitions. This effect is called electrical masking, but analog simulations show that the propagated timing information is often sufficient for classification. All the cells on a propagation path suffer from variations such that a signal may not arrive at a flipflop in the appropriate timing window to be captured. This effect is referred to as timing or latch-window masking. Timing masking is hard to be analyzed, since in general the propagation time of a path has to be modeled in a computationally expensive way by a skewed multi-variable distribution [59].

The effects of electrical and timing masking are illustrated in Fig. 3 for the NAND gate of Fig. 1. The histograms show Monte Carlo analog simulation results both for an isolated cell and using the cell as the initial point of a chain of 16 inverters with varying fanout loads.

At nominal voltage, the delays of the defect-free and defective NAND cells lead to rather separated histograms (Fig. 3a), and the delay could be used already for classification at the cost of some yield loss. If the behavior of the cell is observed at the end of the inverter-chain, the histograms overlap significantly, and in most of the delay bins we find both defective and defect-free cells (Fig. 3b). Figure 2

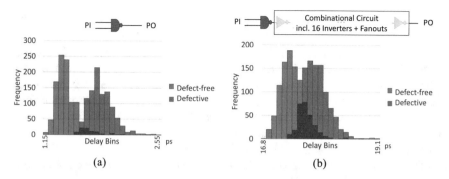

Fig. 3 Delay histograms of a defective and defect-free NAND-cell. (**a**) Direct observation of the cell. (**b**) Observation at the end of an inverter-chain

indicates that this will happen for any feasible voltage, and a more fine-grained analysis requires measurements at multiple voltages.

In today's leading edge systems, adaptive voltage and frequency scaling (AVFS) is used for power and energy optimization, aging reduction, and lifetime increase as well as for allowing robust operation under variations. AVFS designs require to execute both speed binning device characterization and manufacturing test at multiple operation points.

Usually, speed binning is executed as a functional test of the device, in order to determine the maximum frequency $F_{max}^d(v_{dd})$ the device can operate with. In AVFS systems, multiple operation voltages are selected from a set V_{op} and applied during speed binning. Hence, we obtain a set of measurements M_d for each device $d \in D$ which contains the maximum frequency obtainable by some voltages $v_{dd} \in V_{op}$. The tuple

$$M_d = (F_{max}^d(v_{dd}) \mid v_{dd} \in V_{op}) \tag{3}$$

describes the features to be used for supervised training of a machine learning scheme. Each tuple has to be labeled, if it corresponds to a defective die or to a defect-free one. This information is generated during production test for the training samples.

Only defects which create a new or extend an existing critical path have an impact on $F_{max}^d(V_{dd})$. Different random instances may also have different critical paths, and the critical paths may also change for different supply voltages V_{dd}.

In general, defects affecting the critical path occupy only a rather small portion of the device area. If defects on short paths have to be covered, too, so-called faster-than-at-speed-test techniques (FAST) have to be employed [1, 14, 32]. A launch-off-shift or a launch-off-capture technique is used for storing the output values at higher frequencies for a given delay test pattern set T in a scan chain. Equation (3) turns into

$$M_{d,o} = (F_{max}^{d,o}(V_{dd}) \mid V_{dd} \in V_{op}) \tag{4}$$

which describes the maximum frequency where the output cone o of device d is passing the complete test T for a given voltage V_{dd}. Adding the results from diagnosis for FAST [30] provides all the information required for supervised learning. The samples M_d or $M_{d,o}$ can also be generated by timing simulation with high resolution, and the results of fault simulation can be used for supervised learning. The simulation-based data set may serve as an initial training set to be completed by real data during manufacturing later on.

The next subsection describes an appropriate learning scheme for device classification based on the samples M_d or $M_{d,o}$, respectively.

2.4 Random Forests for Resistive Open Classification

In this section, we do not use different notations for the tuples M_d and $M_{d,o}$ but consider them just as a set of instances I generated either by measurements and diagnosis after production or by accurate timing simulation. The maximum frequencies are transformed into the critical path delays, $\tau_i(V_{dd}) = 1/F_{max}^i(V_{dd})$, and the tuples

$$\Theta_i = (\tau_i(V_{dd}) \mid V_{dd} \in V_{op}), i \in I \tag{5}$$

are now subject to further investigations.

Each Θ_i has to be classified if it is probably the response of a defective circuit, or if the timing measurements are better explained by some variations. There is a plethora of machine learning schemes available for classification, and selecting an appropriate one depends on both the computing complexity and the expected accuracy. Below, we first discuss the sources of classification errors and conclude the so-called Random Forests can be an appropriate scheme. Next, we discuss, how decision trees and random forests can be constructed for our classification problem. Finally, in this section we explain the generation of the training sets in detail.

2.4.1 Classification Errors

Let us assume, we have generated a set of training instances $\Theta = \{\Theta_i\}_{i \in I}$ with a label $l(i) = N$ for defect-free circuits and $l(i) = P$ for defective ones, and we have to select a supervised learning scheme for classification. The types of classification errors of such a scheme can be irreducible errors, errors due to bias, and variance errors. *Irreducible errors* have their source in properties of the available data set. In our example, test coverage and diagnosis may not be able to provide an accurate labelling, or the resolution of the timing simulator may be too low for generating realistic data always.

Variance and bias errors are not caused by an imperfect data set, but they are properties of the learning schemes. *Variance* describes the sensitivity of the scheme on the selection of the training set. Drawing different training sets from the same population may result in great changes of the learned model. An example of a scheme with high variance errors is a Classification and Regression Tree [12].

Bias errors are due to the assumption on the structure of the model to be learned. An example is to learn coefficients for linear regression while the underlying problem is a non-linear one. Strong assumptions on the model's structure reduce the variance errors but may result in a large bias. On the other hand, weaker assumptions increase the sensitivity to the training set and hence the variance errors. For our problem, we are looking for a robust learning scheme, which reduces both the variance and the bias errors.

The next section briefly describes the use of decision trees. After this, an ensemble learning technique is introduced to balance the bias and variance errors.

2.4.2 Decision Trees

Decision Trees or in modern terms Classification and Regression Trees (CART) are widely used due to their flexibility and adaptability. Yet this flexibility comes with a rather high variance error. In this section, we briefly describe the construction of CARTs, and in the next section, we describe random forests as one example of so-called ensemble techniques to reduce the variance error.

We restrict the entire presentation to binary classification as it is needed for differentiation between defects and variations. A decision tree corresponds to a binary tree, each non-terminal node n points to a subset $I(n) \subseteq I$ of instances and one voltage $v_{dd}^n \in V_{op}$ determining the feature variable $(\tau_i(v_{dd}^n))_{i \in I}$. At an internal node n, the instance set $I(n)$ is partitioned into

$$I(n_r) \cup I(n_l) = I(n), \ I(n_r) \cap I(n_l) = \emptyset$$

for the left child n_l and the right child n_r by using some heuristics. Each partitioning should provide more homogenous subsets, and at each leaf n_{leaf} we want to observe a set of instances $I(n_{leaf})$ which is as pure as possible, preferably that all the instances assigned to a leaf node have the same label, i.e. $\forall_{i \in I(n_{leaf})} l(i) = P$ or $\forall_{i \in I(n_{leaf})} l(i) = N$.

We create this tree starting from the root node $n = n_{root}$ and $I(n_{root}) = I$, and at each node we have to decide if it will be a terminal node, and otherwise we have to select the feature $v_{dd}^n \in V_{op}$ and the partitioning $I(n_r)$, $I(n_l)$. The heuristic for splitting $I(n)$ has to be a measure for homogeneity, and it is usually the Gini coefficient. For our binary problem it is defined as

$$G(I(n)) = 2 * p_P(I(n)) * p_N(I(n)) \tag{6}$$

where

$$p_P(I(n)) = \frac{|\{i \in I(n)|l(i) = P\}|}{|I(n)|} \tag{7}$$

is the portion of instances i in $I(n)$ with label $l(i) = P$, and $p_N(n)$ is the portion with label N. If $I(n)$ is completely homogeneous, the coefficient is $G(I(n)) = 0$, and if on the contrary it is completely mixed, the value is $G(I(n)) = 0.5$.

In order to partition $I(n)$, we investigate each voltage, search for a delay threshold d_n, put instances slower than d_n at this voltage into one subset, the faster instances form the other subset, and we search the threshold which yields the best two subsets. Finally, the feature for splitting corresponds to the voltage which delivers the overall best results.

In a more formal way, for each voltage $v \in V_{op}$ and each delay d we can split the set of instances into

$$I^+(n, v, d) = \{i \in I(n)|\tau_i(v) \geq d\}$$

and

$$I^-(n, v, d) = \{i \in I(n)|\tau_i(v) < d\}.$$

The best possible split at node n for a voltage $v \in V_{op}$ can be evaluated by

$$\text{split}(v) = \min_d \left(G(I^+(n, v, d)) * \frac{|I^+(n, v, d)|}{|I(n)|} + G(I^-(n, v, d)) * \frac{|I^-(n, v, d)|}{|I(n)|} \right) \tag{8}$$

which can be interpreted as the weighted average of the Gini coefficients of the created two partitioning sets, $d_{v_{dd}^n}$ is the corresponding delay where $I(n)$ is split. We set now

$$v_{dd}^n = \arg \min_{v \in V_{op}} (\text{split}(v)), \tag{9}$$

$$I(n_l) = I^-(n, v_{dd}^n, d_{v_{dd}^n}),$$

and

$$I(n_r) = I^+(n, v_{dd}^n, d_{v_{dd}^n}).$$

We observe that a value $v \in V_{op}$ may occur multiple times along a branch. The algorithm stops, if $I(n)$ is a homogeneous set of instances, however, we may want to stop earlier to avoid overfitting and for efficiency reasons. Possible criteria are the size of $I(n)$ which should not fall below a certain minimum, or the depth of the tree.

The label $l(j)$ of a new instance j is predicted by following the tree along one branch, and at each node n, the value of $\tau_j(v_{dd}^n)$ decides to follow the left or the right child, until the leaf set is reached. The majority of the labels in the leaf set provides the prediction for instance j.

Already from the construction procedure it can be seen that such a tree is very sensitive to the training set and prone to bias errors. The next section provides a method for increasing the robustness of this approach.

2.4.3 Classification with Random Forests

The basic idea of random forests is to overcome the variance errors of a decision tree by using multiple of them and decide about the result by voting. This approach will only succeed, if there is some diversity between these trees, and an ensemble of weak estimators will in general perform better than a few stronger ones. Even if some of the weak estimators suffer from overfitting, their collection may perform quite well.

Bagging is an abbreviation of Bootstrap Aggregation and can be used to generate multiple data sets from a limited original set. For each tree t we create a new set I_t out of the original set I of instances by randomly drawing instances from I with replacement. Usually, we generate sets of the same size as the original one, $|I_t| = |I|$, and we use the new sets I_t for training different trees t.

The second technique for generating diversity applies to Eq. (9) when we split a node n according to the minimum over $v \in V_{op}$. Instead of using the entire feature set V_{op}, we draw randomly a subset $\tilde{V}_{op} \subset V_{op}$ and set

$$V_{dd}^n = \arg \min_{v \in \tilde{V}_{op}} (\text{split}(v)). \tag{10}$$

As a rule of thumb, $|\tilde{V}_{op}| = \lfloor \sqrt{|V_{op}|} \rfloor$ is often used.

The trees generated this way may be weaker than an original decision tree, but they are sufficiently diverse and outperform a single tree if an appropriate number of trees is applied. There is no general rule for pre-computing the optimal number of trees, and usually the number is increased until accuracy saturates.

2.4.4 Generating Training Sets

For generating training sets, two cases have to be considered. Initially, the training sets have to be created by accurate timing simulation, preferably at switch level. Later on, manufacturing data can be added and used as well.

For simulation, standard cell characterization measures and models characteristics like propagation delays, pin transition times, power consumption, leakage or setup/hold constraints. In [44], an industry-standard component model is used, which is calibrated against measurements from industrial 14 nm FinFET technology

[4, 45]. It considers silicon-validated parameter distributions to accurately model process variation in the transistor, including transistor gate length, fin thickness, fin height, gate dielectric thickness, and the impact of MGG on the work function of the gate. To model standard cells under the impact of process variations, Monte Carlo electrical simulation has to be performed while the corresponding model parameters are varied. The simulations have to provide results for each supply voltage $V_{dd} \in V_{op}$ used later on, and eventually a large population of cell instances is generated.

In addition, models for defective cells have to be generated. Based on inductive fault analysis [56] or so-called cell-aware test as the commercial counterpart [28], resistive opens of various sizes are injected and simulated. The models of defective cells are created under variations as well, and eventually defect-free and defective circuit net lists are available for high-resolution timing simulation. In a defective net list, one correct cell is replaced by a defective one, and the defect size is selected large enough to be visible at the circuit output, but small enough to keep the circuit timing still in specification. If the defect size were too large, the device would fail the test or characterization procedures directly, and there would be no need for any further classification. If the defect size were too small, defective and defect-free circuits would behave in the same way, and any characterization would be impossible. Circuit instances $i \in I$ which are generated by the responses of a defect-free net list get the label $l(i) = N$. Instances generated by a defective net list receive the label $l(i) = P$.

Creating the training sets from real test data is much easier. For a defect-free device, Θ_i collects just the measurements of a chip passing the entire test procedure, and it is labeled $l(i) = N$. For a defective chip, the data is taken from the measurements of $F_{max}^{d,o}$, if these values are still in specification but later the chip fails the complete production test due to an error at output o, and the label is $l(i) = P$.

2.5   How Far We Are

We have shown how to create a machine learning model to distinguish the responses from a defect circuit with resistive opens and the responses of a defect-free circuit under process variations. The model can be generated either from precise timing simulation data or from results of manufacturing test and diagnosis.

In the next section, we extend the approach in order to distinguish different fault models at gate level.

3   Neural Networks for Defect Classification

The previous section tried to classify a circuit at gate level and decide, if its timing behavior is explainable by variations or if it indicates the presence of a resistive open defect. This section extends the range of possible defect mechanisms. As already pointed out in the previous chapters, logic diagnosis comprises essentially three tasks:

1. *Fault detection* is implemented by applying a test set created by an appropriate algorithm for automatic test pattern generation (ATPG). The algorithm has to take into account realistic defect models, inductive fault analysis (IFA) mentioned above [56] is able to determine the possible faulty behaviors of a cell and to compute its likelihood based on the critical area of a design. This can be used to create a user-defined fault model for each cell, characterize the cell, and store the relevant patterns at the cell input in the library. In cell-aware testing, ATPG is controlled to generate patterns at the cell inputs and to make the cell responses observable [28].

2. *Fault localization* requires more efforts, since the test pattern sets need not only to detect the faults, but they have to distinguish them. In recent years, quite effective and efficient algorithms have been developed [21, 24, 29–31, 53, 65], some of them are described in Chap. 2.

3. *Fault characterization* or *fault analysis* deals with the extraction of the defect behavior at a victim line, and the analysis of its root cause. Physical failure analysis (PFA) analyzes a device by physical means after a defective region has been identified is rather an expensive task.

Below we present a machine learning technique to extract as much knowledge as possible about a defective cell from the test and diagnosis results of steps 1 and 2, in order to reduce the need for additional information about the defect and the costs of PFA in step 3. The material presented below is based on [25], readers with deeper interest may also see [52].

Figure 4 shows an application scenario for saving PFA effort. Usually, the test data obtained by a failing die is analyzed by the diagnostic procedures described in the previous chapters. The die will be subject of a second test application step with diagnostic patterns and may be passed to expensive PFA (Fig. 4a). The last two steps would be superfluous, if already diagnosis information from the production test data allowed to identify the defect class, and the frequency of this type of defects were still within the expectations. In Fig. 4b, a classifier uses both the test response data and the diagnostic results from volume test for determining the probable defect class, and a second diagnostic test application pass including PFA is only required if the frequency of this class is a reason of concern.

For step 3 above, more general fault models are required like the conditional line flip model [64], the user defined fault model [28], or pattern dependent faults [9] which allow considering timing, indeterminism or layout neighborhoods [21, 65]. Layout-aware diagnostic approaches provide a good first approximation to understand the underlying defect. When a large population is available, they can be successfully combined with yield learning [10].

Next, we present a procedure to match the observed behavior with a fault model based on volume test outcomes and standard logic simulation. This will avoid the

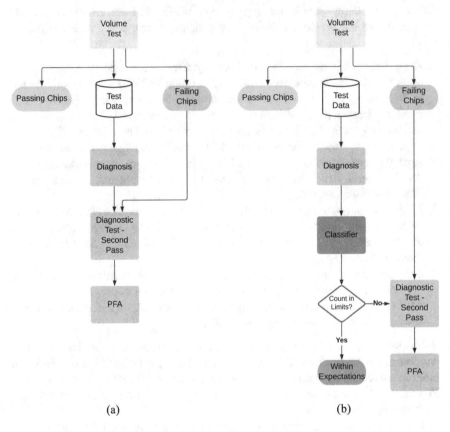

Fig. 4 Test flow. (**a**) Standard test flow. (**b**) Test flow with classifier

need for waiting for PFA results for every faulty chip, while it provides more detailed information and may speed up PFA if required.

3.1 Defect Classification Overview

Defects can be classified according to various fault models as described above, and the goal of the classification procedure is to assign a detected and localized fault to one of the already predefined fault classes. In addition to the widely used stuck-at faults, these classes may include:

- *dominant-and bridge:* due to an unwanted (resistive) bridge, two lines are connected, and the victim line is set to the AND value of the aggressor's and its own value.

- *dominant-or bridge:* due to an unwanted (resistive) bridge, two lines are connected, and the victim line is set to the OR value of the aggressor's and its own value.
- *byzantine bridge:* there are cases where one line is a victim and may flip erroneously, and cases where the other line is a victim and flips erroneously, if the two lines have opposite values.
- *slow-to-rise gates:* the gate output shows a longer than expected rise time.
- *slow-to-fall gates:* the gate output shows a longer than expected fall time.
- *crosstalk delays:* a victim line is delayed because of the switching activity of a neighboring line in the opposite direction.

The classification procedure is based on a preprocessing step during the design phase which consists of ATPG and logic simulation of the generated test set T. The fault coverage obtained by ATPG has an impact on the classification accuracy later on, and appropriate tools are available from both academia and commercial vendors.

For each possible victim line with fault f, ATPG delivers the set $T_f \subseteq T$ of detecting patterns. Below we define certain features, which are extracted from the fault-free behavior of the circuit if the patterns T_f are applied. They can be determined by logic simulation of the patterns T_f and are stored as a feature vector V_f. Finally, an Artificial Neural Network (ANN) is trained to classify the fault based on the vector of features corresponding to T_f. An overview of the training process is depicted in Fig. 5.

During volume test, the ANN is applied as shown in Fig. 6. Now, T_f is the set of the actually failing patterns from the applied test set T. The complete test set has already been pre-simulated, and the failing patterns are used for feature selection and computing the feature vector V_f. The features resulting from the test results are passed to the ANN which delivers a prediction of the corresponding fault class. Since the circuit is subject to variations, the actual detecting test set $\tilde{T}_f$ may be different to any of the test sets used for training.

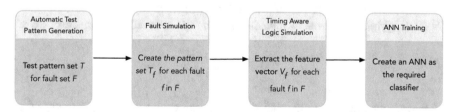

Fig. 5 ANN training for fault classification

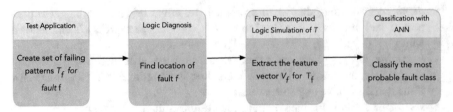

Fig. 6 Classification with volume test results

3.2 Feature Extraction

We follow the feature extraction method used in [25] for a victim line vl and a test pattern set T_f for a fault f at this victim line. The supervised learning scheme targets the fault classes mentioned above which do not include the stuck-at faults as they can be classified easily by matching the observed test outcome with the generated test set under any variations. The method is based on information obtainable from the circuit topology and from logic simulation of both the complete test set T and the test pattern set T_f which detects the fault f. For training, the patterns are generated by ATPG, and for classification, they are observed during test later on. The following values are derived from the fault-free circuit and recorded as features:

1. *Ratio of failing patterns with victim line value 0:* If for all patterns in T_f the victim line value is 0 in the fault-free case, then the fault is detected if it is setting this line to 1, and there is evidence that the victim line may be driven by a dominant-or bridge. Let $T_f^0 \subseteq T_f$ be the subset of observed failing patterns which set the victim to 0 in the fault-free case, and set

$$vl@0 = \frac{|T_f^0|}{|T_f|}.$$

   Then vl@0 is a feature related to dominant-or faults.

2. *Ratio of failing patterns with victim line value 1:* In exactly the same way we set

$$vl@1 = \frac{|T_f^1|}{|T_f|}$$

   which is the ratio of failing patterns where the victim line is 1 in the fault free case among all the failing patterns. It is an indication of a dominant-and bridge.

3. *Ratio of falling transitions at victim line:* This feature is calculated as

$$vlFall = \frac{|T_{fall}|}{|T_f|}$$

   where

$$T_{fall} = \{t_i \epsilon T_f | vl(t_i) = 0 \wedge vl(t_{i-1}) = 1\}.$$

Hence, T_{fall} is the set of failing patterns at which the victim line should have value 0, and a transition from the previous cycle should have taken place. This feature is a strong indicator for slow-to-fall faults.

4. *Ratio of rising transitions at victim line:* Similarly, we define

$$vl\,Rise = \frac{|T_{rise}|}{|T_f|}$$

with

$$T_{rise} = \{t_i \epsilon T_f | vl(t_i) = 1 \wedge vl(t_{i-1}) = 0\}$$

which quantifies the evidence of a slow-to-rise fault.

5. *Driving gate inputs:* Let g be the driving gate of the victim line, and let $g_{in}(t)$ the values at the inputs of gate g if pattern t is applied. Features are added to include information events at the gate inputs. We collect six additional numbers from logic simulation. T_{c0} and T_{c1} are the portions of patterns for which at least one gate input changed but the gate output $g(t)$ remained constant. T_{s0} and T_{s1} are the portions of patterns for which the inputs and the output remained stable with the output value 0 or 1, respectively. T_{tf} and T_{tr} are the portions of patterns for which a transition at the gate inputs caused a falling or rising transition at the gate output, respectively. Their formal definition is:

$$T_{c0} = \frac{|\{t_i \epsilon T_f | g(t_i) = g(t_{i-1}) = 0 \wedge g_{in}(t_i) \neq g_{in}(t_{i-1})\}|}{|T_f|}$$

$$T_{c1} = \frac{|\{t_i \epsilon T_f | g(t_i) = g(t_{i-1}) = 1 \wedge g_{in}(t_i) \neq g_{in}(t_{i-1})\}|}{|T_f|}$$

$$T_{s0} = \frac{|\{t_i \epsilon T_f | g(t_i) = g(t_{i-1}) = 0 \wedge g_{in}(t_i) = g_{in}(t_{i-1})\}|}{|T_f|}$$

$$T_{s1} = \frac{|\{t_i \epsilon T_f | g(t_i) = g(t_{i-1}) = 1 \wedge g_{in}(t_i) = g_{in}(t_{i-1})\}|}{|T_f|}$$

$$T_{tf} = \frac{|\{t_i \epsilon T_f | g(t_i) = 0 \wedge g(t_{i-1}) = 1 \wedge g_{in}(t_i) \neq g_{in}(t_{i-1})\}|}{|T_f|}$$

$$T_{tr} = \frac{|\{t_i \epsilon T_f | g(t_i) = 1 \wedge g(t_{i-1}) = 0 \wedge g_{in}(t_i) \neq g_{in}(t_{i-1})\}|}{|T_f|}$$

6. *Ratio of unexplained patterns:* We define $T_f^u \subseteq T_f$ as those failing patterns which cannot be explained by flipping the victim line. The feature *unexplained* = $|T_f^u|/|T_f|$ reflects the situation that there must be at least one more culprit in the

circuit, one reason of this can be a bridging fault which affects two lines in an arbitrary way.

7. *Maximum ratio of transitions in the neighborhood*: The physical neighborhood $N_h(vl)$ of a victim line should be extracted from layout information, or as a surrogate by topological information from the gate net list. Logic simulation determines for each neighbor $v \in N_h(vl)$ the switching activity when applying the failing patterns T_f, and the maximum switching activity describes the feature:

$$max_{trans} = \frac{\max_{v \in N_h(vl)} |\{t_i \in T_f | v(t_i) \neq v(t_{i-1})\}|}{|T_f|}$$

The feature max_{trans} helps to point out crosstalk faults.

8. *Average ratio of neighbors with a different value than the victim line:* The feature

$$avg_{dist} = \frac{\sum_{v \in N_h} \sum_{t \in T_f} |\{v | v(t) \neq vl(t)\}|}{|N_h||T_f|}$$

computes the average of the ratio of failing patterns in which the neighbors in N_h have a different value than the victim line vl. This feature helps classifying bridges.

3.3   Neural Network Training

Artificial Neural Networks have already been introduced in Chap. 3. The exemplary feature selection described above delivers a vector X with values for the thirteen features resulting from step 1 to 8 above (step 5 delivers six features), if a test set T_f and a fault f are provided. Supervised learning is implemented based on the simulation results for the fault classes specified in Sect. 3.1. In [25] it is reported that a dense, fully connected sequential ANN is sufficient for an accurate classification. The output layer has as many nodes as the number of fault classes $C = \{c_1, \ldots, c_m\}$. The output of the last layer is a vector $F(X)$ of real numbers of length m, which is converted into a probability vector by using the softmax function:

$$P(faultclass = c_i | X) = \frac{e^{F(X)_i}}{\sum_{j=1}^{m} e^{F(X)_j}}$$

and the class c_i for which the corresponding probability is highest, is taken as the result.

3.4 Classification During Volume Test

The classification procedure depicted in Fig. 6 is placed after volume test and fault localization by logic diagnosis, in order to decide if further investigations by PFA is required. This is the case either when the ANN does not provide plausible and sufficiently certain results, or when there is an unexpected increase in the frequency of certain fault classes.

For each detected fault f, the detecting patterns T_f, the set of applied patterns T, and the fault location l are available. The feature values for this fault are extracted from the annotated circuit net list as a result of the precomputed logic timing simulation, and the instance generated this way is evaluated by the ANN. Both, feature extraction and ANN evaluation are extremely fast due to the pre-computations, and [25, 52] report accuracies up to 90% depending on the fault class. The accuracy can be improved further if additional features are included which consider not only the failing patterns T_f but also the set of passing vectors $T \setminus T_f$ [52].

It observed that the classifying ANN has not to be necessarily the same one which is trained for the respective circuit. Accuracy drops only slightly, if a neural network trained for one circuit is applied to a different one.

3.5 How Far We Are

Meanwhile, we learned techniques to distinguish effects caused by benign variations from effects due to circuit marginalities. If these marginal circuits are subject to logic diagnosis, suspect locations will be reported and are analyzed further in this section. A method has been presented to generate feature vectors based on the information of failing test patterns and the probable fault location. An ANN-based classifier has been presented, which matches the fault location with one out of a set of fault models. While the classifier obtains best results, if it is applied to the same circuit as it has been trained with, it can also be transferred to new circuits not yet investigated and provides satisfactory performance.

The next section will investigate, if the occurrence of such a fault is just a transient event or if there is an intermittent mechanism making such a fault a reliability threat.

4 Bayesian Belief Networks for Intermittent Fault Identification

The presence of process variations requires special efforts to ensure the robustness of a design. AVFS as introduced in Sect. 2.3 is only one technique to make systems resilient against variations, fault tolerance and fault masking are needed as well. These techniques complicate test and diagnosis, since not each unexpected behavior indicates the presence of a defect to be sorted out, and benign and maleficent behaviors have to be distinguished. Below we describe a technique for differentiating transient faults already tolerated by the design from permanent and intermittent faults. The material summarizes the results presented in [26], more details and results are found in [52].

Transient faults affecting complex systems are caused by various sources beyond the threat of particle strikes due to radiation. They randomly appear at victim nodes v and usually last for at most one clock cycle. Transient faults can be caused by external noise [8] but even more likely, delay problems occurring at clock boundaries or dynamic parameter variations such as power supply and interconnect noises, electromagnetic interferences and electrostatic discharges can lead to violations of timing safety margins which manifest themselves as transient faults [11, 17]. This type of faults is not caused by defects and they are meant to be tolerated during operation. Increasing parameter variations in nano-scale CMOS have expedited new strategies for "under-designed and opportunistic" computing [22, 27]. To fully exploit the potential of technology scaling, these approaches avoid an overly pessimistic design with large guard bands. Instead, they include mechanisms to detect and compensate a certain amount of transient faults caused by parameter variations or external noise [22, 46]. A structural test, however, will report an error, even if the behavior is benign, and yield loss will result.

Intermittent faults can be traced to unstable or marginal hardware and are activated by specific conditions, like a certain activity in the neighborhood, increasing temperature or decreasing voltage. Their observable behavior is similar to that of transient faults, but they occur regularly under certain conditions, and they may evolve over time such that they become as critical as permanent faults. Unexpected power droop, crosstalk, or temperature dependent effects are further examples which may lead to intermittent faults. They will repeatedly occur at the same location or in its neighborhood. Even though they are not permanent, they may severely affect the system functionality, indicate potential early life failures, or reduce robustness against transient faults.

Distinguishing transient and intermittent faults is particularly challenging. Transient faults are uncritical for appropriately designed "robust" circuits, and test failures due to transient faults cause unnecessary yield loss in this case. Intermittent faults impact quality, but they may lead to similar observations during test as transient faults. To control the trade-off between yield and quality, test and diagnosis procedures must be able to distinguish transient, intermittent, and permanent faults.

4.1 Modeling Intermittent Faults

For permanent faults, a variety of fault models has already been described in previous chapters. These models have both structural and functional aspects. The structural aspects relate to the locations of the defect in the circuits, while the functional aspects cover the erroneous behavior of the defective component. In addition to that, intermittent faults have activation conditions which do not depend only on the logic values and timing behavior of the defective cell, but also on physical, unknown or even random conditions of the component's neighborhood or environment. The conditional line flip (CLF) model provides means to deal with this complex situation [64]. Fault injection according to this calculus follows

$$line \oplus [condition]$$

where *line* represents the boolean value of a victim line, and *[condition]* is an expression which is true when *line* is affected by an error. The expression *[condition]* may cover a variety of circumstances:

- *[condition]* can be a boolean expression with arguments which are inputs of the cell driving *line*, including the timing conditions.
- *[condition]* can describe the instantaneous activities in the neighborhood of *line* for expressing bridge, power droops or crosstalk faults.
- *[condition]* can also describe physical parameters like temperature.
- *[condition]* may also include a random variable for describing events with an unknown or unpredictable cause. An activation rate λ can describe the random aspects of intermittent faults.

Let us consider a line $y = a \,\text{NAND}\, b$ to explain how most of the used fault models can be expressed as a CLF. The stuck-at-0 fault is $y \oplus [y]$, the stuck-at-1 fault is $y \oplus [\bar{y}]$, and a stuck-open fault where y may keep its previous value for the input pattern $(ab) = (01)$ can be expressed as $y \oplus [\bar{a}b\bar{y}_{-1}]$. Depending on the discrete timing resolution of a simulator, the expression y_{-1} may denote the value of y during the previous clock cycle or just during the previous time step, which can also model small delay faults due to resistive opens or bridges.

Static bridges do not consider timing and they usually involve two signal lines which can follow several different fault models. Rousset et al. [53] presents a taxonomy of these models which can be expressed by CLFs. For a bridge between line a and line b, we describe two functions $f_a(b) \in \{0, 1, \bar{b}, b\}$ and $f_b(a) \in \{0, 1, \bar{a}, a\}$ and inject two CLFs:

$$a \oplus [f_a(b)(a \oplus b)]$$

$$b \oplus [f_b(a)(a \oplus b)]$$

Table 1 is taken from [64] and shows the selection of f_a and f_b for relevant static fault models for bridges.

A more complex dynamic behavior can be expressed as well, for instance

$$b \oplus [(a_{-1} \oplus a)(a \oplus b)]$$

Table 1 Static bridge types as CLFs

$f_a(b)$	$f_b(a)$	Bridge type
0	0	Fault free
0	1	a dominates b
0	$\bar{a}$	a AND-dominates b
0	a	a OR-dominates b
1	1	a and b swap values
1	$\bar{a}$	b dominates a and a AND-dominates b
1	a	b dominates a and a OR-dominates b
$\bar{b}$	$\bar{a}$	Wired-AND
$\bar{b}$	a	b AND-dominates a and a OR-dominates b
b	a	Wired-OR

describes a crosstalk fault where aggressor a changes its value and produces a glitch at b. Beyond logic values and timing, physical and environmental parameters can also be considered. The extra-functional properties like temperature, switching activity of cores in the neighborhood or power consumption are usually not completely predictable and form a source of intermittent faults. In addition, there may be unknown and hidden relations so that the activation of these faults appear to be random. This is modeled by a boolean random variable x_λ with the activation rate or mean $E(x_\lambda) = \lambda$ and

$$y' = y \oplus [x_\lambda, condition] = \begin{cases} y & \text{for } x_\lambda = 0 \\ y \oplus [condition] & \text{for } x_\lambda = 1 \end{cases} \quad (11)$$

For testing a combinational circuit, there is a set of test patterns $T_{f_y} \subseteq T$ detecting a stuck-at fault $f_y \epsilon \{y \oplus [y], y \oplus [\bar{y}]\}$. More complex faults like delay faults, opens or crosstalk may require even sequences of patterns in T_{f_y}. It was explained in the previous chapters how the knowledge about T_{f_y} and the corresponding circuit responses are used for locating the victim line y. In case of an intermittent fault at line y with the activation rate λ, the set of failing patterns may be reduced from $T_{f_y, \lambda}$ to $T_{f_y, \lambda} \subseteq T_{f_y}$ which may complicate the localization of the failing line y. Yet robust diagnosis techniques will provide a ranked list of victim candidates, perhaps with reduced resolution, which will be the basis of further classification.

The next section describes briefly the application of the test set T for volume diagnosis and the corresponding responses for fault localization.

4.2 Session-Based Diagnosis

Volume diagnosis should interfere with the test flow as little as possible. The diagnostic procedures must be compatible with standard test architectures and response compaction schemes, as for example the widely accepted STUMPS

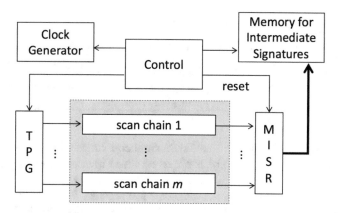

Fig. 7 STUMPS architecture

architecture [6]. For built-in self-test (BIST), a test pattern generator (TPG) and a multiple-input signature register (MISR) are added to the circuit under test (CUT). In each clock cycle a "slice" of a test pattern is loaded into the scan chains as shown in Fig. 7. When a complete test pattern is shifted in, the test response is captured in the flip-flops. While the response is shifted out and compressed by the MISR, the next pattern can be shifted in.

For embedded test, the TPG is replaced by an on-chip decompressor receiving encoded test data from the automatic test equipment (ATE), and the compacted test responses are sent back to the ATE [51]. If a transient fault occurs in the CUT, a failing test response may be obtained and stored in the flip-flops. The fault effect is preserved in the flip-flops during shifting and may lead to a faulty signature. Even if a transient fault lasts longer than one clock cycle, only the fault effect during the capture cycle has an impact on the signature. Therefore, in the sequel transient faults are assumed to have a duration of a single clock cycle only. In contrast to that, intermittent faults can be activated by several patterns over multiple cycles.

Because of the limited bandwidth in embedded test and the limited storage capacities in built-in self-test, the amount of response data to be analyzed is minimized for either test strategy. It is beyond the scope of this chapter to give a complete review of the state of the art in this extensively studied field. Both strategies partition the test into certain test sessions, and evaluate the compacted test response which is the basis of further diagnosis in the presence of an error. The diagnosis scheme of [63] repeats sessions with a failing signature in a special diagnostic mode without compaction, and the approach in [18, 19] can determine the fault location directly from the cumulative signature of a session and provides an additional reduction of the response data. In the sequel, we assume that diagnosis is directly performed either on the compacted output during embedded test or on the signatures, and present a technique using results from [2].

To distinguish between permanent and non-permanent faults, a session is repeated immediately in case of a mismatch between the observed and expected

Table 2 Encoding the outcome of a test session

Observation	Code	Interpretation
$S \neq S' = S_{REF}$	A	Transient or intermittent fault
$S = S' \neq S_{REF}$	B	Permanent or intermittent fault
$S \neq S_{REF}, S \neq S' \neq S_{REF}$	C	Intermittent fault activated by different patterns or multiple transient faults

response data. R_{max} denotes the maximum number of times a specific test session is executed if an erroneous outcome is observed. Since in most cases, the probability of an erroneous signature due to a transient fault is rather low, $R_{max} = 2$ is selected. Let S_{REF} denote the expected outcome of a test session, S the observed outcome and S' the outcome of a repeated session. With these notations, Table 2 shows how possible test outcomes are encoded.

Since multiple transient faults in consecutive test sessions are considered extremely unlikely, both case B and case C of Table 2 indicate the presence of an unacceptable fault, and the die has to be sorted out.

Another criterion can be applied to the remaining dice, in order to sort out those with an unacceptable error rate. Let N be the number of planned test sessions, let N_f be the number of failing sessions which are repeated, then $N - N_f$ is the number of sessions already correct at the first time. A robust design has to assume a certain maximum transient error rate μ and $q = (1 - \mu)^{n_t}$ is the probability that a test session with n_t test patterns is passed without any transient error. The probability of a specific outcome with N_f repeated sessions is $p(q, N_f) = (1 - q)^{N_f} q^{N - N_f}$. After some maximum likelihood estimation, $p(q, N_f)$ is maximum for $q = \frac{N - N_f}{N}$. If the number of failing sessions N_f leads to

$$\delta < (1 - \mu)^{n_t} - \frac{N - N_f}{N},$$

for a given $\delta > 0$, the die has to be considered defective as well. The threshold for δ depends on the user defined quality requirements.

While the techniques discussed above and the codes B or C of Table 2 allow for sorting out some defective dice under test, code A needs deeper investigations, a solution based on Bayesian belief networks is discussed below.

4.3 Some Background on Bayesian Belief Networks

Bayesian Belief Networks (BBN) provide an engineering framework for the analysis of statistical data. They are widely used in classification problems such as medical diagnosis applications. A problem is modeled by a collection of random variables and the dependencies between them, and a Bayesian network provides a graph representation of this model. The structure and inference strategies for fault classification by BBNs described here are following [5, 48]. The interested reader may find deeper information on BBNs in these textbooks.

A BBN is a formal framework to reason about the probabilities of events which are represented by random variables. For our purposes, we can restrict the random variables v, x to the discrete case. The random variable v takes its values from $dom(v)$, and for $a \in dom(v)$, the probability that v takes the value a is $p(v = a)$. We have

$$\sum_{a \in dom(v)} p(x = a) = 1$$

and we use $p(x)$ to describe a distribution which is a function

$$p(v) : dom(v) \rightarrow [0, 1[$$

$$a \mapsto p(v = a)$$

In many cases, $dom(v)$ is just the two valued Boolean Algebra and v is a Boolean random variable. In this case, we use the abbreviation $p(v = 1) = p(v)$. In general we assume that $v, v_1, \ldots, x, x_1, \ldots$ are just discrete random variables, and denote the number of states of variable x_i as $dim(x_i) = |dom(x_i)|$.

The conditional probability $p(v_1 = a | v_2 = b)$ quantifies the probability of the event $v_1 = a$ given that $v_2 = b$. The expression $p(v | x_1, \ldots, x_n)$ is the conditional distribution of v under given $x_1, \ldots, x_n$, hence it is a function with domain

$$dom(v) \times dom(x_1) \times \ldots \times dom(x_n)$$

into $[0, 1]$. Since for each fixed assignment of the variables $x_1, \ldots, x_n$ the values $p(v | x_1, \ldots, x_n)$ will sum up to 1, a brute-force representation of this distribution would consist of

$$(dim(v) - 1) \prod_{i=1}^{n} dim(x_i) \tag{12}$$

table entries, and BBNs allow to reduce this effort.

In a more formal way, a BBN describes a distribution of the form

$$p(x_1, \ldots, x_D) = \prod_{i=1}^{D} p(x_i | pa(x_i)) \tag{13}$$

where $pa(x_i)$ represents the set of parental variables of the variable x_i [5]. This distribution is represented as a directed acyclic graph where the arcs or edges point from a parental variable to a child variable, and the i-th node corresponds to the factor $p(x_i | pa(x_i))$. The arcs (v_1, v_2) usually express a cause-effect relation, in

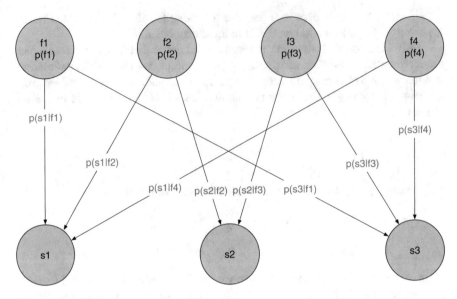

Fig. 8 Example of a Bayesian network for fault diagnosis

which event v_1 is a cause of event v_2, and they are annotated with the conditional probability $p(v_2|v_1)$.

Considering the Bayesian rule

$$p(v_1|v_2) = \frac{p(v_1 v_2)}{p(v_2)} = p(v_2|v_1)\frac{p(v_1)}{p(v_2)} \qquad (14)$$

we notice that reasoning in a BNN does not just follow the direction of the arcs in the graph, but information can be retrieved in both directions. "A priori" probabilities describe the initial probabilities for each node which may come from previous observations. "A posteriori" probabilities are computed by reasoning, after certain values are inserted into any subset of nodes of the BNN. The goal of Bayesian inference is to deduce these "a posteriori" probabilities of faults and use them to guide the diagnosis. Figure 8 shows an example BNN for a diagnosis problem.

The vertices f_1 to f_4 correspond to faults in a system. They are characterized by their "a priori" probabilities of occurrence $p(f_1)$ to $p(f_4)$. The vertices s_1, s_2, and s_3 represent the "symptoms" observed during a test. The edges are labeled with conditional probabilities, where $p(s|f)$ denotes the probability that symptom s is observed under the condition that fault f has occurred. In contrast to that, $p(f|s)$ is the probability that the fault f is a correct diagnosis for the symptom s.

Computing $p(x_i|x_1, \ldots, x_{i-1}, x_{i+1}, \ldots, x_D)$ in a direct way would be computationally extremely expensive, and graphical reasoning on the BNN structure is used to simplify the computation by mainly two techniques:

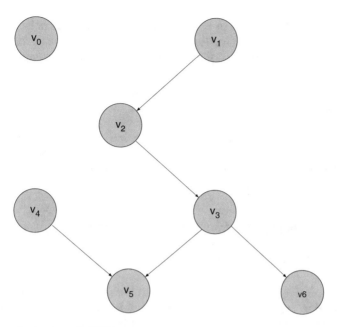

Fig. 9 Example structure of a BNN

1. If there is no path between x_i and some x_j in the undirected graph, the variable x_j does not have any impact on x_i and can be omitted.
2. If there are sets of independent variables, the joint distribution can be computed as the product of the distributions of the two sets, instead of the expensive evaluation as a table following Eq. (12).

Let us illustrate these techniques with the help of Fig. 9. Here, node v_0 is disconnected from v_5, and we have $p(v_5|v_0, \ldots, v_4, v_6) = p(v_5|v_1, \ldots, v_4, v_6)$ which illustrates case 1.

For illustrating case 2, we assume variable v_3 takes a fixed value $v_3 = v_3^*$. For each of such a fixed value, the sets $\{v_5, v_6\}$ and $\{v_1, v_2\}$ are independent, and we have

$$p(v_3 = v_3^*|v_1, v_2, v_5, v_6) = \frac{p(v_1, v_2, v_5, v_6|v_3 = v_3^*)p(v_3 = v_3^*)}{p(v_1, v_2, v_5, v_6)} =$$

$$\frac{p(v_1, v_2|v_3 = v_3^*)p(v_5, v_6|v_3 = v_3^*)p(v_3 = v_3^*)}{p(v_1, v_2)p(v_5, v_6)}$$

which cuts down the tables for four variables in Eq. (12) to two tables of two variables.

It has to be noted that injecting the additional knowledge into the BNN may not only remove dependencies as seen above for case 2, but it may also create new ones. Let us consider variable v_5 in Fig. 9 where v_3 and v_4 seem to be independent without any cause-effect relation. However, setting some fixed value for variable v_5 may create a new dependency. This is easily understood by an example where the BNN above models the random events in a digital circuit, and the nodes are just gate outputs. Assume we have $v_5 = XOR(v_3, v_4)$. By injecting $v_5 = 0$, we get $v_3 = v_4$ which is highly correlated.

This effect can be generalized by using the concepts of *D-separation* and *D-connection* [48] whose details are beyond the scope of this chapter. However, the principal way for asking certain queries, for instance the probabilities of some events or the consequences of some observations, should be clear by now.

4.4 Intermittent Fault Identification with BBNs

4.4.1 BBNs for Diagnosis

The network in Fig. 8 is typical for a diagnosis problem where multiple faults can occur simultaneously. If only a single fault is considered at a time, a simpler network can be used with one multi-valued random variable representing the possible faults (cf. Fig. 10) [50].

In the context of electronic testing, Bayesian networks have already been successfully used for the diagnosis of analog and power circuits, e.g. [3, 35, 38]. Also at the system and board level some pioneering approaches are available for coarse-grained diagnosis [7, 47, 67].

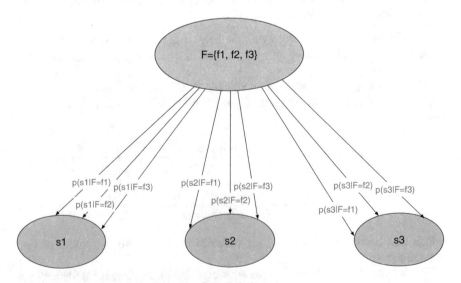

Fig. 10 Bayesian network for diagnosis of single faults

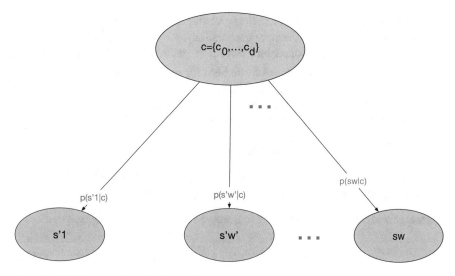

Fig. 11 BBN for intermittent fault classification

In [20] a proof of concept for system diagnosis in the presence of intermittent faults is described. Intermittent faults are represented by pairs of a priori probabilities: the probability of occurrence and the probability of faulty behavior in the presence of the fault. The diagnosis procedure observes the system for several time steps. At each time step, a Bayesian network is updated according to the observed behavior. The diagnosis strategy selects internal observation points, such that the "a posteriori" probabilities of a wrong diagnosis decrease quickly.

4.4.2   Intermittent Fault Classification

We assume, w test sessions are planned, and in case of an erroneous signature, $w' \leq w$ sessions are repeated and provide the correct signature after the second run, otherwise the die would be sorted out.

The $w - w'$ correct signatures and the w' faulty signatures of the first run are passed to logic diagnosis which creates a list of $c_1, \ldots, c_d$ fault candidates. From this, a multi-valued fault variable $c = (c_0, \ldots, c_d)$ is created, here c_0 represents the fault free case which could also be possibly affected by some transient noise. The situation can be modeled by a rather simple BBN of just two layers. The root consists of the fault variable c, and the leafs are the $w + w'$ signatures computed during test. Figure 11 shows the straightforward structure of the corresponding BBN, it has to be noted that the nodes with an erroneous result s' cause a repetition which may not happen in all the sessions, and $w + w' \leq 2w$ is expected. Altogether, the BBN has to evaluate $w - w'$ correct signatures and w' erroneous ones.

Now we want to determine the a priori probabilities annotated in Fig. 11. Let T be the complete test set of a session. For each fault c_j, $j = 1, \ldots, d$, the corresponding test set $T_{c_j} \subseteq T$ is partitioned into the set of patterns $T_{c_j}^0$ where the fault was active, and into $T_{c_j}^1$ which contains those tests for c_j, where the fault was inactive. Let μ be the error rate of transient faults, and let $\lambda > \mu$ the activation rate of intermittent faults. Let s' be an erroneous signature which can be explained by c_j. The probability to observe s' can be computed as a product of three factors:

- the probability that c_j was active during the detecting patterns $T_{c_j}^0$ which is $\lambda^{|T_{c_j}^0|}$,
- the probability that c_j was not active during $T_{c_j}^1$ which is $(1 - \lambda)^{|T_{c_j}^0|}$,
- the probability that a transient fault did not occur which is $(1 - \mu)^{|T|}$.

Putting this together, the probability to observe s' in the presence of fault c_j is

$$p(s'|c_j) = \lambda^{|T_{c_j}^0|}(1 - \lambda)^{|T_{c_j}^0|}(1 - \mu)^{|T|}. \tag{15}$$

If the fault c_j cannot explain the observed signature, then some transient noise has masked it leading to

$$p(s'|c_j) = 1 - (1 - \mu)^{|T|}. \tag{16}$$

If an intermittent fault is not present, we have

$$p(s_{REF}|c_0) = (1 - \mu)^{|T|}. \tag{17}$$

The BBN can be annotated with these conditional probabilities, and if a priori knowledge about the probabilities of each fault class is not available, we assume a uniform distribution setting the probability of c_j to $1/(d + 1)$.

After test application, we obtain an observed signature $s_i^{observed}$ for each session, and the a-posteriori belief of the fault variable is computed

$$p(c|s_i = s_i^{observed}, i = 1, \ldots, w + w') \tag{18}$$

which gives us the conditional probability for each single intermittent fault and for the fault-free or transient case in $c = (c_0, \ldots, c_d)$. Due to the extremely simple structure of the BBN, the values can be computed exactly by the means of [48].

In [52], an accuracy higher than 0.96 for intermittent faults and more than 0.66 for transient faults is reported. The classification of transient faults is obviously much more challenging, due to the much lower number of activating patterns which complicates both logic diagnosis and Bayesian classification.

4.5 How Far We Are

Innovative technologies allow higher performance and denser scaling, but they come also with increased threats of variations and reliability risks. Techniques for statistical learning can help significantly in diagnosis and yield-ramp up. Measuring chips at multiple operating points provides sufficient data to identify a benign behavior or a reliability threat leading to early life failures. After localization of a fault, the best matching fault model can be identified, even if there are uncertainties and indeterminism in the test outcome. Variations make robust designs mandatory which are resilient against certain transient faults. In these cases, distinguishing between tolerable transient faults and maleficent intermittent and permanent faults is essential for keeping the yield.

The plethora of machine learning techniques allows to select for the different challenges specifically tuned approaches. As examples, we introduced the application of random forests, artificial neural networks and Bayesian belief networks.

References

1. Ahmed, N., Tehranipoor, M., Jayaram, V.: A novel framework for faster-than-at-speed delay test considering IR-drop effects. In: Proceedings of the 2006 IEEE/ACM International Conference on Computer-Aided Design (ICCAD'06), pp. 198–203. Association for Computing Machinery, New York (2006). https://doi.org/10.1145/1233501.1233541
2. Amgalan, U., Hachmann, C., Hellebrand, S., Wunderlich, H.J.: Signature rollback - a technique for testing robust circuits. In: 26th IEEE VLSI Test Symposium (VTS 2008), pp. 125–130 (2008). https://doi.org/10.1109/VTS.2008.34. ISSN: 2375-1053
3. Aminian, F., Aminian, M.: Fault diagnosis of analog circuits using Bayesian neural networks with wavelet transform as preprocessor. J. Electron. Test. 17(1), 29–46 (2001). https://doi.org/10.1023/A:1011141724916
4. Amrouch, H., Pahwa, G., Gaidhane, A.D., Dabhi, C.K., Klemme, F., Prakash, O., Chauhan, Y.S.: Impact of variability on processor performance in negative capacitance FinFET technology. IEEE Trans. Circuits Syst. I Regul. Pap. 67(9), 3127–3137 (2020). https://doi.org/10.1109/TCSI.2020.2990672. Conference Name: IEEE Transactions on Circuits and Systems I: Regular Papers
5. Barber, D.: Bayesian Reasoning and Machine Learning. Cambridge University Press, Cambridge (2012)
6. Bardell, P., McAnney, W.: Self-testing of multichip logic modules. Undefined (1982). https://www.semanticscholar.org/paper/Self-Testing-of-Multichip-Logic-Modules-Bardell-McAnney/276f40aa9972e41e5ab1953911eeada4e562c14e
7. Barford, L., Kanevsky, V., Kamas, L.: Bayesian fault diagnosis in large-scale measurement systems. In: Proceedings of the 21st IEEE Instrumentation and Measurement Technology Conference (IEEE Cat. No.04CH37510), vol. 2, pp. 1234–1239 (2004). https://doi.org/10.1109/IMTC.2004.1351288. ISSN: 1091-5281
8. Baumann, R.: Soft errors in advanced computer systems. IEEE Design Test of Computers 22(3), 258–266 (2005). https://doi.org/10.1109/MDT.2005.69. Conference Name: IEEE Design Test of Computers

9. Blanton, R., Chen, J., Desineni, R., Dwarakanath, K., Maly, W., Vogels, T.: Fault tuples in diagnosis of deep-submicron circuits. In: Proceedings of International Test Conference, pp. 233–241 (2002). https://doi.org/10.1109/TEST.2002.1041765. ISSN: 1089-3539

10. Blanton, R.D., Tam, W.C., Yu, X., Nelson, J.E., Poku, O.: Yield Learning through physically aware diagnosis of IC-failure populations. IEEE Des.Test Comput. 29(1), 36–47 (2012). https://doi.org/10.1109/MDT.2011.2178587. Conference Name: IEEE Design Test of Computers

11. Borkar, S.: Designing reliable systems from unreliable components: the challenges of transistor variability and degradation. IEEE Micro 25(6), 10–16 (2005). https://doi.org/10.1109/MM.2005.110. Conference Name: IEEE Micro

12. Breiman, L., Friedman, J.H., Olshen, R.A., Stone, C.J.: Classification and Regression Trees. Routledge, New York (2017). https://doi.org/10.1201/9781315139470

13. Brown, A.R., Idris, N.M., Watling, J.R., Asenov, A.: Impact of metal gate granularity on threshold voltage variability: a full-scale three-dimensional statistical simulation study. IEEE Electron Device Lett. 31(11), 1199–1201 (2010). https://doi.org/10.1109/LED.2010.2069080. Conference Name: IEEE Electron Device Letters

14. Chakravarty, S., Devta-Prasanna, N., Gunda, A., Ma, J., Yang, F., Guo, H., Lai, R., Li, D.: Silicon evaluation of faster than at-speed transition delay tests. In: 2012 IEEE 30th VLSI Test Symposium (VTS), pp. 80–85 (2012). https://doi.org/10.1109/VTS.2012.6231084. ISSN: 2375-1053

15. Champac, V., Gervacio, J.G.: Timing Performance of Nanometer Digital Circuits Under Process Variations. Springer, Berlin (2018). https://doi.org/10.1007/978-3-319-75465-9. https://link.springer.com/content/pdf/10.1007/978-3-319-75465-9.pdf. Type: BOOK

16. Cheng, B., Dideban, D., Moezi, N., Millar, C., Roy, G., Wang, X., Roy, S., Asenov, A.: Statistical-variability compact-modeling strategies for BSIM4 and PSP. IEEE Des. Test Comput. 27(2), 26–35 (2010). https://doi.org/10.1109/MDT.2010.53. Conference Name: IEEE Design Test of Computers

17. Constantinescu, C.: Trends and challenges in VLSI circuit reliability. IEEE Micro 23(4), 14–19 (2003). https://doi.org/10.1109/MM.2003.1225959

18. Cook, A., Elm, M., Wunderlich, H.J., Abelein, U.: Structural in-field diagnosis for random logic circuits. In: 2011 Sixteenth IEEE European Test Symposium, pp. 111–116 (2011). https://doi.org/10.1109/ETS.2011.25. ISSN: 1558-1780

19. Cook, A., Hellebrand, S., Indlekofer, T., Wunderlich, H.J.: Diagnostic test of robust circuits. In: 2011 Asian Test Symposium, pp. 285–290 (2011). https://doi.org/10.1109/ATS.2011.55. ISSN: 2377-5386

20. De Kleer, J.: Diagnosing multiple persistent and intermittent faults. In: Twenty-First International Joint Conference on Artificial Intelligence (2009)

21. Desineni, R., Poku, O., Blanton, R.D.: A logic diagnosis methodology for improved localization and extraction of accurate defect behavior. In: 2006 IEEE International Test Conference, pp. 1–10 (2006). https://doi.org/10.1109/TEST.2006.297627. ISSN: 2378-2250

22. Ernst, D., Das, S., Lee, S., Blaauw, D., Austin, T., Mudge, T., Kim, N.S., Flautner, K.: Razor: circuit-level correction of timing errors for low-power operation. IEEE Micro 24(6), 10–20 (2004). https://doi.org/10.1109/MM.2004.85. Conference Name: IEEE Micro

23. Garcia-Gervacio, J.L., Nocua, A., Champac, V.: Screening small-delay defects using inter-path correlation to reduce reliability risk. In: Microelectronics Reliability. Elsevier, Amsterdam (2015). https://www.sciencedirect.com/science/article/pii/S0026271415000578. Publisher: Elsevier

24. Girard, P., Landrault, C., Pravossoudovitch, S.: A novel approach to delay-fault diagnosis. In: [1992] Proceedings 29th ACM/IEEE Design Automation Conference, pp. 357–360 (1992). https://doi.org/10.1109/DAC.1992.227778. ISSN: 0738-100X

25. Gómez, L.R., Wunderlich, H.J.: A neural-network-based fault classifier. In: 2016 IEEE 25th Asian Test Symposium (ATS), pp. 144–149 (2016). https://doi.org/10.1109/ATS.2016.46. ISSN: 2377-5386

26. Gómez, L.R., Cook, A., Indlekofer, T., Hellebrand, S., Wunderlich, H.J.: Adaptive bayesian diagnosis of intermittent faults. J. Electron. Test **30**(5), 527–540 (2014). https://doi.org/10.1007/s10836-014-5477-1
27. Gupta, P., Agarwal, Y., Dolecek, L., Dutt, N., Gupta, R.K., Kumar, R., Mitra, S., Nicolau, A., Rosing, T.S., Srivastava, M.B., Swanson, S., Sylvester, D.: Underdesigned and opportunistic computing in presence of hardware variability. IEEE Trans. Comput.-Aided Des. Integr. Circuits Syst. **32**(1), 8–23 (2013). https://doi.org/10.1109/TCAD.2012.2223467. Conference Name: IEEE Transactions on Computer-Aided Design of Integrated Circuits and Systems
28. Hapke, F., Redemund, W., Glowatz, A., Rajski, J., Reese, M., Hustava, M., Keim, M., Schloeffel, J., Fast, A.: Cell-aware test. IEEE Trans. Comput.-Aided Des. Integr. Circuits Syst. **33**(9), 1396–1409 (2014). https://doi.org/10.1109/TCAD.2014.2323216. Conference Name: IEEE Transactions on Computer-Aided Design of Integrated Circuits and Systems
29. Holst, S., Wunderlich, H.J.: Adaptive debug and diagnosis without fault dictionaries. J. Electron. Test **25**(4), 259–268 (2009). https://doi.org/10.1007/s10836-009-5109-3
30. Holst, S., Schneider, E., Kochte, M.A., Wen, X., Wunderlich, H.J.: Variation-aware small delay fault diagnosis on compressed test responses. In: 2019 IEEE International Test Conference (ITC), pp. 1–10 (2019). https://doi.org/10.1109/ITC44170.2019.9000143. ISSN: 2378-2250
31. Huang, Q., Fang, C., Mittal, S., Blanton, R.D.: Improving diagnosis efficiency via machine learning. In: 2018 IEEE International Test Conference (ITC), pp. 1–10 (2018). https://doi.org/10.1109/TEST.2018.8624884. ISSN: 2378-2250
32. Kampmann, M., Kochte, M.A., Liu, C., Schneider, E., Hellebrand, S., Wunderlich, H.J.: Built-in test for hidden delay faults. IEEE Trans. Comput.-Aided Des. Integr. Circuits Syst. **38**(10), 1956–1968 (2019). https://doi.org/10.1109/TCAD.2018.2864255. Conference Name: IEEE Transactions on Computer-Aided Design of Integrated Circuits and Systems
33. Karel, A., Comte, M., Galliere, J.M., Azais, F., Renovell, M.: Resistive bridging defect detection in bulk, FDSOI and FinFET technologies. J. Electron. Test **33**(4), 515–527 (2017). https://doi.org/10.1007/s10836-017-5674-9
34. Kim, Y.M., Kameda, Y., Kim, H., Mizuno, M., Mitra, S.: Low-cost gate-oxide early-life failure detection in robust systems. In: 2010 Symposium on VLSI Circuits, pp. 125–126 (2010). https://doi.org/10.1109/VLSIC.2010.5560326. ISSN: 2158-5636
35. Krishnan, S., Doornbos, K.D., Brand, R., Kerkhoff, H.G.: Block-level bayesian diagnosis of analogue electronic circuits. In: 2010 Design, Automation Test in Europe Conference Exhibition (DATE 2010), pp. 1767–1772 (2010). https://doi.org/10.1109/DATE.2010.5457100. ISSN: 1558-1101
36. Lee, J., McCluskey, E.J.: Failing frequency signature analysis. In: 2008 IEEE International Test Conference, pp. 1–8 (2008). https://doi.org/10.1109/TEST.2008.4700561. ISSN: 2378-2250
37. Li, X., Qin, J., Bernstein, J.B.: Compact modeling of MOSFET wearout mechanisms for circuit-reliability simulation. IEEE Trans. Device Mater. Reliabil. **8**(1), 98–121 (2008). https://doi.org/10.1109/TDMR.2008.915629. Conference Name: IEEE Transactions on Device and Materials Reliability
38. Liu, F., Nikolov, P., Ozev, S.: Parametric fault diagnosis for analog circuits using a Bayesian framework. In: 24th IEEE VLSI Test Symposium, pp. 6, pp.–277 (2006). https://doi.org/10.1109/VTS.2006.54. ISSN: 2375-1053
39. Martins, M., Matos, J.M., Ribas, R.P., Reis, A., Schlinker, G., Rech, L., Michelsen, J.: Open cell library in 15nm FreePDK technology. In: Proceedings of the 2015 Symposium on International Symposium on Physical Design, ISPD'15, pp. 171–178. Association for Computing Machinery, New York (2015). https://doi.org/10.1145/2717764.2717783
40. Mesalles, F., Villacorta, H., Renovell, M., Champac, V.: Behavior and test of open-gate defects in FinFET based cells. In: 2016 21th IEEE European Test Symposium (ETS), pp. 1–6 (2016). https://doi.org/10.1109/ETS.2016.7519305. ISSN: 1558-1780
41. Moreno, J., Renovell, M., Champac, V.: Effectiveness of low-voltage testing to detect interconnect open defects under process variations. IEEE Trans. Very Large Scale Integr. (VLSI) Syst. **24**(1), 378–382 (2016). https://doi.org/10.1109/TVLSI.2015.2397934. Conference Name: IEEE Transactions on Very Large Scale Integration (VLSI) Systems

42. Najafi-Haghi, Z.P., Wunderlich, H.J.: Resistive open defect classification of embedded cells under variations. In: 2021 IEEE 22nd Latin American Test Symposium (LATS), pp. 1–6. IEEE, Piscataway (2021)

43. Najafi-Haghi, Z.P., Hashemipour-Nazari, M., Wunderlich, H.J.: Variation-aware defect characterization at cell level. In: 2020 IEEE European Test Symposium (ETS), pp. 1–6 (2020). https://doi.org/10.1109/ETS48528.2020.9131600. ISSN: 1558-1780

44. Najafi-Haghi, Z.P., Klemme, F., Amrouch, H., Wunderlich, H.-J.: On extracting reliability information from speed binning. In: 2022 IEEE European Test Symposium (ETS) (2022)

45. Natarajan, S., Agostinelli, M., Akbar, S., Bost, M., Bowonder, A., Chikarmane, V., Chouksey, S., Dasgupta, A., Fischer, K., Fu, Q., Ghani, T., Giles, M., Govindaraju, S., Grover, R., Han, W., Hanken, D., Haralson, E., Haran, M., Heckscher, M., Heussner, R., Jain, P., James, R., Jhaveri, R., Jin, I., Kam, H., Karl, E., Kenyon, C., Liu, M., Luo, Y., Mehandru, R., Morarka, S., Neiberg, L., Packan, P., Paliwal, A., Parker, C., Patel, P., Patel, R., Pelto, C., Pipes, L., Plekhanov, P., Prince, M., Rajamani, S., Sandford, J., Sell, B., Sivakumar, S., Smith, P., Song, B., Tone, K., Troeger, T., Wiedemer, J., Yang, M., Zhang, K.: A 14nm logic technology featuring 2nd-generation FinFET, air-gapped interconnects, self-aligned double patterning and a 0.0588 μm2 SRAM cell size. In: 2014 IEEE International Electron Devices Meeting, pp. 3.7.1–3.7.3 (2014). https://doi.org/10.1109/IEDM.2014.7046976. ISSN: 2156-017X

46. Nicolaidis, M.: GRAAL: a new fault-tolerant design paradigm for mitigating the flaws of deep-nanometric technologies. In: IEEE International TEST Conference (ITC'07). IEEE Computer Society, Santa Clara (2007). https://hal.archives-ouvertes.fr/hal-00226406. ISBN: 1-4244-1128-9/07

47. O'Farrill, C., Moakil-Chbany, M., Eklow, B.: Optimized reasoning-based diagnosis for non-random, board-level, production defects. In: IEEE International Conference on Test, pp. 7–179 (2005). https://doi.org/10.1109/TEST.2005.1583974. ISSN: 2378-2250

48. Pearl, J.: Probabilistic Reasoning in Intelligent Systems: Networks of Plausible Inference. Morgan Kaufmann, San Francisco (1988)

49. Polian, I., Engelke, P., Becker, B., Kundu, S., Galliere, J.M., Renovell, M.: Resistive Bridge fault model evolution from conventional to ultra deep submicron. In: 23rd IEEE VLSI Test Symposium (VTS'05), pp. 343–348 (2005). https://doi.org/10.1109/VTS.2005.72. ISSN: 2375-1053

50. Przytula, K., Thompson, D.: Construction of Bayesian networks for diagnostics. In: 2000 IEEE Aerospace Conference. Proceedings (Cat. No.00TH8484), vol. 5, pp. 193–200 (2000). https://doi.org/10.1109/AERO.2000.878490. ISSN: 1095-323X

51. Rajski, J., Tyszer, J., Kassab, M., Mukherjee, N.: Embedded deterministic test. IEEE Trans. Comput.-Aided Des. Integr. Circuits Syst. 23(5), 776–792 (2004). https://doi.org/10.1109/TCAD.2004.826558. Conference Name: IEEE Transactions on Computer-Aided Design of Integrated Circuits and Systems

52. Rodríguez Gómez, L.: Machine learning support for logic diagnosis. Ph.D. Thesis (2017). http://elib.uni-stuttgart.de/handle/11682/9216. Accepted: 2017-08-15T12:44:56Z

53. Rousset, A., Bosio, A., Girard, P., Landrault, C., Pravossoudovitch, S., Virazel, A.: DERRIC: a tool for unified logic diagnosis. In: 12th IEEE European Test Symposium (ETS'07), pp. 13–20 (2007). https://doi.org/10.1109/ETS.2007.16. ISSN: 1558-1780

54. Sauer, M., Kim, Y.M., Seomun, J., Kim, H.O., Do, K.T., Choi, J.Y., Kim, K.S., Mitra, S., Becker, B.: Early-life-failure detection using SAT-based ATPG. In: 2013 IEEE International Test Conference (ITC), pp. 1–10. IEEE, Piscataway (2013)

55. Schneider, E., Wunderlich, H.J.: Switch level time simulation of CMOS circuits with adaptive voltage and frequency scaling. In: 2020 IEEE 38th VLSI Test Symposium (VTS), pp. 1–6 (2020). https://doi.org/10.1109/VTS48691.2020.9107642. ISSN: 2375-1053

56. Shen, J.P., Maly, W., Ferguson, F.J.: Inductive fault analysis of MOS integrated circuits. IEEE Des. Test Comput. 2(6), 13–26 (1985). https://doi.org/10.1109/MDT.1985.294793. Conference Name: IEEE Design Test of Computers

57. Tehranipoor, M., Peng, K., Chakrabarty, K.: Test and Diagnosis for Small-Delay Defects. Springer, New York (2012). https://doi.org/10.1007/978-1-4419-8297-1. http://link.springer.com/10.1007/978-1-4419-8297-1

58. Turakhia, R., Daasch, W.R., Ward, M., Slyke, J.V.: Silicon evaluation of longest path avoidance testing for small delay defects. In: 2007 IEEE International Test Conference. IEEE Computer Society, Washington (2007). https://doi.org/10.1109/TEST.2007.4437564. https://www.computer.org/csdl/proceedings-article/test/2007/04437564/12OmNzdoMKh. ISSN: 1089-3539

59. Wagner, M., Wunderlich, H.J.: Probabilistic sensitization analysis for variation-aware path delay fault test evaluation. In: 2017 22nd IEEE European Test Symposium (ETS), pp. 1–6 (2017). https://doi.org/10.1109/ETS.2017.7968226. ISSN: 1558-1780

60. Wang, X., Brown, A.R., Cheng, B., Asenov, A.: Statistical variability and reliability in nanoscale FinFETs. In: 2011 International Electron Devices Meeting, pp. 5.4.1–5.4.4 (2011). https://doi.org/10.1109/IEDM.2011.6131494. ISSN: 2156-017X

61. Wang, X., Brown, A.R., Idris, N., Markov, S., Roy, G., Asenov, A.: Statistical threshold-voltage variability in scaled decananometer bulk HKMG MOSFETs: a full-scale 3-D simulation scaling study. IEEE Trans. Electron Devices **58**(8), 2293–2301 (2011). https://doi.org/10.1109/TED.2011.2149531. Conference Name: IEEE Transactions on Electron Devices

62. Wang, L., Brown, A.R., Nedjalkov, M., Alexander, C., Cheng, B., Millar, C., Asenov, A.: Impact of self-heating on the statistical variability in bulk and SOI FinFETs. IEEE Trans. Electron Devices **62**(7), 2106–2112 (2015). https://doi.org/10.1109/TED.2015.2436351. Conference Name: IEEE Transactions on Electron Devices

63. Wohl, P., Waicukauski, J.A., Patel, S., Maston, G.: Effective diagnostics through interval unloads in a BIST environment. In: Proceedings of the 39th annual design automation conference, DAC'02, pp. 249–254. Association for Computing Machinery, New York (2002). https://doi.org/10.1145/513918.513984

64. Wunderlich, H.J., Holst, S.: Generalized fault modeling for logic diagnosis. In: Wunderlich, H.J. (ed.) Models in Hardware Testing. Lecture Notes of the Forum in Honor of Christian Landrault, Frontiers in Electronic Testing, pp. 133–155. Springer, Dordrecht (2010). https://doi.org/10.1007/978-90-481-3282-95

65. Xue, Y., Poku, O., Li, X., Blanton, R.D.: PADRE: physically-aware diagnostic resolution enhancement. In: 2013 IEEE International Test Conference (ITC), pp. 1–10 (2013). https://doi.org/10.1109/TEST.2013.6651899

66. Yan, H., Singh, A.: Experiments in detecting delay faults using multiple higher frequency clocks and results from neighboring die. In: International Test Conference, 2003. Proceedings. ITC 2003, vol. 1, pp. 105–111 (2003). https://doi.org/10.1109/TEST.2003.1270830. ISSN: 1089-3539

67. Zhang, Z., Wang, Z., Gu, X., Chakrabarty, K.: Board-level fault diagnosis using Bayesian inference. In: 2010 28th VLSI Test Symposium (VTS), pp. 244–249 (2010). https://doi.org/10.1109/VTS.2010.5469569. ISSN: 2375-1053

Machine Learning in Logic Circuit Diagnosis

Ronald D. (Shawn) Blanton, Chenlei Fang, Qicheng Huang, and Soumya Mittal

1 Introduction

The objective of logic diagnosis is to uncover the cause of failure within a logic circuit contained within an integrated circuit. It is the first step in understanding the underlying reason for circuit failure. Implemented in software, the diagnosis process typically uses design descriptions (netlist and layout), the test vectors applied to the circuit to detect failure, and the test response measured by the ATE (automatic test equipment) to characterize failure of a logic circuit, where characterization may include, for instance, (i) localization, (ii) failure behavior identification, and (iii) root cause determination.

The accomplishment of the aforementioned objectives of diagnosis accelerates the overall process of yield learning. For example, the onus of defect characterization is on cost- and time-intensive physical failure analysis (PFA) which is mitigated by a less-expensive and time-effective diagnosis. An ideal diagnosis would thus minimize the need for PFA, thereby, further enabling rapid yield analysis. Numerous methods have been recommended in the literature to enhance the quality of diagnosis. Diagnosis quality is improved in three ways—better algorithms, better tests and/or better design [35, 41]. Diagnosis methods that develop new algorithms are further differentiated based on the type of fault model used at the logic level (i.e., whether a binary error or an unknown value is assumed at a candidate location), the scoring technique employed (i.e., deterministic versus statistical), how precisely a defect is localized (i.e., whether a defect candidate is reported at the logic level, within a cell, or in the interconnect), and if multiple defects affecting a single circuit can be analyzed/identified adequately.

R. D. (Shawn) Blanton (✉) · C. Fang · Q. Huang · S. Mittal
Department of ECE, Carnegie Mellon University, Pittsburgh, PA, USA
e-mail: rblanton@andrew.cmu.edu; chenleif@alumni.cmu.edu; qichengh@alumni.cmu.edu; soumyami@alumni.cmu.edu

© The Author(s), under exclusive license to Springer Nature Switzerland AG 2023
P. Girard et al. (eds.), *Machine Learning Support for Fault Diagnosis of System-on-Chip*, https://doi.org/10.1007/978-3-031-19639-3_5

135

Over the last decade, machine learning (ML) has become an essential tool for improving the effectiveness of logic diagnosis along the three dimensions of localization, behavior identification, and root cause determination. ML uses existing data to form models that predict missing properties for new data. For example, ML has been successfully used to identify bridge defect behavior among a group of failures where failure behavior is not consistent with existing fault models [1].

To the best of our knowledge, the work in [1] is the first published paper on the use of ML for diagnosis. Since then, there has been many publications discussing the use of ML in diagnosis. These publications focused on a wide variety of diagnosis topics that ranged from reordering test vectors to optimizing the outcomes of diagnosis, to data collection prediction for identifying the minimal amount of test-response data for optimizing diagnosis, to the use of existing diagnosis outcomes (from both actual failures and simulation) to improve failure localization and behavior identification, just to name a few.

In this chapter, we examine representative work of the use of ML in diagnosis from the Advanced Chip Testing Laboratory of Carnegie Mellon University. We partition the work into three categories, namely, pre-diagnosis, during-diagnosis, and post-diagnosis. Pre-diagnosis is concerned with any activities that are performed before diagnosis is deployed. Examples of pre-diagnosis activities includes classic work such as diagnostic ATPG, and DFT for increasing testability. Pre-diagnosis activities that utilize ML include test reordering and optimizing test response data collection, for example. In addition, there is work that predicts the outcome of diagnosis so that the use of compute resources can be optimized. During-diagnosis involves the algorithms (e.g., cause-and-effect, path tracing, etc.) and the underlying technologies used by the algorithms such as fault simulation. The use of ML in during-diagnosis activities generally involves learning while diagnosis executes. For example, in [2], a k-nearest neighbor model is created, evolved, and used during on-chip diagnosis to improve diagnosis outcomes. Finally, post-diagnosis includes all the activities that occur after diagnosis execution. These approaches usually involve volume diagnosis (i.e., using the outcome results of many diagnoses) to improve diagnostic resolution, that is, the number of possible failure locations (within the netlist and/or the layout). The use of ML in post-diagnosis activities is a perfect application for ML given the abundance of well-structured, labeled data from prior diagnoses, physical failure analyses, and precise fault simulations. The bulk of the work conducted by ACTL lies in post-diagnosis and focusses on localization and behavior identification improvement.

In the following chapter subsections, details of ACTL work in pre-, during-, and post-diagnosis are described.

2 Pre-Diagnosis

This chapter section describes the use of ML to improve the outcomes of diagnosis before diagnosis is executed. Specifically, in Sect. 2.1, an approach for learning to predict the outcome of diagnosis as "successful" or "too expensive" is described. Such a capability is invaluable for optimizing diagnostic outcomes when there are limited compute resources. In Sect. 2.2, an approach for re-ordering test patterns for optimizing the data collected for diagnosis is described. Such a method is appropriate when it is not cost-feasible to apply a conventional diagnostic ATPG test set. Finally, in Sect. 2.3, a method that complements test re-ordering is described. Specifically, given that the amount of data collected from a failing logic circuit is limited due to ATE memory, it is prudent to consider which data should be collected to maximize the likelihood of a successful diagnosis.

2.1 Diagnosis Outcome Preview Through Learning

Given the fabrication perturbations inherent to integrated circuits (IC), yield learn-ing is a crucial process to identify and mitigate sources of yield loss. Due to the virtually unlimited layout patterns that are created by new standard-cell libraries and the techniques employed by evolving place-and-route algorithms, yield learning for logic faces unique challenges for both process development and high-volume manufacturing. Logic-circuit diagnosis serves as a key first step in yield learning under both scenarios. Diagnosis is a software-based process that identifies possible locations and/or behaviors of defects in a failing chip. The output of a diagnosis tool often serves as a guide for physical failure analysis (PFA) to identify the root cause of the failure. Diagnosis results from a population of failing chips also serve as input to various volume-diagnosis techniques to reveal important statistics.

While diagnosis plays an important role in yield learning, various uncertainties in the fabrication process bring a number of challenges, resulting in diagnosis with undesirable outcomes or low efficiency. For example, real defects may behave in an unexpected manner different from the fault models implicitly used by diagnosis tools. In addition, fault equivalence/dominance relationships and the existence of multiple defects further challenge the effectiveness of diagnosis tools. Such challenges may result in diagnosis outcomes characterized by low quality (e.g., poor resolution or inaccuracy), or even diagnosis failure (i.e., diagnosis exits without reporting any actionable information). Also, diagnosis computation time may be very long, which aggravates time-to-market pressure.

According to one of our surveys from a dozen industrial practitioners from foundries, integrated device manufacturers, and fabless design houses, the number of diagnoses performed per week per design during technology development can be 10,000, and one indicated that number exceeds 100,000. Given the amount of diagnoses that can lead to no actionable information, it would be very beneficial to know beforehand which fail logs would result in actionable outcomes and which would result in either logic circuit or chain failure since different yield learning techniques can be brought to bear. In a high-volume scenario, practitioners that are handling a large number of diagnoses indicate that sampling must be employed, implying that there are limited resources for diagnosis. Like technology development, resources can also be easily wasted if diagnosis results are not meaningful or not actionable (e.g., diagnosis failure, poor resolution and/or large defect count). It would therefore again be very beneficial to know beforehand which fail logs would result in actionable information. Moreover, for a given set of fail logs, a constraint on diagnosis resources, and with all other things being equal, it is prudent to choose fail logs that have short runtime. The above observations motivate the building of a diagnosis previewer, which is able to predict the diagnostic outcomes (diagnosis success, resolution, runtime, etc.) before diagnosis even starts, so that the practitioners can prioritize the fail logs for best allocation of diagnosis resources.

To predict diagnostic outcomes directly from fail-log information, however, is not trivial. Fortunately, it is found that there indeed exists **complex non-linear** correlations between fail-log features and diagnostic outcomes in a high-dimensional space. Due to the complexity of the correlation and the need to build models for different kinds of chips, machine learning (ML) is employed for its advantage in uncovering high-dimensional correlations to develop separate models for different IC designs [3, 4].

Specifically, the ML-based system developed can provide a preview of diagnosis outcomes. The focus is on several informative aspects indicating diagnosis quality and efficiency, including (i) diagnosis success, i.e., whether a diagnosis run successfully reports defect candidate(s); (ii) failure type, i.e., whether the reported candidate is located in a scan-chain or in the logic; (iii) existence of multiple defects (referred to as defect count); (iv) whether the diagnosis resolution is acceptable, and (v) the order of magnitude of diagnosis execution time, e.g., $1\sim10$ s, $10\sim100$ s, etc. The previewer predicts these five aspects of diagnosis outcome by four separate predictors. These aspects are the key metrics indicating the quality and efficiency of a diagnosis run, and provide a comprehensive preview of the diagnosis outcome.

- **Diagnosis success and defect count:** the two aspects are predicted by the same classifier C_1, with three possible labels: unsuccessful (indicated by "0"), single defect ("1") and multiple defects ("2").
- **Failure type:** Failure type is predicted by classifier C_2 with a binary outcome: either logic ("0") or scan-chain ("1").
- **Resolution:** The third classifier C_3 aims to predict whether a fail log would result in an diagnosis outcome with an acceptable resolution ("0" for yes and "1" for no). Typically, an acceptable resolution is not larger than three.
- **Runtime magnitude:** Users usually only need an approximation of the order of runtime magnitude. Different from the previous classification problems, runtime prediction is a regression problem and the regression model R outputs a real number.

The previewer is trained on a subset of fail logs, which are already diagnosed with all the information concerning diagnosis outcome available. The whole system is then applied to the remaining fail logs which have not been diagnosed yet. As outputs of the previewer, the different predicted aspects can be used as evaluation metrics to order diagnosis priority of the fail logs. These metrics can be freely selected and combined according to practitioners' needs.

For the three classification tasks and the regression task, the *Random Forest* (RF) algorithm [5] is used due to its good performance in many ML applications. RFs are capable of both classification and regression tasks, so we use *Random Forest Classification* (RFC) for predictors $C_1 \sim C_3$, and *Random Forest Regression* (RFR) for the predictor R.

With such information in hand, practitioners are able to evaluate the effectiveness of a diagnosis *a priori*, and decide its priority to be executed accordingly, which helps allocate the diagnosis resources in a smarter way and expedites yield learning at minimal cost. In experiments on a 28 nm test chip and a high-volume 90 nm part, the predictors can provide accurate prediction results for each aspect. In a virtual application scenario, the overall previewer can result up to 9X speed-up for the test chip and 6X for the high-volume part.

The works in [3, 4, 6] have demonstrated that there exist high-dimensional nonlinear correlations between fail-log features and diagnosis outcome, and ML techniques can be used to uncover such correlations and predict diagnosis outcome solely based on fail logs. While the employed RF model does not require a large training dataset to derive parameters, the performance of RF degrades significantly with *very little data*. For example, for a 28 nm test chip, 2000 training data samples are crucial to guarantee a 80%+ recall for logic failure prediction; the value drops to around 70% with fewer than 500 samples [6]. Collecting sufficient training data, unfortunately, can be time-consuming. For example, one training sample may require several hours of compute time for performing diagnosis. Also, limited data inherently occurs during yield ramp-up. That is, as a new IC is being produced, the number of failing chips is likely large but not sufficiently large for accurate prediction. This situation is quite problematic because new IC ramp is the most crucial period for yield learning. Therefore, it is necessary and beneficial to explore new frameworks to improve prediction performance when the amount of data is limited.

It has been observed that many datasets generated during testing and diagnosis processes are highly correlated. For example, diagnosis data of different partitions in the same design, diagnosis data from different rounds of tape-outs, or fault simulation results and real testers responses. The correlated datasets could provide some prior knowledge as a supplement to the small dataset being focused upon. Utilization of prior knowledge from correlated data has been widely used in electronic design automation (EDA) [7–9]. Similar applications for testing and diagnosis have not however been similarly explored. We therefore focus on how to make full use of prior knowledge from a correlated diagnosis dataset to improve diagnosis outcome prediction, when only limited diagnosis data is available for the design of interest.

Towards this goal, a comprehensive flow that uses the "transferring RF" technique [10] is developed to incorporate prior knowledge for more efficient diagnosis-outcome prediction. Specifically, a model is trained from the fail-log features and diagnosis results from a correlated dataset, and then combine the prior knowledge encoded in this model with limited data from the design of interest. This leads to a new RF that predicts diagnosis outcomes more accurately. Experiments on

Fig. 1 Transfer learning
process

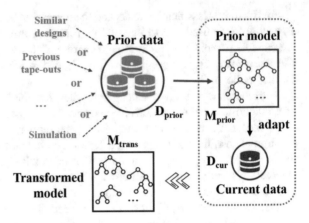

industrial designs demonstrate that in the "limited data" scenario, the transferred RF can significantly improve prediction performance (e.g., a **6:1** training sample reduction for the same performance) when suitable prior knowledge exists.

Figure 1 gives an illustration of the transfer learning flow. It is assumed that there are two components for diagnosis preview—a prior RF model M_{prior} and a few training data D_{cur} for the current design of interest. M_{prior} is trained from some prior data D_{prior} which are (i) correlated to D_{cur} and (ii) sufficient in size to train a well-performed model M_{prior}. Several possible sources for D_{prior} include similar designs, previous tape-outs or fault-simulation data.

Since D_{cur} and D_{prior} are correlated, we expect that M_{prior} should be able to provide useful side-information for building a model using D_{cur}, especially when the size of D_{cur} is very small (e.g., with fewer than 100 samples). Therefore, beginning with the architecture of M_{prior}, we adapt it based on D_{cur} to generate a transferred RF model M_{trans}. By combining these two sources of knowledge from both M_{prior} and D_{cur}, M_{trans} is expected to have better performance than a model solely constructed using D_{cur}.

2.2 Adaptive Test Pattern Reordering for Accurate Diagnosis

The quality of a diagnosis is generally measured in terms of its resolution and accuracy. Resolution refers to the number of possible defect locations, and a diagnosis is deemed accurate if the failing location is among the sites reported. Two different approaches exist for performing diagnosis: cause-effect and effect-cause. Cause-effect method simulates all possible faults (typically single stuck-at faults), stores the faulty outputs as a fault dictionary, and compares the outputs to circuit under test (CUT) responses. Effect-cause, on the other hand, starts from "back-coning" from the erroneous outputs and identifies the candidate faulty locations, as shown in Fig. 2. Most state-of-the-art diagnosis tools adopt a combination of these

Fig. 2 An example of back-coning a CUT with two failing patterns; T_1 with output O_1 failing, and T_2 with output O_2 failing

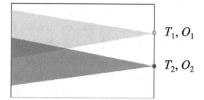

T_1, O_1

T_2, O_2

two approaches[11]. That is, the tool first uses effect-cause to find a rather small portion of the CUT, then simulates faults in the portion in a cause-effect manner.

In general, the more tester-response collected, a more precise diagnosis can be obtained. For example, as shown in Fig. 2, suppose we have a chip failing for pattern T_1 at output O_1, and T_2 at output O_2. If only the tester response for pattern T_1 is at hand, then the entire back-cone of O_1 will be possible candidates for the effect-cause step. Even if fault simulation is executed for all these locations to rule out some candidates, many candidates will still remain, resulting in poor resolution. However, if the responses from pattern T_2 is also included for diagnosis, the possible candidate area can shrink to the overlap of the two back-cones, leading to a potentially better resolution (assuming only one defect exists). The need to collect sufficient fail data is stated in the user manual of commercially-available tools[1], and is often verified in practice.

Despite the benefit brought by additional fail data, collecting more data is not always feasible. For example, collecting more tester data increases cost due to increased tester time, and may even increase tester cost if more memory is needed to store fail data. For the purpose of saving testing time, stop-on-first-fail is widely adopted in industry, which stops applying more patterns once the first failure is observed for each chip. Such a strategy is enough for cases where detecting faulty chips is the only requirement, but is not enough for high-quality diagnosis. Furthermore, the limited size of ATE memory also restricts the amount of data that can be collected and stored—patterns applied after the memory being filled up will be of no use.

An adaptive test pattern reordering algorithm is described in [13], which collects sufficient fail data to ensure good diagnosis outcomes with fewer applied test patterns. The approach used is based on the personalized movie/book recommending systems that are widely adopted. These systems usually build a "user-item" model that is based on the following assumption: the preference of one individual can be predicted based on preferences of other similar individuals. Once a user creates an account, and indicates several preferences, the system will automatically recommend items that are popular with similar users.

---

[1] The user manual for Tessent, a diagnosis tool from Siemens, states the following "Include failures from at least 30 failing scan test patterns to achieve good logic diagnosis resolution. Mentor Graphics recommends 100 failing scan test patterns for optimal results." [12].

In the testing and diagnosis scenario, the goal is to dynamically rearrange a fixed set of test patterns to produce an optimal diagnosis outcome on a per chip basis. That is, each chip, we want to apply the patterns that the chip is more likely to fail before other patterns that are more likely to pass. A common property found in semiconductor fabrication is exploited to achieve the goal: defects occurring for a given design and fabrication process will not be completely random, but instead also include so-called systematic defects that are similar in location and behavior. This property implies that chips with similar defects will fail for similar test patterns. In this way, the chips and test patterns act as the "users" and "items", respectively, in a recommending system. The similarity between these techniques motivates the use of recommending system approaches for test reordering.

In the first step, a set of failing chips that have been fully tested are required for the training set. Next, a CUT has a small set of test applied to obtain the data for making a recommendation of the next test pattern. Specifically, the pattern reordering algorithm will use the data to recommend test patterns based on similar failing chips found within the training set. The failing patterns of these similar chips are applied to the CUT.

The k-Nearest Neighbors (kNN) algorithm is implemented to accomplish this task. kNN makes no assumptions on the underlying data distribution, and does not require a training process typical of other machine learning algorithms, because there is no model to build. kNN is based merely on comparing the feature similarity between existing and new data instances. A failing test pattern signature (i.e., the number of failing outputs for each test pattern) is adopted as the feature for similarity comparison.

This approach does not require extensive computation for fault simulation and model training. In addition, it directly accounts for the characteristics of the relevant defects because it utilizes historical data from the same design/fabrication environment for making test-pattern recommendations. It is also assumed that the reordering algorithm is conducted online, which means the CUT does not have to be taken off the tester for offline computation. This assumption is reasonable because modern testers have the computation capability to perform ML inference, so the algorithm is able to be conducted on the tester with proper settings. Therefore, the test patterns can be reordered dynamically on the tester, which means applying the algorithm does not introduce extra test cost. Experiments using three industrial chips demonstrate the efficacy of the proposed methodology; specifically, the recommended test pattern order led to a 35% reduction, on average, while maximizing the amount of failure data collected.

The aforementioned work demonstrates a significant reduction in the number of tests needed to collect sufficient fail data. However, this approach has to be conducted on the tester to achieve maximal test cost reduction. In order to execute this algorithm on the tester, all the pattern signatures stemming from the training samples have to be stored in tester memory. In addition, because of the adoption of online learning, the amount of data to be stored in memory further increases, which brings the feasibility of this approach into question. For example, for an industrial design with M test patterns and N failing chips, a matrix with size $M \times N$ must be stored on the tester. An alternative approach is to adopt a pattern-centered view, that is, identifying similarity among test patterns, which resembles the item-centered methods in recommendation systems [14].

In modern ICs, systematic defects dominate over random defects, implying that the patterns that detect these systematic defects occur more often than the remaining test patterns. The work in [15] takes advantage of this observation by (i) tabulating the number of times that each pattern has failed, and (ii) creating a corresponding reordered pattern set that is sorted based on the most-frequently failing patterns. In other words, the algorithm orders the patterns by their "popularity" with respect to failing chips. The pattern order obtained is called the *most-frequent order*. It is assumed that by using the most-frequent order, a CUT is more likely to fail earlier. However, the most-frequent order is not always beneficial. Analysis has shown that chips tend to form "clusters" instead of failing for the same set of patterns. Each CUT is more likely to fail for test patterns that are common to its cluster of chips. In other words, chips in the same cluster are similar, and thus fail for similar patterns. Instead of finding a most-frequent order for all chips, finding the most-frequent order for each cluster is more likely to achieve superior results. An algorithm based on this idea is described in [14]. Compared to the kNN-based pattern reordering, this algorithm only requires storage for the K pattern orderings, where K is the number of clusters identified. The space complexity is $O(KM)$, where K is the number of clusters. Because $K \ll N$, where N is the number of failing chips, this method significantly saves tester memory compared to the kNN-based method. Experiments conducted on several fabricated chips have demonstrated efficacy. The algorithm achieves similar performance but only requires as little as $1/500$ memory to store failing data on tester compared to previous work.

2.3 Test-Data Volume Optimization for Diagnosis

As discussed in the previous section, generally, failure diagnosis achieves its best possible outcome when a large amount of test measurement data from a failing IC is available for analysis. However, the cost of collecting test measurement data can be significant due to the extra test time incurred for going beyond the first failing pattern. The data storage cost can also be significant, especially for high-volume products. Also, even when it is desirable to collect ample amounts of test data, it may not be possible due to the limited memory that modern ATE (automatic test equipment) has for storing test response data. To understand the incongruence between test-execution cost reduction and diagnosis further, consider a "stop-on-first-fail" test strategy. For this case, test data is minimized but the negative impact on diagnosis is likely quite significant since there is minimal information for distinguishing the various sources that could be responsible for failure. Moreover, once a subset of tests is selected from the original set, it is fixed and little consideration is given to chip to chip variation. This means that different failing chips may require varying amounts of test-data for an accurate diagnostic analysis. A dynamic method can therefore be quite useful and could counter any negative effects on diagnosis due to test-cost reduction.

The work in [16] focuses on reducing the cost of test-data collection from a different perspective. In the analyses, we have observed, for a substantial number of failing ICs, that their diagnosis results do not change with an increasing amount of test data. It is even possible that test data from the first few failing test patterns is sufficient for obtaining an accurate diagnosis result with optimal resolution. If the testing procedure can be terminated after a sufficient amount of data is collected for diagnosis, test-data collection can be reduced.

In this work, the objective is to predict the minimal amount of test data necessary to produce both a precise and accurate diagnosis. A method is proposed which dynamically predicts test termination for producing an amount of test data that is sufficient for obtaining a quality diagnosis result. The prediction model is learned from the test data produced by a sample of fully-tested ICs that have failed. The learned model is then deployed in production to predict the termination point when testing future ICs of the same type. Specifically, when an IC is tested, for each new failing pattern observed, the prediction model is invoked to determine if sufficient data has been collected for producing a quality diagnosis. The method is dynamic since test termination can vary from one failing IC to next rather than being fixed. In this way, the variation in individual ICs is taken into account in order to produce an optimized test-data volume for each, ensuring that the overall cost from collecting test data is minimized without sacrificing diagnosis quality. The test-termination prediction model is based on analysis of the failing tests and outputs of the ICs, which exhibit certain patterns that can be interpreted as a signal for judging the termination point of testing. Several statistical learning techniques are employed to discover these patterns from full-fail test data.

A defect in an IC, once activated, produces one or more errors that may be propagated to one or more outputs. Since each output has a fan-in logic cone, back-cone tracing is employed during diagnosis to locate possible defect sites. Consequently, determining the test-termination point for a given IC under test is based on simple observations made from the failing outputs. It is difficult however to use this raw data directly to discover underlying patterns and trends that suggest a termination point for testing. Therefore, the raw test data is processed and then organized into a set of features that are more readily usable for forming a test-termination prediction model.

Predicting test termination can be formulated as a classification task in statistical learning: for each failing test pattern, a decision is made about whether to terminate testing, or to continue applying further test patterns, by observing and analyzing the test data thus far collected. A binary classification problem of this nature can be handled by various statistical learning techniques, one of which is chosen based on effectiveness. The authors of [16] have evaluated the performance of kNN, linear regression (LASSO), SVM (Support Vector Machine), and decision tree with experiments, and decision tree performs the best. Experiment results from both industrial and simulated data sets demonstrate that the amount of test data needed to obtain an accurate diagnosis can vary from IC to IC, and the prediction method proposed in [16] allows the test-termination points to be accurately predicted.

3 During Diagnosis

This section describes the use of ML *during* diagnosis. Specifically, Sect. 3.1 describes a physically-aware diagnosis methodology called LearnX [17, 18] that focuses on accurate defect localization. In Sect. 3.2, a diagnosis methodology called MD-LearnX [18, 19] is described; it builds upon LearnX to cope with multiple defects within a failing logic circuit. Finally, Sect. 3.3 critiques an on-chip test and diagnosis methodology to ensure the lifetime reliability of an IC. Specifically, this section focuses on describing methods that employ machine learning to improve the quality of diagnostics via designing a better algorithm. Approaches that manipulate a test set or alter the design itself complement them. Work in [20] and [21] that specialize in accurate physical localization and behavior identification of back- and front-end defects further strengthen these methods.

3.1   LearnX: A Deterministic-Statistical Single Defect Diagnosis Methodology

An enhanced diagnosis procedure is extremely important for improving design, test and manufacturing of a chip. It makes volume diagnosis, and subsequently, PFA more effective in their ability to pinpoint and verify the most probable cause of yield loss. LearnX is a step in that direction. There are three salient features of LearnX.

First, it is a physically-aware diagnosis approach that utilizes layout information to identify not only the defect location but also its physical defect type (e.g., interconnect open, interconnect bridge and cell-internal defect). It should be noted that LearnX complements/strengthens approaches like [20, 21] where physical resolution is improved using layout neighborhood analysis.

Second, instead of just using a deterministic fault model where an erroneous value of either 0 or 1 is assumed, it employs the X-fault model as well because it is immune to error masking. As a result, it allows an error to propagate from a defect location conservatively, which likely avoids removing a correct candidate. Outputs with an X are assumed to be potentially erroneous.

The X-value simulation of a candidate is said to explain a failing pattern observed on the tester if the erroneous circuit outputs are subsumed by the set of simulated outputs that possess an X value.

LearnX is thus applicable to defects exhibiting arbitrary misbehaviors, a feature that is not present in state-of-the-art commercial tools but becoming increasingly significant to include with advancing process nodes.

Third, it applies machine learning to generate a scoring model to uncover the hidden correlations between a candidate and the tester response for identifying the candidate that best represents a defect.

Prior scoring methods use various candidate-ranking heuristics to identify the best candidates. They are independent of the type of the defect, and hence cannot capture the characteristics specific to each defect type. Additionally, prior ranking expressions are created by intuition and domain knowledge, and hence are not guaranteed to work for every defect type, design and/or process node.

Conversely, instead of using an ad-hoc scoring technique, a data-driven scoring model can discover (or learn) latent correlations between the correct candidate and the tester response, and possibly achieve better diagnosis accuracy and resolution. The scoring model can be built separately for each defect type to capture its unique characteristics.

It is thus not surprising that machine learning (ML) has been applied before to improve the quality of diagnosis. For instance, a multi-class support vector machine is employed in [22, 23] to locate a defect in a failing chip, where each class represents a fan-out free region in the design. A large design can be divided into smaller partitions to reduce memory and runtime overhead. However, the resolution is limited to a fanout-free region and hence inadequate for subsequent failure analysis.

In [1], a random forest is used to predict whether a failing chip is affected by a bridge defect. In that work, several features for each candidate are extracted from the design and the test data. The features identified from a logic description of the design aim to differentiate a bridge from other defect types. For example, one feature checks if the bridged nets drive a parity cell; a short between the inputs of a parity cell is equivalent to a stuck-at fault at its output, and thus adds ambiguity to bridge defect prediction. The features extracted from the test data indicate how well the simulation response of a bridge candidate matches with the observed tester response.

The approach proposed in [24] accomplishes a similar objective of defect classification by training a neural network. However, the methods presented in [1, 24] predict the defect type but do not pinpoint the defect location, which means they do not improve candidate-level resolution.

To address the shortcomings associated with prior scoring methods, LearnX uses machine learning to identify the defect type as well as defect location. Specifically, it employs a supervised learning algorithm to distinguish the correct candidate from incorrect ones.

Figure 3 shows the overview of LearnX. LearnX is a two-phase diagnosis methodology. The first phase aims to identify a defect that mimics the behavior of classic fault models such as the stuck-at, the bridge (specifically, the AND-type, the OR-type and the dominating bridge) and the open fault model (where a net is assumed to be stuck at the opposite value of the expected value for each pattern) through a set of strict rules. These strict rules must ensure that (a) the actual defect behavior and location are accurately captured by one of the identified candidates and (b) the minimum number of candidates are reported.

Fig. 3 An overview of the proposed diagnosis methodology, LearnX. The sequence of steps in each phase are marked with a different color

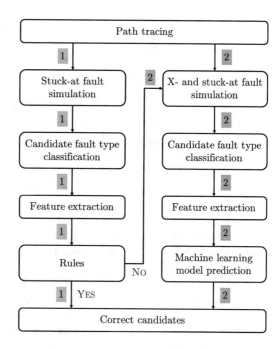

The defects that do not satisfy these rules are diagnosed in the second phase of the methodology. Such defects are identified to have a non-trivial behavior and cannot be modeled using the fault models employed in Phase 1. The two main steps in Phase 2 are (a) fault simulation using the X-fault model, which aims to capture the complex behavior of a defect with relatively high accuracy, and (b) machine learning classification to distinguish between the correct and the incorrect candidates.

To build the ML model, numerous fail logs are created by injecting and simulating realistic defect behaviors. Each fail log is then diagnosed by LearnX (up to the step named "feature extraction" in Fig. 3) to obtain an initial set of candidates. Forty-four features are extracted for each candidate by comparing its simulation response with the observed tester response (i.e., the simulation response of the injected defect). Because the location of an injected defect is known *a priori*, each candidate is then labeled either as correct or incorrect.

A supervised learning algorithm examines the training dataset to build a prediction model. Given an undiagnosed fail log, a set of candidates is obtained by following the flow of Fig. 3 up to the step named "feature extraction". The generated prediction model classifies each candidate as correct/incorrect. A commonly used machine learning method called a random forest is utilized in this work.

Note that the training dataset is highly imbalanced. For each injected fault, there is only one correct candidate. Therefore, it is entirely possible that the default classification/decision threshold of 0.5 (which corresponds to majority voting for a random forest algorithm) is not optimum. Therefore, a precision-recall (PR) curve is used here to find the optimum decision threshold for a random forest.

The learned model inherently acts like a scoring framework, and assigns a probability (or a "score") to each defect candidate. The score assigned to each candidate represents the likelihood of a candidate being correct. Any candidate whose score is more than the decision threshold is deemed a correct candidate.

Several experiments are conducted to evaluate the performance of LearnX. Simulation-based experiments are carried out for four different designs with 6000 faulty circuits each (that are created using a variety of realistic defect behaviors including byzantine bridges and opens). The experiments demonstrate the potential of LearnX. LearnX achieves an average accuracy of 97.3%, while two state-of-the-art commercial diagnosis tools achieve 88.1% and 76.0%. Thus, LearnX identifies the correct candidate for at least 9.2% more fail logs than commercial diagnosis.

In addition, LearnX returns a single candidate for 70.1% fail logs, an improvement of 46.7% and 41.5% over commercial tools. When the number of diagnoses with high resolution (i.e., when at most three candidates are reported) are analyzed, it is observed that LearnX returns a resolution of three or less for a majority of fail logs (specifically, 89.3%). More importantly, LearnX hits a home run (i.e., when a single candidate is identified correctly) significantly more often than commercial diagnosis. LearnX delivers a home run for 67.9% of fail logs, which is 67.5% better than one commercial tool and almost double the second commercial tool.

Moreover, significance of LearnX is substantiated by diagnosing 2400 silicon fail logs from a design fabricated in an advanced process technology. It is revealed that LearnX returns an ideal resolution for 46.9% fail logs, which is almost twice as often as commercial diagnosis. Additionally, LearnX returns 7.1 fewer candidates per fail log, on average. LearnX enhances the resolution for 71.5% of fail logs, and achieves an average resolution improvement of 40.8%.

The effectiveness of LearnX is further corroborated by inspecting 19 failing chips that are PFA'ed. **LearnX is able to correctly locate a defect in each failing chip, while reporting fewer candidates than state-of-the-art commercial diagnosis.** LearnX returns fewer defect candidates than commercial diagnosis for 52.6% fail logs. On average, LearnX identifies 2.4 fewer candidates per fail log. Furthermore, LearnX delivers a home run for 52.9% of fail logs, which is 12.6% better than commercial diagnosis.

3.2   MD-LearnX: A Deterministic-Statistical Multiple Defect Diagnosis Methodology

Results presented in Sect. 3.1 demonstrate the superior performance of LearnX for single-defect diagnosis when compared with leading-edge commercial diagnosis. However, with decreasing feature sizes, and increasing interconnect density and manufacturing complexity, more chips are failing due to multiple defects, particularly when systematic defects (that arise from unforeseeable process-design interactions) are the dominant yield limiters (either in the early stages of yield

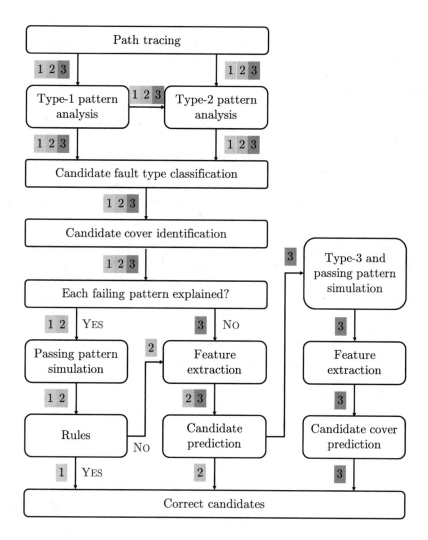

Fig. 4 Overview of the proposed three-phase diagnosis methodology, MD-LearnX. The sequence of steps in each phase are marked with a different color

learning or due to yield excursion). To that end, a single-chip diagnosis methodology called MD-LearnX is described here. It builds on LearnX (hence the name "MD-LearnX", where "MD" indicates multiple defects) to effectively tackle the task of locating more than one defect, and in turn, aid in accelerating the design and process development.

Figure 4 shows the overview of MD-LearnX. It is a three-phase diagnosis methodology. The sequence of analysis steps involved in each phase are marked with a different color.

The first phase focuses on finding defects that mirror the behavior of well-established fault models. Such defects are henceforth referred to as class-1 defects. Specifically, a set of strict rules are constructed for each candidate fault type to identify the correct cover of candidates. A defect candidate of a cover is deemed correct if it satisfies the rules.

Defect candidates of a cover that do not comply with the rules are further analyzed in Phase 2, which is especially geared towards the diagnosis of class-2 defects. It should be noted that the covers with the least number of class-2 defects are passed on to Phase 2 (by virtue of Occam's Razor).

Phase 2 aims to identify a defect whose behavior deviates from that of each fault model considered in Phase 1 and whose error propagation path does not interfere with errors stemming from other defect locations for each failing pattern. Such defects are henceforth referred to as class-2 defects.

The primary objective of Phase 2 is to apply machine learning to discern the correct candidate for each defect. The ML procedure is similar to Phase 2 of LearnX, albeit with one notable difference. It should be recalled that Phase 2 of MD-LearnX identifies multiple defects whose fault effects do not intervene with each other. Thus, for each defect candidate, only the outputs that are reachable from its location are considered during feature extraction. The remaining steps remain the same as LearnX Phase 2.

Finally, any failing chip left undiagnosed is analyzed in Phase 3. Such a failing chip is affected by multiple defects where errors manifesting from at least two defects interact with each other such that no single candidate can explain the observed response for at least one failing pattern.

In Phase 3, machine learning classification is applied at two different levels; first, at a defect level, where defect-type specific learning models built using the approach outlined in Phase 2 classify each defect candidate as correct/incorrect, and, second, at a candidate-cover level, where the entire cover is predicted as correct/incorrect.

However, compared to Phase 2, there are a few differences when defect-level classification is applied in Phase 3. First, a candidate cover, up to this point, is only simulated for the failing patterns explained by a single or non-interacting X faults, and not the remaining failing or passing patterns. Thus, only the relevant features that correspond to the failing patterns are used to train a defect-level ML model in Phase 3. Second, the training data is created from analyzing virtual fail logs that are produced by injecting and simulating **multiple** defects in the circuit. Third, in order to be conservative and avoid eliminating a correct candidate cover, a cover is analyzed further if at least one of its component defect candidates is predicted correct by the machine learning model built for the corresponding fault type. Each cover is then simulated for the remaining patterns.

Next, machine learning is utilized at a candidate cover level to predict a cover of defect candidates as correct or incorrect. A single random forest is trained using the steps outlined in Sect. 3.1 with the difference being that each training instance is a candidate cover here. A cover is then said to represent actual defects when it is predicted correct by the machine learning model.

A comprehensive simulation-based experiment is conducted to assess MD-LearnX, where a total of 28,000 faulty circuits with varying defect multiplicities and behaviors are created and analyzed. Three metrics, namely, diagnosability (i.e., proportion of injected defects that are correctly located), precision (i.e., proportion of reported defects that are correctly identified) and *home run* (i.e., when a single correct candidate is reported for each injected defect), are employed to measure the performance of MD-LearnX. The effectiveness of MD-LearnX is also compared with two leading-edge commercial diagnosis tools (that are referred to as Tool 1 and Tool 2).

The proposed methodology achieves an average diagnosability of 0.78, which is 15.4% and 58.2% better than Tool 1 and Tool 2, respectively. The average precision for MD-LearnX is 0.7, and is an improvement of 57.0% (2X) over Tool 1 (Tool 2). Additionally, MD-LearnX hits a home run for 40.0% of fail logs, which is twice as often as commercial diagnosis.

The efficacy of MD-LearnX is impressive for large values of defect multiplicity. Specifically, when the number of injected defects is at least five, the diagnosability of MD-LearnX is 22.8% higher than Tool 1 and 2.3X times Tool 2, on average. MD-LearnX returns a correct candidate for each reported defect 2.4X (6.2X) more often than Tool 1 (Tool 2). Moreover, MD-LearnX delivers a home run for 6.8% of fail logs; however, commercial diagnosis returns a perfect diagnosis for less than 0.3% of fail logs.

The capability of MD-LearnX is further demonstrated with a silicon experiment, where 36 fail logs whose PFA results are available are diagnosed. It is seen that MD-LearnX returns fewer candidates than commercial diagnosis for 69.4% of the fail logs, without sacrificing accuracy. Moreover, MD-LearnX reports 5.3 fewer candidates per fail log, on average.

3.3   Improving On-Chip Diagnosis Accuracy via Incremental Learning

Ensuring the lifetime reliability of integrated systems has become a central concern. A robust system should be able to continue acceptable operations over its intended lifetime even in the presence of failures. Although manufacturing tests are performed to help ensure reliability, a chip may still degrade and even fail in the field due to various locations of failure. Early-life failure, also called infant mortality, is caused by the defects that are not exposed during manufacturing tests. However, electrical and thermal stress during in-field use will eventually degrade the defect to a significant failure in functionality. Wear-out, also called aging, manifesting as progressive performance degradation, is induced by various mechanisms, e.g., negative-bias temperature instability (NBTI) and hot-carrier injection (HCI).

Various methods have been proposed to detect and avoid failures. First, forward error control (FEC) method uses error correction codes (ECC) to detect and correct data faults during transmission by adding data redundancy to the packets [25–28]. However, FEC does not target permanent failures, e.g, early- life and wear-out [29]. On the other hand, most error detection/correction methods incur significant power, performance and area penalties [30]. Second, failure-prediction schemes provide an early warning of circuit aging before errors appear. Specifically, an aging sensor periodically checks the slack of a critical path in order to avoid the occurrence of delay faults. This approach requires an aging sensor and a stability checker for each flip-flop which however incurs a large area overhead. Third, on-chip self-test schemes test the system periodically for failure detection [2].

To enable failure localization, diagnosis is performed when on-chip self-test detects a failure. On-chip effect-cause techniques need a complete model of the design, test-vector set, and the test response from the failing circuit to identify possible fault locations, and hence require significant memory and runtime. On-chip cause-effect techniques are ineffective for modern designs because of the size of the fault dictionary.

Work in [2] performs diagnosis at a module level (e.g., sub-core/sub-uncore) to reduce the dictionary size. In addition, instead of storing the entire simulation response, it only stores a single bit per test that indicates the pass/fail status of a subset of faults. However, it sacrifices diagnostic resolution because it only narrows down the defect location to a module.

To address the drawbacks of prior work, [31] describes a novel incremental-learning algorithm called a dynamic k-nearest-neighbor (DKNN) to improve the accuracy of on-chip, module-level diagnosis. Here, accuracy is defined as the probability that the identified module is the one with the failure, assuming that a single module is faulty. It is assumed that on-chip testing is performed with a test clock that has a higher frequency than the system clock because it allows failure sources that slowly degrade system timing to be tracked over time. For example, delay degradation due to NBTI can be monitored, enabling system adjustments (e.g., task scheduling, sleep scheduling, etc.) that mitigate adverse effects on system lifetime. Another consequence of using a faster clock for test, means failures will be much more frequent, thus creating sufficient data for learning a model for improving diagnostic accuracy. Different from the conventional KNN, DKNN employs online data to update the learned classifier, enabling the classifier to evolve as new data becomes available. Consequently, DKNN is able to track the fault distribution, especially when the fault distribution is non-stationary.

Pattern recognition methods have been used in the past to make diagnostic decisions in analog circuits [32]. In particular, possible circuit defects are identified through inductive fault analysis. Then a set of classifiers, trained offline using data from fault simulation, is used to map each defect to a score according to its likelihood of occurrence. In [33], an incremental KNN algorithm that also revises the composition of data set by exploiting a "correct-error" teacher is proposed. In their approach, each instance in the data set is associated with a dynamically-evolving weight, that requires additional overhead if implemented on chip. Moreover, the memory required increases as more data is collected, making it difficult to determine the required amount of memory a priori. Compared to the

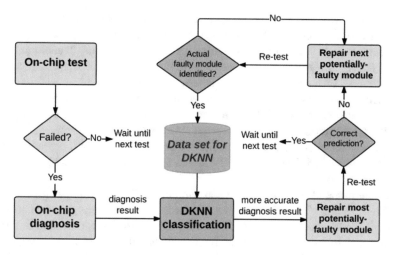

Fig. 5 An on-chip diagnosis result is fed into the on- chip DKNN classifier. The module predicted by the DKNN classifier is repaired. If the predicted label proved to be incorrect, the DKNN data set is updated

incremental KNN in [33], DKNN is more hardware-friendly because it maintains a fixed-size data set, and only requires little additional logic for data replacement.

There are two main features of the described work [31]:

- DKNN copes with non-stationary distributed data by updating the classifier incrementally using online data.
- DKNN can be implemented on chip for improving diagnostic accuracy, using little additional logic and a fixed-size data set.

The flow for improving the accuracy of on-chip diagnosis is depicted in Fig. 5. On-chip test/diagnosis generates a coarse diagnosis result which is then refined, if necessary, by the DKNN classifier. DKNN takes as input the diagnosis result from the on-chip diagnosis scheme described in [2], performs classification using the labeled data that is stored in the on-chip buffer, and updates the data set if the classification is later deemed to be incorrect.

The DKNN data set is initialized using fault simulation data, i.e., injecting faults into the gate-level netlist and then generating a fault-count for each module using simulation tools.

A periodic test/diagnosis starts with testing a core/uncore. If the test fails, then diagnosis is applied to identify which module of the core/uncore is most likely the reason for the observed test failure. The diagnosis result is reported as an n-dimensional vector of integers, where n is the number of modules. For each diagnosis result, a check for ideal resolution (i.e., only one fault-count is non-zero) is performed. If the resolution is ideal, then prediction via DKNN is obviously not needed. If the resolution is non-ideal, then DKNN is performed to predict which of the n modules is the faulty one. Specifically, the module is labeled using the outcome of a majority vote provided by its k nearest neighbors, and then data replacement is performed if the predicted label is incorrect. It is noted that only relevant data is considered when the classifier is searching for nearest neighbors. In addition, if a tie occurs during majority vote (i.e., two or more modules contribute the same number of neighbors), then either one module is selected as the label.

When a module is predicted to be faulty, it will be repaired and the system will be re-tested to determine if it was the actual faulty module. If the system passes, then the prediction is assumed to be correct; otherwise, the second-potentially faulty module will be repaired. Test, diagnosis, and repair continue until the actual faulty module is identified. Finally, after the actual faulty module is identified, the data replacement is performed.

To summarize, DKNN described here employs online data to update the learned classifier dynamically, so that the learned classifier can evolve as new data becomes available. Specifically, if the classification for a data instance is proved to be wrong, the nearest neighbor that is responsible for predicting the wrong label is replaced by the data instance itself.

Results from a simulation experiment conducted on two benchmark circuits demonstrate the efficacy of DKNN. Compared to select-max (where the module with the maximum fault count is deemed as faulty [2]), DKNN improves the accuracy up to 2X. Compared to KNN, DKNN improves the accuracy by 5%.

4 Post Diagnosis

In this section, we describe methodologies that use diagnosis results for a variety of activities. Many of these methodologies use the results of a large set of diagnoses for a given objective, all of which are typically referred to as volume diagnosis. In Sect. 4.1, two techniques (PADRE and LAIDAR) are described. Both use ML-based volume diagnosis to improve diagnostic resolution without negatively impacting diagnostic accuracy. Section 4.2 uses a classical expert system for scan chain diagnosis. Specifically, using initial tester results, this technique iteratively applies rules from the expert system to identify the location of a failing scan cell. Section 4.3 is the first-ever diagnostic technique that uses ML. For diagnostic results that do not exactly match the behavior of a conventional bridge fault, the method ABC (Automatic Bridge Classification) uses a decision tree classifier to predict whether a failure is a bridge defect. The last two sections (Sects. 4.4 and 4.5) use the results of volume diagnosis, not to improve diagnosis itself, but instead to achieve other objectives, namely, DFM rule/guideline evaluation and systematic defect identification.

4.1   Diagnosis Quality Enhancement Through Learning

Much effort has been dedicated to improving diagnosis because it is an important part of the yield learning process for understanding the root cause of failure. Diagnosis can be followed by physical failure analysis (PFA), a time-consuming and destructive approach for exposing the defect physically in order to characterize the failure mechanism [34]. Due to the high cost and destructive nature of PFA, the accuracy and resolution of diagnosis is of critical importance.[2]

In addition to being an integral part of PFA, diagnosis results from a population of failing chips also serve as input for a variety of analyses besides PFA [35]. For instance, volume diagnosis can reveal important statistics including the defect distribution or the primary yield detractors [36, 37], and provide useful feedback for evaluating and improving the quality of manufacturing test [37–39].

In practice, diagnosis tends to be non-ideal for a variety of reasons. Two such reasons include the limitation on test set size, and the equivalent circuit I/O behavior that inherently exists among candidates. Because there is a trade-off between the time needed to both create and apply tests, and the cost of test, it is always the case that not all possible defects are fully exposed when they are detected by the production test set. Even if a comprehensive test set is economically viable, there still can be candidates that have equivalent behavior among the many locations that are specific to the standard cells used and their interconnections. Also, the fault models adopted for both test and diagnosis are not perfect either, meaning it is quite likely that the actual defective behavior cannot be fully explained by the fault model(s) adopted [40]. The overall result is an imperfect diagnosis that typically produces an accurate result but a non-ideal resolution.

Physically-aware diagnostic resolution enhancement (PADRE) [41, 42] improves diagnostic resolution through the use of ML. Specifically, defect- and candidate-specific features are derived to characterize and distinguish the diagnosis candidates. Some of the features are well established such as TFSP, TPSF, and TFSF [41, 42]. Some others are new, established in [41, 42] for characterization of more sophisticated candidate properties. The feature data from a population of candidates are used to learn a classifier that separates good candidates from bad ones.

Unlike other ML-based techniques, PADRE constructs good-labeled and bad-labeled training sets from unlabeled candidates that are originally available, instead of relying on historical data or simulation data. As shown in Fig. 6, a two-level classifier is deployed by PADRE. The first level is a simple rule-based check of the neighborhood consistency of each candidate, which is found by past work [1, 5, 11–13] to be very adept at identifying some types of bad candidates. The second level uses an support vector machine (SVM) classifier that is learned from some of the candidates that pass the first level to predict which candidates are

---

[2] Here, a diagnosis is *accurate* if the true defect location is included in the reported candidate list, whereas the *resolution* is the total number of candidates reported for each defect. The diagnosis outcome is considered ideal if and only if the resolution equals to one and it is accurate.

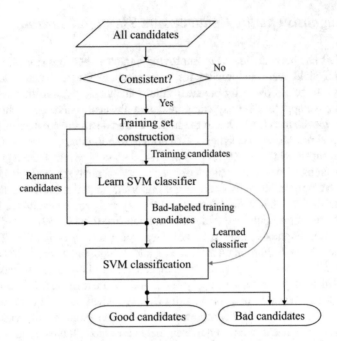

Fig. 6 PADRE takes all the candidates through a two-level classifier, in which bad candidates are eliminated and good candidates are identified. By reducing the total number of candidates while maintaining accuracy, the resolution of the diagnosis is improved

good. To characterize the candidates, PADRE incorporates not only well-established features used by conventional diagnosis, but also many newly-established features. These features are analyzed for their effectiveness and are shown to boost overall classification accuracy. In the experiments, PADRE is shown to improve resolution of failing chips significantly for three different data sets, including one virtual data set and two silicon data sets. In addition to improving resolution, PADRE is shown to maintain accuracy, as verified in the virtual data set experiment and on a few PFA'ed chips in silicon data set experiment.

PADRE demonstrates that it is possible to better distinguish the diagnosis candidates by exploiting the logical and physical characteristics related to the candidates as well as the its neighborhood under different activation and sensitization conditions. Also, as an add-on resolution improvement technique, PADRE can be incorporated easily with any existing diagnosis technique without much additional cost.

As with any classification algorithm, the accuracy of PADRE can be further improved with the addition of high quality training data. Such training data can stem from PFA, which reveals the true labels of all the candidates associated with a defect. However, not only are the PFA resources typically limited, but the results from PFA may not always improve classification accuracy. Therefore, it becomes important to identify which defects should be PFA'ed so that these limited resources can be efficiently tapped for ensuring and improving classification accuracy in addition to the conventional purpose of revealing the root-cause of failure.

In [43], active learning (AL) helps to mitigate the conflict between limited PFA resources and the large amount of failed defects by querying the most informative defects for improving the classification accuracy of PADRE. In each AL iteration, from a population of diagnosed defects that are qualified for PFA by other necessary criteria (e.g., defects that have no more than five candidates), the most valuable defect is identified by one of two selection methods; namely discrepancy check and within-margin. Then the labels for the candidates of the queried defect are obtained through a successful PFA, which are in turn fed back to PADRE as training data for updating the classifier. This iterative process continues until PFA resources are exhausted, or when no more defects satisfy the selection criteria

AL PADRE is validated by both simulation-based experiment and silicon experiment. Simulation-based experiments show that by using AL PADRE, the number of PFAs required for increasing the accuracy to a stable level of 90% is reduced by more than 6 compared to baseline approach, and AL PADRE consistently outperforms the baseline approach for accuracy improvement in various scenarios. In the silicon experiment, by using AL PADRE, the number of chips needed to undergo PFA was reduced by more than 6X in order to increase diagnosis accuracy by more than 20%.

While PADRE has the valuable characteristic of "learning from the current data and applying the knowledge back," its prediction performance is largely restricted by the limited labeled data and some incorrect labels inherent to its approach. As a result, the SVM classifier learned from limited and partially correct data can make mispredictions that compromise accuracy (when good candidates are predicted as bad) and resolution (when bad candidates are predicted as good). For example, the work of [41] reports that 476 out of 3441 good candidates are incorrectly predicted as bad by PADRE for a synthesized dataset. To improve the accuracy of PADRE, the work of [43] adds active learning with more informative labeled samples from PFA. The approach is limited however given the small number of PFAs performed due to cost.

On the other hand, we have observed many potential sources of information that can be exploited for distinguishing good and bad candidates, which are not used by previous techniques like PADRE. For example, while a large portion of data samples cannot be labeled beforehand, their correlations to the labeled samples can facilitate the creation of a better classification model. Also, besides candidate-level information, defect-level information such as which candidates correspond to the same defect, the similarity between defects, etc. can be exploited as additional knowledge.

Given the limitation of previous methods and the new opportunities that manifest in various aspects of diagnosis data, a new methodology named *LAIDAR* (Learning for Accurate and Ideal DiAgnostic Resolution) [44] is proposed to enhance diagnostic outcome towards the ideal case—accurate with ideal resolution. LAIDAR adopts a similar labeling procedure and the similar candidate features as PADRE, but aims to fully exploit available information from the dataset using more advanced

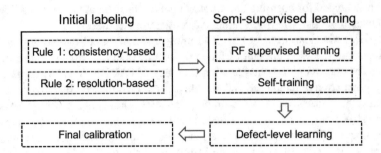

Fig. 7 The overall flow of LAIDAR consists of four stages

ML techniques. Specifically, LAIDAR deploys a semi-supervised learning (SSL) algorithm to consider not only the labeled data but also their correlation with unlabeled data. In addition, unlike PADRE which considers each candidate to be independent, LAIDAR incorporates defect-level information to further improve resolution. Experiments with both virtual and silicon datasets show that LAIDAR significantly outperforms PADRE in terms of both accuracy and resolution (e.g., 6.4X increase in occurrences of perfect diagnosis).

As shown in Fig. 7, the overall flow of LAIDAR is divided into four stages. LAIDAR takes as input the original diagnosis results from a population of failing chips. Like PADRE, each sample in the input dataset contains the extracted features of a certain candidate but also includes the corresponding defect information.

Initially, all the samples are unlabeled. In the initial labeling stage, two rules are adopted from PADRE with some modification. After this stage, some samples are labeled and used for the follow-on semi-supervised learning (SSL) stage. One SSL algorithm, namely self-training, which is based on a random forest (RF) supervised learning (SL) model is deployed. The SSL model predicts a label for each sample in the dataset. Then the defect-level learning process follows to further improve the prediction by leveraging defect-level information. Finally, a calibration procedure is employed to ensure diagnostic accuracy.

The overarching goal for LAIDAR is to first guarantee accuracy, and then improve resolution as much as possible. The motivation behind this goal is that an inaccurate diagnostic result can be misleading, and can directly lead to a an unsuccessful PFA. On the other hand, high resolution is not so detrimental because one can choose to perform PFA on other chips with low-resolution diagnostic results. Towards this goal, each step has its own focus on either guaranteeing accuracy or improving resolution. For example, the optimization goal of the RF model during self-training maximizes the recall of the good class, so that accuracy is given priority. For the defect-level learning process, the aim is to eliminate bad candidates that are previously predicted as good for resolution improvement. Finally, in the calibration process, each defect must have at least one good candidate, so if a defect has all of its candidates predicted as bad, the labels for all candidates will be reversed to good—this is a measure that trades resolution improvement for better accuracy.

4.2 Automated On-ATE Debug of Scan Chain Failures

As the complexity and size of modern integrated circuit design continue to increase, so does the complexity and volume of tests required for adequately screening for failures. SoCs will require a 20% increase in year-over-year test data volume [45]. Though pre-silicon verification of these complex tests is performed, its scope is limited by the massive number of tests, the runtime and compute requirements of the verification techniques, and the aggressive tapeout-to-first silicon timelines. Consequently, the first-silicon bring-up phase of a new product is extremely important, where each test must be verified for all of the cores and performance modes.

A test that meets the required specifications can be included in a production test program. However, when a test does not, it must be debugged to identify the underlying issue, followed by the search for a resolution. While improvements in test generation methodologies are constantly being made, the increasing volume and complexity of tests has still resulted in an increasing number of tests that do not work. Non-working tests stem from sources that span across all involved domains, including the physical setup of the Automated Test Equipment (ATE), potential errors in the test setup which configures the design-for-test (DFT) logic, misalignment between on-silicon observations and pre-silicon modeling, etc. [46]. This range of sources usually requires a set of engineers that have knowledge of all the domains, or more often, expert debug engineers that have significant debug experience to call upon. Expert debug engineers have already seen a number of these issues, and over these experiences have obtained knowledge on the various domains, enabling them to connect dots that are not immediately apparent. As a result, these experts can more easily and accurately construct a hypothesis on the source of the failure, and quickly develop targeted experiments that prove or disprove the hypothesis. Debug expertise offers significant acceleration of test-failure analysis, driving a case toward a hypothesis with targeted experimentation, where it may otherwise meander or stall altogether. However, given the high volume of non-working tests across concurrent products, this limited debug expertise is often unable to scale and meet debug needs.

An Automated, on-ATE AI (AAA, pronounced "triple A") framework is proposed to address this challenge [47]. By representing the thought process of a debug expert within an AI expert system of if-then rules, an inference engine can be used to obtain knowledge to satisfy a desired objective. It is demonstrated that an expert system tightly coupled with dynamic control of the ATE can perform hypothesis formation, experiment implementation, test execution, and analysis in a fully automated fashion.

An expert system revolves around a set of rules and facts [48] . Rules are in the "if-then" form, where if the antecedents (statement(s) under "if") are all true, the consequent (statement under "then") are deemed facts. Facts are the statements which have been asserted or inferred as true for the current case under examination. These base elements are key to the typical three main components of an expert system:

- The **Knowledge Base** consists of the complete set of rules for the given domain. When the rule antecedents are all facts, it follows that the consequent must also be a fact.
- The **Database of Facts** represents all of the true statements being used for evaluation in the current case. The facts within the expert system may be standard for all use cases, input by the user to represent the active case, or facts inferred using the rules in the knowledge base.
- An **Inference Engine** performs the logical processing, tracing through the knowledge rules with the use case information in the fact database. The engine interacts with the expert system user through an interface, accepting objectives, asking questions, and providing results.

As shown in Fig. 8, the overall automated debug process is initiated by the user, who provides an objective to the Inference Engine of the expert system. The Inference Engine will then query the Knowledge Base, which is complete with expert rules identified from the Knowledge Acquisition process. When a rule-required fact exists within the Database of Facts, that fact is used to satisfy (or negate) the rule. In addition to rules for inference, there also exist a set of information gathering rules, which when reached by the inference engine, request execution of a function external to the system. Most commonly, an information gathering rule will request the results of a test execution, either in pass/fail form or as set of failing cycles. With this request, a call is made to a connected, live, online ATE with the die-under-test loaded and active. Other sources of information include relevant DFT information and commercial diagnosis tools.

AAA provides a number of benefits, most significantly the savings of human resources for many debug tasks. AAA can significantly accelerate initial data collection by performing classification and reporting of the main failure objectives. Automation of these tasks frees up debug expert engineers from this data collection, allowing them to instead focus on those particularly complex failure cases that require extra attention. In addition, with a common debug procedure, a systematic set of steps is followed and a familiar result is produced. This makes the summarization and communication of debug status significantly more efficient as increasingly many projects are being executed and managed at once. Finally, as compared with other potential methods of artificial intelligence, expert systems provide a high degree of explainability. This framework enables a step-by-step reporting of all rules that were encountered as the system aimed to satisfy the objective. A study of the triggered rules represents the thought process of an expert, from which other engineers can become debug experts as well. Experimental evaluation on a set of industrial chips demonstrates the speed-up in extracting the objectives of failure

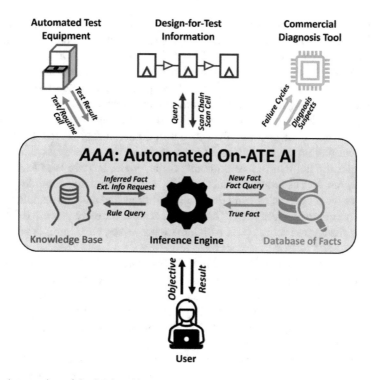

Fig. 8 An overview of the AAA architecture

debug, in turn accelerating time-to-resolution of failing tests during new product bringup. AAA provides these significant benefits at this particularly critical point in the overall product life-cycle.

4.3   Automatic Classification of Bridge Defects

Any investment that advances processing techniques must be supported by an investment in yield learning technologies. Inline inspection tools that optically examine wafers between key processing steps present a potential bottleneck as the number of steps increases. As on-chip dimensions decrease, so does the effectiveness of optical inspection techniques, leading to reports of a glut of potential defects that ultimately have no effect on yield. Discrete test structures provide excellent resolution and accuracy of defects, but at the precious cost of silicon area, forcing a significant tradeoff between the ability to learn about defects and satisfy product-volume requirements. Late in the process lifecycle, discrete test structures become relegated to the scribe lines, significantly decreasing the number of defects they observe.

The natural evolution of yield learning techniques is to use the product itself as a characterization vehicle in conjunction with failing test results generated by defective chips. Early in the process lifecycle, SRAMs will no doubt continue to be the learning vehicle of choice, but even at this stage where baseline process parameters are established, new, advanced techniques to leverage test results can provide valuable insight into both systematic and random yield detractors.

Using the product as a yield learning vehicle has three significant advantages over other approaches. First, it covers the entire wafer, maximizing defect observability. Second, working products can be sold for profit, whereas nonproduct characterization vehicles are not themselves profitable if they yield, but can later lead to yield improvements for customer products. Finally, the product itself provides the best diversity of layout geometries and the exact conditions that must be manufacturable. While the extreme regularity of SRAMs is convenient for early process debugging, ultimately the entire product (i.e., ASIC, microprocessor, GPU, etc.) is what needs to yield.

A method is proposed in [1] to aid product-test-based yield learning by providing an automatic way to determine the defect type that leads to chip failure. In particular, this work emphasizes the automatic classification of chips as either failing due to a bridge defect, or failing due to some non-bridge defect. This is a key enabling technology for related work that aims to measure defect density and size distributions (DDSDs) of bridge-causing defects using only product test results [49, 50]. It should be noted that it is not necessary that every bridge defect be identified, nor is it necessary that there be zero misclassification. Some level of imperfection is tolerable in many applications (e.g., DDSD measurement has been shown to tolerate a mis-classification rate of 10% [51]). Having improved methods to automatically group failing chips by defect type could open up new opportunities for automatic Pareto generation as well as drive decisions about which chips should undergo costly and time-consuming PFA.

Although many logic diagnosis techniques exist, they traditionally focus on localizing a defect. The emphasis in this work is on determining, with some degree of confidence, whether the defect causing a chip to fail is a bridge or some other defect type. Figure 9 shows the overall flow of the bridge classification method. Logic diagnosis is performed on scan-test results to identify candidate nets that could potentially be affected by a defect within a failing chip. Candidate nets resulting from diagnosis serve as the starting point for this work. This is a much better choice than initiating an analysis that uses all nets from the entire design. The bridge classification method allows for many different types of diagnosis at this stage. That is, a precise diagnosis technique can be employed to narrow the set of initial candidates to a very small number, or a more conservative diagnosis approach can be used that allows a large set of candidate nets to be analyzed.

For each candidate net pair, several features are identified from the logic model that specifically relate to that net pair. Examples of logic features include signal correlation and whether or not a feedback path exists. Features of the failing test response are also extracted for the net pair. Examples of test response features include stuck-at fault model performance and various bridge fault model

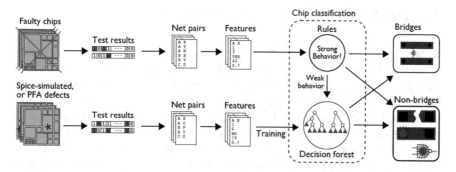

Fig. 9 An overview of the classification method

performances. Bridges and other defects can have complex behavior that is not captured well by logic models of the defect. Thus, several logic models are considered collectively in addition to features that are selected based on their ability to help predict whether the defect affecting a faulty chip is a bridge. There are two steps to the proposed classification technique. The first applies "rules" designed to identify defects that comprehensively mimic all the characteristics of known bridge behaviors with additional constraints, or alternatively exhibit behavior that cannot be caused by a bridge defect. Bridges identified in this step exhibit not only bridge behavior, but provide strong evidence that it is not some other defect type masquerading as a bridge. Non-bridge defects identified in this step have behavior that strongly indicates the defect cannot be a bridge. These rules must identify bridges with very little chance for defect-type misidentification. Defects that do not satisfy the bridge rules may still be bridges. For these subtle bridge defects, a machine-learning approach known as a decision forest is used [5]. The decision forest classifier is built using test results from thousands of SPICE-simulated defects.

The two-step approach to classification identified 36% of all bridges using the model-based rules. Once the second step was applied, an average of 80% of all bridges were identified correctly with a final data set consisting of an average of over 91% bridges. Classification involving net pairs that were formulated using only actual defect sites led to 91% of all bridges, on average, being correctly identified. The final data set consisted of an average of 97% bridges. This particular result shows that very good diagnosis will lead to outstanding classification results. However, the resolution offered by today's diagnostic tools are more than adequate to produce significantly large populations of failing chips that are nearly homogeneous. Overall, these results indicate that the method presented to classify defects can identify a significant portion of net bridges, while preventing non-bridges from being misclassified. The ability to automatically create a set of chips that fail due to bridges will enable product-test-based yield learning techniques such as automatic measurement of defect density and size distributions (DDSDs)

4.4 DFM Evaluation Using IC Diagnosis Data

Design for manufacturability (DFM) rules are constraints placed on the physical characteristics of a design (i.e., the layout) that are intended to improve yield or some other desired design property. The cost of DFM for a given design can be measured in terms of the additional die area, power consumption, and design time resulting from imposing the rules on the design layout. The payoff in terms of yield is extremely difficult to measure however, especially when rules are applied in varying degrees and in an ad hoc fashion. For example, it is not uncommon to hear from designers that top-priority rules are applied or imposed at a 0.90 adherence rate, second-priority rules are imposed at a 0.60 adherence rate, and low-priority rules are imposed at a 0.40 rate, where adherence rate is the fraction of layout locations where the rule is applicable and actually imposed. More often than not, the applicable layout locations a rule is imposed or not imposed is somewhat arbitrary in nature, not taking into account, for example, if rule adherence at a site A is more advantageous than at site B.

Unlike previous approaches which are employed for evaluating DFM rules prior to high-volume manufacturing [1–3], DFM rule evaluation using manufactured silicon (DREAMS) [52], has the goal of measuring DFM rule effectiveness using information extracted from actual failed ICs. The DFM rule-violation database is generated pre-silicon and is a tabulation of all the layout locations where some DFM rule/guideline has been violated. Failed-IC diagnosis is applied post-silicon to fabricated instances of the design (i.e., chips) that have failed testing. The outcome of diagnosis (especially layout-based diagnosis) are the suspected layout locations where the diagnosis tool/procedure believes failure could have occurred ([4–6]). Ideally, diagnosis reports one location for each actual failure site, but in most cases, several possible locations are reported. DREAMS correlates the locations reported by diagnosis for a population of failed ICs with the DFM rule-violation database to identify those rules that are effective in preventing failure.

DREAMS take into account the type of failure that a given rule is meant to guard against, along with precise information concerning the frequency of violations within the design and among the locations reported by diagnosis. These additional insights, along with a custom formulation of the expectation-maximization (EM) algorithm [8], improve accuracy dramatically and also, as a by-product, improve diagnostic resolution [9].

Experiments involving both virtual and silicon/design data have demonstrated the efficacy of DREAMS. In addition, it has been shown that DREAMS is superior to prior work and applicable to various scenarios characterized by differing degrees of rule violation. Application of DREAMS to an industrial design and diagnosis data has revealed that two of the twenty rules examined have an overall importance level that significantly higher than the values of the remaining rules. A by-product of DREAMS is a significant improvement in diagnostic resolution. Specifically, by understanding which rule is most likely responsible for failure of a given chip allows one to further localize the failure to a more precise location. For the design and test data examined in [52], resolution is improved by 69%.

There are several trends in the chip industry that make DREAMS and other similar methodologies worthwhile:

1. Many design houses and foundries commonly fabricate product-like test chips at volumes that sufficiently meet the sample-size requirements of DREAMS. This means that a DFM rule deck can be evaluated using actual product-like layout features, with results on rule importance being fed back to designers for deployment on actual customer designs.
2. For a given technology, it is typically the case that many subsequent designs are launched after the lead product. This is certainly true in the automotive industry where up to a dozen follow-on designs are launched after the lead product. Under this scenario, DFM rule evaluation can be continuously applied to chips 1 through i with the resulting learning applied to chip design $i + 1$.
3. DREAMS can also be used as on-going monitor of the fabrication process. Specifically, monitoring rule-failure rate gives insight into what part of the process has to be tuned since a given rule is concerned with particular features fabricated by specific steps of the manufacturing process (e.g., a via type between two adjacent layers i and $i + 1$).

4.5 LASIC: Layout Analysis for Systematic IC-Defect Identification using Clustering

Defects caused by random contaminants were the main yield detractor in IC manufacturing before the nanoscale era. However, as CMOS technology continues to scale, the process complexity increases tremendously, which results in design-process interactions that are difficult to predict and control. This, in turn, translates into an increased likelihood of failure for certain layout features that are sensitive to particular process corners. Unlike random contaminants, the defects that result from design-process interactions are systematic in nature, i.e., they can lead to an increase in likelihood of failure in locations with similar layout geometries (note that it does not mean that they must lead to a drastic yield loss). Since a product IC typically contains a diverse set of layout features, it is difficult to use test structures to completely characterize the design-process interactions because of their limited volume.

To address these issues, volume diagnosis is increasingly deployed to improve/supplement yield learning. Because it uses data from failing chips, it does not intrude into the fabrication and test process. It only consumes CPU cycles and thus does not consume extra silicon real estate. Another benefit of volume diagnosis is that it is performed on actual product ICs which contain the diverse geometries that may render conventional test structures inadequate.

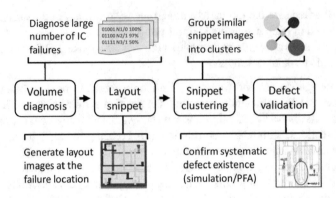

Fig. 10 Overview of the LASIC methodology

LASIC (Layout Analysis for Systematic IC-defect identification using Clustering) [53] identifies systematic defects for yield improvement using volume diagnosis. By clustering layout images of diagnosis-implicated regions, layout-feature commonalities (if any) that underlie the failures are identified. Features present in large clusters can then be analyzed further to confirm the existence of a systematic issue, or alternatively physical failure analysis (PFA) can be performed on one or more of the actual failing chips that have the failing feature.

Figure 10 overviews LASIC. LASIC consists of four steps: (1) volume diagnosis, (2) layout snippet extraction, (3) snippet clustering, and (4) validation.

Volume diagnosis is simply applying diagnosis to a sufficiently-large number of IC failures. The outcome of diagnosis is a set of candidates, where each candidate consists of the suspect net or cell where the defect is believed to reside.

A layout region (i.e., a region involving one or more layers in the layout that contains the features of interest, for example, a net and its surrounding nets) is extracted for each candidate. Different candidates consist of different polygons that reside in various layers, thus making their comparison non-trivial. Therefore, to standardize the scale and the dimension of the layout region, the layout region implicated by a candidate is split into smaller regions of fixed scale and dimension that are called layout snippets.

The layout snippets are then saved as images in the png format. After the images are generated, the discrete cosine transform (DCT), which is a widely used technique in image processing, is performed on each image for dimensionality reduction and accelerating subsequent analyses.

Clustering is then applied to group similar snippet images together to identify any commonalities. The goal of clustering is to discover structure in the data by exploring similarity within the data. Employing clustering involves the following:

1. Representation of the object, which can involve feature extraction or selection,
2. A measure of similarity between objects,
3. Selection and application of a clustering algorithm such as K-means [54].

The method of employing K-means in this work is novel as well. It (1) performs K-means on the snippet images for each candidate first, (2) selects representative snippet images from each candidate using the resulting clusters based on the cluster centers, and (3) performs K-means again on the representative snippet images from all the candidates. This approach identifies unique representative layout features from each candidate so that repeating/similar features from the same candidate will not bias the overall clustering process. In addition, this approach has the advantage that the number of images to be clustered (in the second pass) is substantially reduced, thereby achieving a faster runtime.

Unfortunately, the distance metric adopted and many other commonly used distance metrics (e.g., cosine distance) are not rotation- or reflection-invariant with respect to the image orientation. Thus, LASIC first generates the clusters using the current distance metric, and then merges the resulting clusters using the cluster centers only, taking into account all possible orientations of the cluster centers. This is advantageous since the cluster centers are only used for the more expensive distance metric calculation. The number of cluster centers is typically far less than the number of images. To merge the clusters, the clustering outcome is represented using a graph, where each vertex represents a cluster center, and an edge exists between two vertices, if and only if the minimum distance between the cluster centers is less than a certain threshold. (This threshold should be chosen to have a similar value as the threshold used in the second K-means pass.) Clearly, two clusters should be merged if a path exists between their corresponding vertices in the graph. Thus, the cluster-merging problem is equivalent to identifying the connected components of the graph, i.e., clusters in the same connected component should be merged. This problem can be easily solved by using a depth-first search (DFS) in linear time.

Finally, the identified layout feature is simulated using a process simulator (such as the lithography tool Optissimo [55]) to confirm the existence of a systematic defect. Validation could take on other forms such as physical failure analysis (PFA) which are more comprehensive and conclusive.

Silicon and simulation based experiments are conducted to evaluate LASIC. In the silicon study, 738 failing ICs from LSI corporation are successfully diagnosed resulting in 2168 suspect nets. LASIC is applied to all diagnosis candidates using a 2um-by-2um area for snippet image extraction. The rationale for choosing this dimension stems from the fact that this dimension should capture all optical interactions caused by sub-wavelength lithography, which is the dominant cause of systematic defects. To ensure accuracy, each snippet image is chosen to have a resolution of 100-by-100 pixels. Figure 11 clearly shows that the layout features are sharply represented with this resolution. Figure 11 shows example images from two different metal-1 clusters after LASIC has been applied. Figure 11a shows four snippet images from a cluster in metal-1, while Fig. 11b shows four snippet images from another metal-1 cluster. Figure 11 shows that geometries in the same cluster resemble each other but are not exactly the same while geometries in different clusters exhibit substantial differences. This example clearly demonstrates

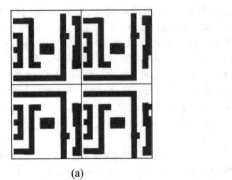 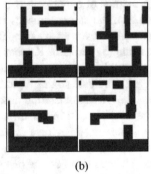

(a)   (b)

Fig. 11 Illustration of clustered snippet images (**a**) from one cluster and (**b**) a second cluster

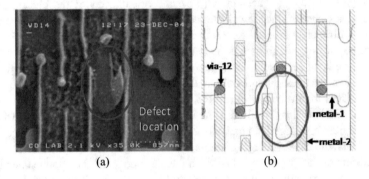

(a)   (b)

Fig. 12 (**a**) SEM image of a bridge in the metal-2 layer, and (**b**) its corresponding layout snippet with its lithography simulation result

that LASIC is able to extract from diagnosis data images that have similar layout features.

To further validate LASIC, the clustering outcome of an IC that was PFA'ed is also examined. The PFA result is a bridge between two nets in metal-2, and its SEM image and layout region are shown in Fig. 12. The snippet image corresponding to this defect belongs to a cluster that is in the top 10.6% of all clusters. These results are strong evidence that LASIC is able to identify systematic defects.

5 Conclusions

In this chapter, the use of machine learning in diagnosis by the Advanced Chip Testing Laboratory (ACTL) has been reviewed. Work in ACTL has been partitioned into three categories, namely, pre-diagnosis, during-diagnosis and post-diagnosis to indicate the applicability of a given methodology with respect to diagnosis execution. The collective work demonstrates that ML is a valuable technology for optimizing diagnostic results and the use of diagnostics for follow-on activities such as DFM rule/guideline evaluation and systematic defect identification. Even after a decade of using ML in diagnosis, there remains a great deal of ongoing and future application of ML in diagnosis.

References

1. Nelson, J.E., Tam, W.C., Blanton, R.D.: Automatic classification of bridge defects. In: International Test Conference (2010)
2. Beckler, M., Blanton, R.: On-chip diagnosis for early-life and wear-out failures. In: 2012 IEEE International Test Conference, pp. 1–10. IEEE (2012)
3. Fang, C., Huang, Q., Mittal, S., Blanton, R.S.: Diagnosis outcome preview through learning. In: VLSI Test Symposium, pp. 1–6. IEEE (2019)
4. Huang, Q., Fang, C., Mittal, S., Blanton, R.: Towards smarter diagnosis: a learning-based diagnostic outcome previewer. ACM Trans. Des. Autom. Electron. Syst. **25**(5), 1–20 (2020)
5. Breiman, L.: Random forests. Mach. Learn. **45**(1), 5–32 (2001)
6. Huang, Q., Chenlei, F., Mittal, S., Blanton, R.D.: Improving diagnostic efficiency via machine learning. International Test Conference (2018)
7. Zeng, W., Zhu, H., Zeng, X., Zhou, D., Liu, R., Li, X.: C-yes: an efficient parametric yield estimation approach for analog and mixed-signal circuits based on multicorner-multiperformance correlations. IEEE Transactions on Computer-Aided Design of Integrated Circuits and Systems (2016)
8. Fang, C., Yang, F., Zeng, X., Li, X.: BMF-BD: Bayesian model fusion on bernoulli distribution for efficient yield estimation of integrated circuits. In: 2014 51st Design Automation Conference (DAC) (2014)
9. Huang, Q., Fang, C., Yang, F., Zeng, X., Li, X.: Efficient multivariate moment estimation via bayesian model fusion for analog and mixed-signal circuits. In: 2015 52nd Design Automation Conference (DAC) (2015)
10. Segev, N., Harel, M., Mannor, S., Crammer, K., El-Yaniv, R.: Learn on source, refine on target: a model transfer learning framework with random forests. IEEE Transactions on Pattern Analysis and Machine Intelligence (2016)
11. Abramovici, M., Breuer, M.A., Friedman, A.D. et al.: Digital Systems Testing and Testable Design, vol. 2. Computer Science Press, New York (1990)
12. Tessent Diagnosis User's Manual: Siemens (2022)
13. Fang, C., Huang, Q., Blanton, R.: Adaptive test pattern reordering for diagnosis using k-nearest neighbors. In: International Test Conference in Asia (ITC-Asia), pp. 59–64. IEEE (2020)
14. Fang, C., Huang, Q., Blanton, R.S.: Memory-efficient adaptive test pattern reordering for accurate diagnosis. In: VLSI Test Symposium (VTS), pp. 1–7. IEEE (2021)
15. Madge, R., Benware, B., Turakhia, R., Daasch, R., Schuermyer, C., Ruffler, J.: In search of the optimum test set-adaptive test methods for maximum defect coverage and lowest test cost. International Test Conference, pp. 203–212 (2004)
16. Wang, H., Poku, O., Yu, X., Liu, S., Komara, I., Blanton, R.D.: Test-data volume optimization for diagnosis. Design Automation Conference (2012)
17. Mittal, S., Blanton, R.D.: LearnX: a hybrid deterministic-statistical defect diagnosis methodology. In: IEEE European Test Symposium (ETS) (2019)
18. Blanton, R.D., Mittal, S.: Integrated circuit defect diagnosis using machine learning. US Patent App. 16/986,963 (2021)
19. Mittal, S., Blanton, R.D.: MD-LearnX: a deterministic-statistical multiple defect diagnosis methodology. In: IEEE VLSI Test Symposium (VTS) (2020)

20. Mittal, S., Blanton, R.D.: PADLOC: physically-aware defect localization and characterization. In: IEEE Asian Test Symposium (ATS) (2017)
21. Mittal, S., Blanton, R.D.: NOIDA: noise-resistant intra-cell diagnosis. In: IEEE VLSI Test Symposium (VTS) (2018)
22. Wang, S., Wei, W.: Machine learning-based volume diagnosis. In: Design, Automation Test in Europe Conference (DATE), pp. 902–905. (2009)
23. Wang, S.: Machine learning based volume diagnosis of semiconductor chips. US Patent App. 12/269,380 (2010)
24. Gómez, L.R., Wunderlich, H.: A neural-network-based fault classifier. In: IEEE Asian Test Symposium (ATS), pp. 144–149 (2016)
25. Gizopoulos, D., Psarakis, M., Adve, S.V., Ramachandran, P., Hari, S.K.S., Sorin, D., Meixner, A., Biswas, A., Vera, X.: Architectures for online error detection and recovery in multicore processors. In: 2011 Design, Automation & Test in Europe, pp. 1–6. IEEE (2011)
26. Shamshiri, S., Ghofrani, A., Cheng, K.-T.: End-to-end error correction and online diagnosis for on-chip networks. In: 2011 IEEE International Test Conference, pp. 1–10. IEEE (2011)
27. Zhang, M., Mitra, S., Mak, T., Seifert, N., Wang, N.J., Shi, Q., Kim, K.S., Shanbhag, R.N., Patel, S.J.: Sequential element design with built-in soft error resilience. IEEE Trans. Very Large Scale Integr. VLSI Syst. **14**(12), 1368–1378 (2006)
28. Mitra, S., Zhang, M., Seifert, N., Mak, T., Kim, K.S.: Built-in soft error resilience for robust system design. In: 2007 IEEE International Conference on Integrated Circuit Design and Technology, pp. 1–6. IEEE (2007)
29. Ghofrani, A., Parikh, R., Shamshiri, S., DeOrio, A., Cheng, K.-T., Bertacco, V.: Comprehensive online defect diagnosis in on-chip networks. In: 2012 IEEE 30th VLSI Test Symposium (VTS), pp. 44–49. IEEE (2012)
30. Li, Y., Kim, Y.M., Mintarno, E., Gardner, D.S., Mitra, S.: Overcoming early-life failure and aging for robust systems. IEEE Design Test Comput. **26**(6), 28–39 (2009)
31. Ren, X., Martin, M., Blanton, R.D.: Improving accuracy of on-chip diagnosis via incremental learning. In: 2015 IEEE 33rd VLSI Test Symposium (VTS), pp. 1–6. IEEE (2015)
32. Huang, K., Stratigopoulos, H.-G., Mir, S., Hora, C., Xing, Y., Kruseman, B.: Diagnosis of local spot defects in analog circuits. IEEE Trans. Instrum. Meas. **61**(10), 2701–2712 (2012)
33. Förster, K., Monteleone, S., Calatroni, A., Roggen, D., Troster, G.: Incremental KNN classifier exploiting correct-error teacher for activity recognition. In: 2010 Ninth International Conference on Machine Learning and Applications, pp. 445–450. IEEE (2010)
34. Wagner, L.C.: Chemical analysis. In: Failure Analysis of Integrated Circuits, pp. 195–203. Springer, Berlin (1999)
35. Chung, K.Y., Nicholson, S., Mittal, S., Parley, M., Veda, G., Sharma, M., Knowles, M., Cheng, W.-T.: Improving diagnosis resolution with population level statistical diagnosis. In: ISTFA 2021, pp. 388–393. ASM International (2021)
36. Huisman, L.M., Kassab, M., Pastel, L.: Data mining integrated circuit fails with fail commonalities. International Test Conference (2004)
37. Tam, W.C., Poku, O., Blanton, R.S.: Precise failure localization using automated layout analysis of diagnosis candidates. In: Proceedings of the 45th Annual Design Automation Conference, pp. 367–372. (2008)
38. Lin, Y.-T., Blanton, R.D.: Test effectiveness evaluation through analysis of readily-available tester data. In: 2009 International Test Conference. IEEE (2009)
39. Yu, X., Blanton, R.D.: Estimating defect-type distributions through volume diagnosis and defect behavior attribution. In 2010 IEEE International Test Conference. IEEE (2010)
40. Aitken, R.C.: Finding defects with fault models. In: Proceedings of 1995 IEEE International Test Conference (ITC), pp. 498–505. IEEE (1995)
41. Xue, Y., Poku, O., Li, X., Blanton, R.D.: PADRE: physically-aware diagnostic resolution enhancement. International Test Conference (2013)
42. Xue, Y., Li, X., Blanton, R.D.: Improving diagnostic resolution of failing ICs through learning. Transactions on Computer-Aided Design of Integrated Circuits and Systems (2016)

43. Lim, C., Xue, Y., Li, X., Blanton, R.D., Amyeen, M.E.: Diagnostic resolution improvement through learning-guided physical failure analysis. International Test Conference (2016)

44. Huang, Q., Fang, C., Blanton, R.S.: LAIDAR: learning for accuracy and ideal diagnostic resolution. In: 2020 IEEE International Test Conference (ITC). IEEE (2020)

45. I.T.R. for Semiconductors: Executive Summary (2009). http://www.itrs.net

46. Molyneaux, R.: Debug and diagnosis in the age of system-on-a-chip. In: Proceedings of the International Test Conference, 2003 (ITC 2003), pp. 1303–1303. IEEE Computer Society (2003)

47. Nigh, C., Bhargava, G., Blanton, R.D.: AAA: automated, on-ATE AI debug of scan chain failures. In: IEEE International Test Conference (ITC), pp. 314–318. IEEE (2021)

48. Hayes-Roth, F., Waterman, D.A., Lenat, D.B.: Building Expert Systems. Addison-Wesley Longman Publishing (1983)

49. Nelson, J.E., Zanon, T., Brown, J.G., Poku, O., Blanton, R.D., Maly, W., Benware, B., Schuermyer, C.: Extracting defect density and size distributions from product ICs. IEEE Design & Test of Computers (2006)

50. Nelson, J.E., Zanon, T., Desineni, R., Brown, J.G., Patil, N., Maly, W., Blanton, R.: Extraction of defect density and size distributions from wafer sort test results. In: Proceedings of the Design Automation & Test in Europe Conference, vol. 1, pp. 1–6. IEEE (2006)

51. Nelson, J., Maly, W., Blanton, R.: Random defect characterization sensitivity to test, diagnosis, and volume. GRC Techcon (2008)

52. Blanton, R.D.S., Wang, F., Xue, C., Nag, P.K., Xue, Y., Li, X.: DFM evaluation using IC diagnosis data. IEEE Transactions on Computer-Aided Design of Integrated Circuits and Systems (2016)

53. Tam, W.C.J., Blanton, R.D.S.: LASIC: Layout analysis for systematic IC-defect identification using clustering. IEEE Trans. Comput. Aided Des. Integr. Circuits Syst. **34**(8), 1278–1290 (2015)

54. MacQueen, J.: Some methods for classification and analysis of multivariate observations. In: Proceedings of the 5th Berkeley Symposium on Mathematical Statistics and Probability, p. 281 (1965)

55. Optissimo: The Optissimo User Manual. PDF Solutions (2001)

Machine Learning Support for Cell-Aware Diagnosis

Aymen Ladhar and Arnaud Virazel

1 Introduction

The usage of Cell-Aware (CA) methodology becomes mandatory for semiconductor industry, especially for designs with the highest product quality. This is because fault models like stuck-at, transition, as well as layout-aware fault models are not accurate enough to achieve very low Defective Part Per Million (DPPM) rates and to resolve the underlying systematic yield detractors for a successful and fast yield ramp-up. Previous works on CA defect test and diagnosis can be classified into two categories. Techniques in the first category extend the application of logic algorithms to deal with transistor defects [1, 2]. The main weakness of these techniques is the quality of the logic fault models that do not properly describe the behavior of potential transistor defects. These methods are limited to cell-level diagnosis and cannot be used during test pattern generation. The second category of cell-internal defect test and diagnosis techniques relies on the realistic assumption that the excitation of a defect inside a cell is highly correlated with the logic values at the input pins of the cell [3, 4]. For this category, a cell-internal-fault dictionary or *CA model* (also referred to as *CA fault model* or *CA test model in the literature*), describing the detection conditions of each potential defect affecting a cell, is used [5, 6]. These techniques are more efficient and can be used to guide the test pattern generation and CA diagnosis phases. However, the main limitation of these techniques is the generation effort needed to characterize all the standard cells per technology in terms of run time, number of SPICE simulator licenses, CPU requirements and disk

A. Ladhar (✉)
Crolles, France
e-mail: aymen.ladhar@st.com

A. Virazel
LIRMM, University of Montpellier / CNRS, Montpellier, France
e-mail: arnaud.virazel@lirmm.fr

© The Author(s), under exclusive license to Springer Nature Switzerland AG 2023
P. Girard et al. (eds.), *Machine Learning Support for Fault Diagnosis of
System-on-Chip*, https://doi.org/10.1007/978-3-031-19639-3_6

usage. The second limitation of this technique is the usage of simple fault models to describe silicon failures, which is not always possible for actual silicon failures. In addition, the assumption that the excitation of a defect inside a cell is highly correlated with the logic values at the input pins of the cell is not always correct mainly for unmodeled cell internal defect.

With the fast development and vast application of Machine Learning (ML) in recent years, ML-based techniques have been shown to be quite valuable for diagnosis purpose, especially in the volume diagnosis scenario [5, 6]. In this chapter, the the usage of machine-learning algorithms is extended to generate the cell-aware models in a first step, then to diagnose cell internal defects.

The remainder of this chapter is organized as follows. Section 2 gives some background on conventional cell-aware generation, test, and diagnosis. Section 3 explains the machine learning cell-aware flow. Section 4 summarizes a machine learning technique to diagnose cell-aware defect. Section 5 presents some industrial case studies performed with latest ML-based diagnosis techniques. Section 6 concludes the chapter and draws some conclusions.

2 Background on Conventional Cell-Aware Generation, Test and Diagnosis

To compete in the fast-growing market of automotive ICs as well as 3D transistor era, semiconductor industries need to address new challenges across the entire design flow. In addition, the requirement to be compliant with the highest standards like the ISO 26262 goal of zero Defective Parts Per Million (DPPM), has an impact on the test and diagnosis phase of nowadays silicon failures [7]. Cell internal defects are one of most critical defects that should be targeted since they are increasingly occurring in latest technology nodes. In addition, this defect type is hard to be targeted with conventional ATPG solution and fault models. Test and diagnosis of cell internal defects is possible thanks to the realistic assumption that the excitation of a defect inside a cell is highly correlated with the logic values at the input pins of the cell [3]. To perform CA test and diagnosis, a CA model is needed. It is obtained through the characterization of each standard cell library with regards to all possible cell-internal defects. These defects can be either transistor defects or inter-transistor defects like open and shorts [8–11]. Electrical simulations are performed to generate CA models for each cell in the library. These models include information about the behavior of each defect within the cell with regards to the stimuli applied at the cell inputs.

One bottleneck of the CA flow deployment in industry is the generation effort in terms of run time and flow complexity. Indeed, it requires extensive computational efforts to characterize all standard cells of a library [12]. Figure 1 represents a typical CA model generation flow. It starts with a SPICE netlist representation of a standard cell which is usually derived from a layout description, e.g., a GDSII file. The DSPF

Fig. 1 Conventional cell-aware model generation flow

(Detailed Spice Parasitic Format) contains information about potential shorts and open defect, as well as their coordinates within the cell. This DSPF cell netlist is then used by an electrical simulator to simulate each potential cell internal defect against an exhaustive set of stimuli containing static and dynamic patterns. Once the simulation process is completed, all cell-internal defects are classified into defect equivalence classes with their detection information (required input values for each defect within each cell) and are synthetized into a CA model. Two tables or matrix exist in each CA model, the first one targets static defects whereas the second one targets dynamic defects. As standard cells may have several inputs, and thousands of cells with different complexities are used for a given technology, the generation time of CA models for complete standard cell libraries of a given technology may reach up to several months, thus drastically increasing the library characterization process cost.

As mentioned, cell-aware models can be either used for test purpose or for fault diagnosis of silicon failures:

- *ATPG usage*: the ATPG engine identifies for each cell in the Circuit Under Test (CUT) the minimum set of stimuli targeting the entire cell internal defects, then it generates test patterns exercising this test stimuli at the input pins of the cell under test and ensures the fault propagation to an observation point.
- *Fault diagnosis usage*: the diagnostic tool extracts the failing and passing logic values at the input pins of the defective cell in the Circuit Under Diagnosis (CUD). This information is then matched with the CA model of the defective cell to identify the suspect internal defect.

3 Learning-Based Cell-Aware Model Generation

The reason behind the ML usage for defective cell characterization is the result of several observations made while performing electrical simulation of several cell internal defects on different standard cell libraries and technology nodes:

- Several cell internal defects are independent of the technology and transistor size, and their defective behavior depends mainly on the location of the defect and test stimuli applied on the cell under characterization [13, 14].
- For the same function, cell schematics are usually quite similar whatever is the technology node.

- Detection tables for static and dynamic defects, in the form of binary matrices describing the detection patterns for each cell-internal defect, are ML friendly.
- CA models may change with respect to test conditions and Process Voltage Temperature (PVT) corners. In fact, CA model generation for the same cell with different test conditions may exhibit slight differences. Few defects can be of different types (i.e., static, or dynamic) or may have different detection patterns. Since CA models are generated for specific test conditions and can be used with different ones, it may lead to inaccurate characterization. This inaccuracy is usually allowed in industry since it is marginal. This indicates that one can also tolerate few error percentages in our ML-based prediction.
- Very simple CA models are used to emulate short and open defects, for which resistance values are often identical for all technologies.
- A large database of CA models is usually available and can be used to train a ML algorithm.

All these above-mentioned reasons intuitively revealed that CA model generation through ML could be feasible. However, the first tricky task is to be able to describe cell transistor netlist as well as corresponding cell-internal defects in a uniform (standardized) manner, so that a ML algorithm can learn and infer from data irrespective of their incoming library and technology. Indeed, similar cells (e.g., cells with same logic function, same number of inputs and same number of transistors) may be described differently in transistor-level (SPICE) netlists of various libraries (e.g., a transistor label does not always correspond to the same transistor in two similar cells coming from two different libraries), and it is therefore mandatory to standardize the description of cells and corresponding defects for the ML-based defect characterization methodology. Heuristic solutions developed to this purpose are described in Sects. 3.1 and 3.2. The second challenging task is to find a solution to represent all these information / input data so that they can be ML friendly. A matrix description of cells and corresponding defects can be chosen to this purpose. Section 3.3 describes how to adapt the ML method to deal with sequential cells.

3.1 Generation of Training and New Data

The learning-based defect characterization methodology proposed in [15] is used to predict the behavior of a cell affected by cell internal defects, hence avoiding costly and heavy electrical defect simulations. The proposed flow is presented in Fig. 2. It is based on supervised learning that takes a known set of input data and known responses (*labeled data*) used as training data, trains a model to classify those data, and then uses this model to predict (*infer*) the class of new data.

Training data are made of various and numerous CA models formerly generated by relying to brute-force electrical defect simulations. For each cell in a library, the

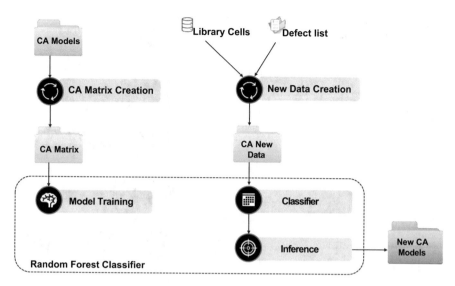

Fig. 2 Generic view of the ML-based defect characterization flow

CA model is transformed into a so-called *CA-matrix* and filled in with meaningful information.

An overview of the CA-matrix flow is described in Fig. 3. It begins by rewriting the CA model in a ML friendly description. Then, it categorizes the activation conditions of each transistor with respect to input stimuli. Once the activation conditions for each transistor have been identified, transistors are renamed in a uniform way. This is a critical step in the proposed flow as it allows the usage of the training data across different libraries and technologies. Finally, the CA-matrix is created with the above information.

Table 1 shows an example of a training dataset for a NAND2 cell. It is composed of four types of information:

- *Cell patterns and responses*. This information represents an extended representation of a NAND2 truth table (A, B: inputs, Z: output). As it can be seen, the test pattern sequence provides all the possible input stimuli that can be applied to the cell. It takes into consideration all the possible transitions that can be applied on the inputs of this cell. These stimuli must also be efficient to detect sequence-depending defects like stuck-open defects. For this reason, a four-valued logic algebra made of 0, 1, R and F is used to represent input stimuli in the CA-matrix. R (resp. F) represents a Rising (resp. Falling) transition from 0 to 1 (resp. from 1 to 0).

- *Active / Passive Transistor Identification*. This indicates the activation conditions of each transistor in the cell schematic. Each transistor can be in the following state: active, passive, switching to active state, switching to passive state.

Fig. 3 Generic view of the
CA-matrix creation flow

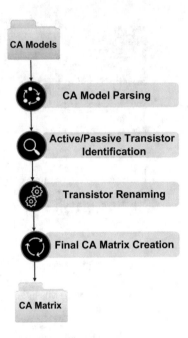

- *Defect description.* This gives information about defect locations in the cell transistor schematic. At least one logic value is used to indicates the name and the transistor port impacted by the defect.
- *Defect detection.* This is the class of the data sample (the output of ML classifier). A value '1' ('0') means that the defect is detected (undetected) by the input pattern.

The first three types of information constitute the inputs of the ML algorithm.

New data represent the cells to be characterized and are obtained for each standard cell from the cell description, corresponding list of defects and cell patterns. The format of a new data instance is like that of the training data, except that the class (label) of the new data instance is missing. The ML classifier is used to predict that class. As for training data, new data are grouped together according to their number of cell inputs and transistors, so that inference can be done at the same time for cells with the same number of inputs and transistors.

As highlighted in Fig. 2, a Random Forest Classifier is used for predicting the class of each new data instance. This choice comes from the results obtained after testing several learning algorithms (k-NN, Support Vector Machine, Random Forest, Linear, Ridge, etc.) and observing their inference accuracies. So, the first main step of our CA diagnosis flow consists in generating a Random Forest model and to train it by using the training dataset. A Random Forest Classifier is composed of several Decision Tree Classifiers, which are models predicting class of samples by applying simple decision rules. During training, a Decision Tree tries to classify data samples and its decision rules are modified until it reaches a given quality criterion.

Table 1 Example of training dataset for a NAND2 cell

Cell inputs & responses			Transistor switching activity				Defect description				About defect		Defect detection
A	B	Z	N0	N1	P0	...	N1_D	N1_G	N1_S	...	Name	Type	fZ
0	0	1	0	0	1	...	0	0	0	...	Free	Free	0
0	1	1	0	1	1	...	0	0	0	...	Free	Free	0
0	F	1	0	F	1	...	0	0	0	...	Free	Free	0
...	...	...	...	...	...	...	...	...	...	...	...	...	...
0	1	1	0	1	1	...	1	0	1	...	D15	Short	1
1	1	0	1	1	0	...	1	0	1	...	D15	Short	0
...	...	...	...	...	...	...	...	...	...	...	...	...	...

Then, the Forest averages the responses of all Trees and outputs the class of the data sample. The second main step consists in using the Random Forest Classifier to make prediction (or inference) when a new data instance must be evaluated. Prediction for a new data instance amounts to answer to the question: "Does this stimulus detects this defect affecting this cell?". Answering to this question allows obtaining a new CA model for a given standard cell.

3.2   Cell and Defect Representation in the Cell-Aware Matrix

This section describes the used method to represent a transistor-level (SPICE) netlist in a CA-matrix, which is a standardized and ML friendly description. This representation must be accurate enough to clearly identify each transistor and each net of the cell transistor schematic. Furthermore, this description must also be able to correlate each transistor to its sensitization patterns and to report the output response for each pattern. Therefore, the cell description process needs several successive operations that are detailed below. Note that this process is applied to all cells in a library to be characterized.

3.2.1   Identification of Active and Passive Transistors

The first step consists in identifying active and passive transistors in the cell netlist with respect to an input stimulus. This information is used to represent the transistor netlist in the matrix format. To this end, a single defect-free (golden) electrical stimulation of each cell to be characterized is performed to identify active and passive transistors for each input stimulus (test pattern) and to measure the cell response on the output pin. A NMOS (resp. PMOS) transistor is considered active if a logic-1 (resp. logic-0) value is measured on its gate terminal. Similarly, a NMOS (resp. PMOS) transistor is considered passive if a logic-0 (resp. logic-1) value measured on its gate terminal. With this information, each cell pattern can be associated to the list of active transistors in the cell. After this step, the CA-matrix contains the columns:

- Cell inputs & responses columns. They contain all input stimuli (test patterns) that can be applied to the cell, and the corresponding responses.
- Transistor switching activity columns. They contain four possible values indicating if the transistor is active (1), passive (0), switching from an active state to a passive one (F) or switching from a passive state to an active one (R). Since PMOS and NMOS transistors are activated in opposite way, the '-' character is used before the PMOS values.

Figure 4 represents an example of a NAND2 cell with its CA-matrix representation. In the partial representation of the CA-matrix of the cell shown in Fig. 4b, columns A and B list all the possible input stimuli for this cell. These columns also define

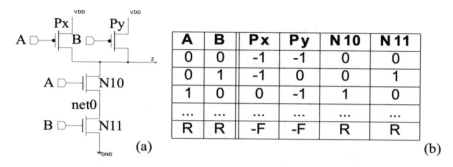

Fig. 4 Example standard cell NAND2: (**a**) cell transistor schematic and (**b**) partial CA-matrix representation

the length of the CA-matrix, which is equal to $2^n 2^n + (4^n - 2^n) 2^n.(2^n - 1)$, n being the number of cell input pins (2^n is the number of static stimuli, $2^n.(2^n - 1)$ is the number of dynamic stimuli). For each stimulus, active and passive information about each transistor of the cell is entered in the CA-matrix. For example, $AB = 00$ leads to two active PMOS transistors (Px and Py) and two passive NMOS transistors N10 and N11).

This matrix description of the cell netlist is clearly dependent on the transistor names and the order they are specified in the SPICE netlist. In fact, two similar cell schematics may have different transistor naming and the order of transistors in the SPICE netlist may differ from cell to cell and from one technology to another. Without an accurate naming convention of each cell transistor in the CA-matrix, any ML algorithm will fail to predict the behavior of the cell in presence of a defect. To prevent this problem, a second step in the cell description process is needed. This step consists in renaming all cell transistors independently of their initial names and order in the input SPICE netlist. The algorithm built to this objective is explained in the next subsection.

3.2.2 Renaming of Transistors

In the CA-matrix, two cells with the same *transistor structure* will have the same transistor names irrespective of their incoming library and technology. A *transistor structure* is a virtual SPICE netlist without specification of the connections between transistor gates, i.e., only source and drain connections between transistors are listed. The transistor-renaming algorithm consists of the following two steps:

- *Determination of branch equations*: Since the transistor gate connections are not considered, the transistor structure is composed of one or more branches. A branch is a group of transistors connected by their drain and source terminals. The entry of each branch is the set of transistor gates, and the exit is the connection net between the NMOS and PMOS transistors. A branch is connected to a power

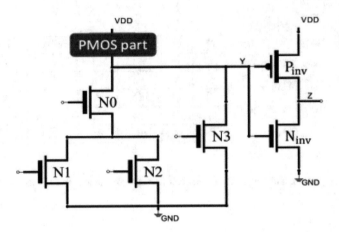

Fig. 5 Schematic example

and/or a ground net. A branch equation is a Boolean-like equation describing how the transistors of the branch are connected, using Boolean-and (symbolized by '&') for serial transistors or serial groups of transistors, and Boolean-or (symbolized by '|') for parallel transistors or parallel groups of transistors.

In Fig. 5, the structure is composed of two branches. The two-transistors output-inverter is the simplest branch whose input is net Y and output is net Z. The inverter creates two paths between branch output and power nets, so its branch equation is $(N_{inv}|P_{inv})$. The equation of the second branch (NMOS branch driving net Y) is $((N0\&(N1|N2))|N3)$. To do not rely on any name present in the SPICE netlist, the branch equations are anonymized, i.e., a NMOS is described by '1n' and a PMOS by '1p'. The anonymized equation of the NMOS branch driving net Y in Fig. 5 is therefore $((1n\&(1n|1n))|1n)$.

- *Sorting of branch equations*: Once all the branch equations for the considered cell have been determined, they are sorted by using deterministic criteria:

  - *Level of each branch.* It is defined in ascending order with respect to the cell output (level-1 branches drive the cell output, level-2 branches drive the gates of transistors in level-1 branches, and so on and so forth),
  - *Number of transistors in each branch* – in ascending order,
  - *Anonymized branch equation* defined in alphabetical order.

3.2.3 Identification of Parallel Transistors

Identifying branch equations is not enough to rename all transistors of a given cell. The problem comes from parallel transistors. In fact, two or more parallel transistors share the same source and drain, which makes their identification quite difficult. For example, in Fig. 5, transistors N1 and N2 can be either presented as "N1|N2" or as

Table 2 Activity values for a NAND2 cell

			old names			
A	B	Comment	Px	Py	N10	N11
0	0	**MSB**	1	1	0	0
0	1		1	0	0	1
1	0		0	1	1	0
1	1	**LSB**	0	0	1	1
Activity value			12	10	3	5
			↓ **Renaming** ↓			
			P1	P0	N0	N1

"N2|N1". As the order of transistors in each branch will determine how transistors will be renamed, such confusing situation cannot by accepted. A solution consists in sorting transistors inside their branch according to their activity with respect to the input stimuli. The algorithm developed to this purpose proceeds as follows. For each transistor, an *activity value* is computed. For a cell with n inputs, the activity value is a 2^n-bit integer that represents the accumulative activation state of a transistor for all possible stimuli applied to the cell. The input stimuli range from $(0,0 \ldots ,0)$ to $(1,1 \ldots ,1)$. For each of these stimuli, the transistor is either active (1) or passive (0). The activity value is defined as a binary number, whose MSB is the activity of the transistor under input stimulus $(0,0 \ldots ,0)$ and LSB is the activity of the transistor under input stimulus $(1,1 \ldots ,1)$, with a bit significance decreasing for increasing binary value of input stimulus.

To compute the activity value, one need to know whether the transistor is active or passive for each input stimulus. This information is already presents in the CA-matrix as described in Section III.A. To illustrate this process, activity values for the transistors of the NAND2 cell in Fig. 4a are given in Table 2.

Finally, transistors of each branch are sorted by increasing activity value to give the final description of the cell in the CA-matrix. For the example in Fig. 4, this leads to: first NMOS transistor of the first branch is named N0, second NMOS transistor of the first branch is named N1, first PMOS transistor of the second branch is named P0, and second PMOS transistor of the second branch is named P1.

3.2.4 Defect Representation

This section details the cell-internal defects representation in the CA-matrix in a standardized and ML friendly manner. Cell internal defects are classified into:

- *Intra-transistor defects.* These defects affect transistor terminals (source, drain, gate and bulk) and can be either an open defect or a short. To describe these defects, all transistor terminals are listed as a column in the CA-matrix (cf. Table I). For an open defect, a value '1' indicates that this transistor terminal is affected by the defect, '0' otherwise. For a short, a value '1' on two transistor terminals indicates that a short exists between these two terminals, '0' otherwise.

Table 3 Defect representation for a NAND2 cell

P0_S	P0_D	P1_S	P1_D		N0_S	...	N1_D	Comment
0	0	1	1		0	...	0	source-drain short on P1
1	0	0	0		1	...	1	net0 & P0-source short

- *Inter-transistor defects.* These defects affect a connection(s) between at least two different transistors. Though these defects are not considered in this work, the matrix representation is flexible enough to represent them. For these defects, the same representation mechanism is used as for intra-transistor defects.

Table 3 is an example of defect description in the CA-matrix of the NAND2 cell example presented in Fig. 4. Row with red cells describes the intra-transistor short defect between drain and source terminals of transistor P1 (formerly Px). Row with blue cells describes the inter-transistor short defect between P0-source and "net0" (net0 connects N0-source and N1-drain).

3.3 Support of Sequential Cells

This section describes the different modifications that must be considered during the creation of the CA matrix in case of a sequential cell:

- *Cell inputs and outputs:* For sequential cells it is important to know the stored value inside the cell before applying the test stimuli. To this end, a new column is included in the CA matrix. It includes the output value of the cell before applying the new test stimuli.
- *Identification of Active, Passive, and pulsing transistors:* One characteristic of sequential cells is the presence of a clock signal, which has an impact on the value applied on each transistor. In fact, clock-signal-controlled transistors can be pulsing (resp. anti-pulsing), which means a 0-1-0 (resp. 1-0-1) sequence appears on the transistor gate terminal during application of the test pattern.
- *Determination of branch equations:* Sequential cells tend to integrate transmission gates to separate the latches from each other and from the inputs. A transmission gate is a transistor configuration acting as a relay that can conduct or block depending on the control signal. It is composed of one PMOS and one NMOS transistors in parallel (i.e., sharing drain and source), and the control signal applied to the gate of the NMOS transistor is the opposite (i.e., NOT-ed) of the signal applied to the gate of the PMOS transistor. A transmission gate directly connects the exit of a branch to the entry of another branch. As such, a transmission gate is considered as an autonomous branch of the transistor structure. The entry of such a branch is the set of transistors gates plus the exit of the previous branch, and the exit is the entry of the following branch. Each time a transmission gate is identified, the branch equation will be considered as

a transmission gate and not as a PMOS with a NMOS transistors. The symbol t will be used into the branch equation.

4 Advanced Cell-Aware Diagnosis Based on Machine Learning

Although conventional diagnosis techniques such as "Cause-Effect" [16] and "Effect-Cause" [17] can achieve a good resolution, in some cases (e.g., complex cells, complex failure mechanisms) the number of candidates may be too high to allow an efficient PFA (Physical Failure Analysis). This problem will be exacerbated in the future with the advent of very deep submicron (i.e., 5 nm and beyond) technologies. Improving diagnosis efficiency at the transistor level (i.e., CA diagnosis) is therefore mandatory.

Previous works on Cell-Aware (CA) fault diagnosis focusing on logic cells can be classified into three approaches. The first approach converts a transistor-level netlist into an equivalent gate-level netlist by means of complex transformations rules [18]. Then, on the equivalent gate-level netlist, any classical fault diagnosis approach can be applied. The main drawback of this approach is that the set of transformation rules depends on the targeted defect and, thereby, the non-modeled defects may not be diagnosed. The second approach is based on the "Cause-Effect" paradigm [18, 19]. The transistor-level netlist of a cell is exploited, in order to inject the targeted defects. Therefore, a defect dictionary is created by transistor-level simulations and the defect signatures of all the defects affecting the cells in the library are stored in this defect dictionary. Then, during fault diagnosis the defect signature of all defects affecting a suspected cell is compared with the observed failures to obtain a list of candidates inside the cell. These approaches can be further classified depending on the "accuracy" of the injected defects and the simulation "precision". In [19], a large number of defects are simulated at transistor-level using SPICE. For a given defect, different resistance values are simulated, in order to be as accurate as possible. This approach leads to more precise results but it requires a huge simulation time. To reduce the simulation time and the fault dictionary size while keeping a high resolution, authors in [18] propose to exploit layout information, in order to consider only realistic defects. For example, for each cell, only the realistic, potential net bridging defects and via open defects are extracted and then simulated. Then, the identified set of realistic defects is simulated at transistor-level. The third intra-cell fault diagnosis approach is based on the "Effect-Cause" paradigm [20]. All the existing diagnosis techniques depend on the targeted fault models or defects. In [20], the main goal is to achieve a resolution close to the transistor-level. However, instead of explicitly considering defects at transistor-level, the idea is to exploit the knowledge of the faulty behavior induced by the defects.

Unfortunately, CA fault diagnosis resolution is typically far from the ideal due to circuit complexity. A mean to achieve this goal is to use supervised learning

Table 4 Test scenarios considered in [21–26]

Fault & Defect	Test Scenarios		
	Static (Low speed)	Dynamic (At-speed)	Static + Dynamic
Stuck-at & Static CA	[22, 24]	[26]	[21, 25]
Transition & Dynamic CA		[23, 24, 26]	[21, 25]

algorithms to determine suspected defects. Supervised learning is now used in numerous classification problems where the knowledge on some data can be used to classify a new instance of such data. This section summarizes the latest developments in the field of CA diagnosis based on supervised learning.

4.1 Preliminaries and Test Scenarios

Several learning-guided solutions for CA diagnosis have been proposed recently in [21–26]. All solutions are based on a Bayesian classification method for accurately identifying defect candidates in combinational standard cells of a customer return. Choosing one solution over another depends on the test scenario (test sequence, test scheme, test conditions) considered during the diagnosis phase and selected according to the types of targeted defects and failure mechanisms.

The test scenarios in [21–26] are sketched in Table 4. In [22, 23], two **distinct** processes were developed to diagnose static and dynamic defects **separately**. In [22], a basic scan testing scheme used to apply static CA test sequences is considered, so that stuck-at faults plus static intra-cell defects are targeted during diagnosis. In [23], a fast sequential testing scheme used to apply dynamic CA test sequences is considered, so that transition faults plus dynamic intra-cell defects are targeted during diagnosis. *Note that* [24] *is just a combination of* [22, 23], *i.e., two testing schemes, one static and one dynamic, and two CA diagnosis flows, one static and one dynamic, are considered independently.* The main limitation of the solutions in [22–24] is the required a priori knowledge of the type of targeted defects in the CUT. In other words, a test engineer needs to know what type of defects is screening before choosing between [22] or [23].

To deal **concurrently** with all types of defects that may occur and without any a priori knowledge of the targeted defect type, a new implementation of the CA diagnosis flow was proposed in [21–25]. *Note that* [21] *is a fully extended version of* [25]. Authors assume a test scenario in which **two** test sequences (static and dynamic) are used successively, each one considering a dedicated testing scheme, i.e., basic scan and fast sequential. First, a static CA test sequence generated by a commercial cell-aware ATPG tool is applied to the CUD. This sequence targets all cell-level stuck-at faults plus cell-internal static defects, considering that these defects are not covered by a standard stuck-at fault ATPG. A standard (low speed) scan-based testing scheme is used to this purpose. Next, another option of the cell-aware ATPG is used to generate a dynamic CA test sequence that targets cell-level transition faults plus intra-cell dynamic defects not covered by a standard transition fault ATPG. In this case, an at-speed *Launch-On-Capture* (LOC) scheme (also called *fast sequential*) is used during test application.

To construct the comprehensive flow described in [21], a new framework was set up in which specific rules were defined to achieve a high level of effectiveness in terms of diagnosis accuracy and resolution. The proposed method was based on a Gaussian Naive Bayes trained model to predict good defect candidates. This method is summarized in the next subsection.

In [26], a new version of the CA diagnosis flow was proposed, assuming a test scenario in which **both** static and dynamic defects can be diagnosed owing to a **single** dynamic CA test sequence applied at-speed. This scenario may happen when such a test sequence has been generated to target transition faults plus cell-internal dynamic defects and appears to also cover the required percentage of stuck-at faults plus cell-internal static defects (or, more generally, satisfies the test coverage specifications). In this case, note that only one (dynamic) datalog is generated after test application and can further be used for diagnosis purpose. Nevertheless, both static and dynamic defects are considered in this scenario. As only dynamic instance tables are manipulated, the representation of training and new data is simplified, i.e., a single type of feature vector is used, without no loss of information and hence without decreasing the quality of the training and inference phases.

4.2 Learning-Based Cell-Aware Diagnosis Flow

Figure 6 is a generic view of the learning-based CA diagnosis flow utilized in [21]. It is based on supervised learning that takes a known set of input data and known responses (*labeled data*) used as training data, trains a model, and then implement a classifier based on this model to make predictions (*inferences*) for the response to new data.

After investigating several ML algorithms and observing their inference accuracies in [22], a Bayesian classification method has been chosen for the learning and inference phases in [22–26]. So, the first main step of the CA diagnosis flow consists in generating a Naive Bayes (NB) model and to train it by using a training dataset. In this step, training data are used to incrementally improve the model's ability to make inference. The training dataset is divided into mutually exclusive and equal subsets. For each subset, the model is trained on the union of all other subsets. Some manipulations, such as grouping data by considering equivalent defects or removing data instances of undetectable defects, are also done during this phase. Once training is complete, the performance (accuracy) of the model is evaluated by using a part of the dataset initially set aside. More details about performance evaluation as done in this framework can be found in [24].

The second main step consists in implementing the NB classifier by using a Gaussian distribution to model the *likelihood* probability functions and use this

Fig. 6 Generic view of the cell-aware diagnosis flow used in [21]

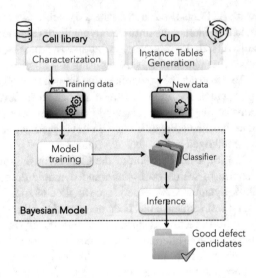

classifier to make prediction when a new data instance has to be evaluated. The next subsections detail the various steps of the CA diagnosis flow, which is able to deal with any type of cell-internal defect (i.e., static and dynamic) that may occur.

4.2.1 Generation of Training Data

Training data are generated for each type of standard cell existing in the CUD during an off-line characterization process done only once for a given cell library. These data are extracted from CA views provided by a commercial CAD tool that contain all characterization results for a given cell type. These results are provided in the form of a fault dictionary containing, for each defect within a cell, the cell input patterns detecting (or not) this defect. An example of training dataset, as used in [21–26] and containing six instances for an arbitrary two-input cell, is shown in Fig. 7. Each instance is associated to a static defect (D_1, D_2, D_3) or a dynamic defect (D_{11}, D_{12}, D_{13}). A 1 (0) indicates that defect D_i is detectable (not detectable) at the output of the cell when the cell-level test pattern P_j is applied at the inputs of the cell. Cell-level test patterns (called *cell-patterns* in the sequel) are static (one input vector – P_1 to P_4 in Fig. 7) or dynamic (two input vectors – P_5 to P_{16} in Fig. 7 in which R (F) indicates a rising (falling) transition at the cell input respectively). For an n-input cell, there exists 2^n static cell-patterns and $2^n.(2^n-1)$ dynamic cell-patterns.

Dynamic defects can be detected not only by dynamic patterns, but also by static patterns applied using a basic scan testing scheme, provided that: i) at least one transition has been generated at the cell inputs between the next-to-last scan shift cycle and the launch cycle, and ii) the delay induced by the defect is large enough to be detected (*these are the detection conditions of a dynamic defect modeled by*

P1	P2	P3	P4	P5	P6	P7	P8	P9	P10	P11	P12	P13	P14	P15	P16	Pattern
00	01	10	11	OR	OF	RO	RR	RF	R1	FO	FF	FR	F1	1R	1F	Defect
1	0	0	0	0	1	0	0	0	0	1	1	0	0	0	0	D1
1	0	0	1	0	1	0	1	0	1	1	1	0	0	1	0	D2
0	1	0	0	1	0	0	0	0	0	0	0	1	1	0	0	D3
0.5	0.5	0.5	0.5	0	0	0	0	0	0	1	0	0	0	0	0	D11
0.5	0.5	0.5	0.5	1	0	0	0	0	0	0	0	1	0	0	0	D12
0.5	0.5	0.5	0.5	0	0	0	0	1	0	0	0	0	0	0	1	D13

Fig. 7 Example of training dataset for all defect types in a two-input cell as used in [21–25]

a stuck-open or a gross delay fault). For this reason, the value '0.5' is assigned to each dynamic defect (D_{11}, D_{12}, D_{13}) for all related static cell-patterns, meaning that such a defect is detectable or not depending on whether or not the above conditions are satisfied.

As only dynamic test sequences are considered in [26], the representation of training data as used in [21–25] could be simplified without losing information and decreasing the quality of the training phase. This comes from the observation that a static defect is a particular case of dynamic defect (e.g., a full open is a resistive open with an infinite value of the resistance), and that all static cell-patterns for a given defect are embedded in its whole set of dynamic cell-patterns. Indeed, a dynamic defect requires a two-vector test pattern ($V_1 V_2$) in which the values of V_1 and V_2 have to be properly defined for the defect to be detected. Conversely, only the value of V_2 is significant for a static defect to be detected by such pattern, irrespective of the value taken by V_1. When looking at Fig. 7, one can see that $P_1 = \{00\}$ is embedded in $P_6 = \{0F\}$, $P_{11} = \{F0\}$ and $P_{12} = \{FF\}$, and the same for P_2, P_3 and P_4. Similarly, one can see that static defect D_2 is detectable by P_1 and P_4, and hence by P_6, P_8, P_{10}, P_{11}, P_{12}, and P_{15}. So, by "compacting" a training dataset as shown in Fig. 8, in which only dynamic cell-patterns are considered, one can see that all meaningful information is still contained in this set, while redundant ('0' and '1' values in the first four columns of Fig. 7) or insignificant ('0.5' values in the same columns for dynamic defects) information is removed. More generally, such compact format for training data makes so that only one type of feature vector (dynamic) is used for both types of defects.

As the goal with training data is to provide a distinct feature vector for each data (defect), it is important to be able to distinguish between static and dynamic defects with such a new format of the training dataset. Let us consider two defects D_1 and D_{11} where D_1 is static and detectable by $\{00\}$ and D_{11} is dynamic and detectable by $\{F0\}$ (note that $\{00\}$ is the second vector of $\{F0\}$). As can be seen in Fig. 8, these two defects can easily be distinguished since their training data instances (or *feature vectors*) are different. The consequence of using such a new format for training data (and hence for new data as will be shown later on) is not an improved accuracy or resolution, but rather a simplified manipulation of feature vectors.

P1'	P2'	P3'	P4'	P5'	P6'	P7'	P8'	P9'	P10'	P11'	P12'	Pattern
OR	OF	RO	RR	RF	R1	FO	FF	FR	F1	1R	1F	Defect
0	1	0	0	0	0	1	1	0	0	0	0	D1
0	1	0	1	0	1	1	1	0	0	1	0	D2
1	0	0	0	0	0	0	0	1	1	0	0	D3
0	0	0	0	0	0	1	0	0	0	0	0	D11
1	0	0	0	0	0	0	0	1	0	0	0	D12
0	0	0	0	1	0	0	0	0	0	0	1	D13

Fig. 8 Example of training dataset for all defect types in a two-input cell as used in [26]

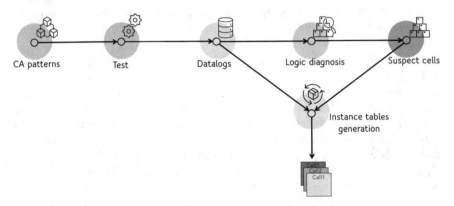

Fig. 9 Generation flow of instance tables

4.2.2   Generation of Instance Tables

An instance table is a failure mapping file generated for each suspected cell by using information contained in the tester datalog. It describes the behavior (pass / fail) of the cell for each cell-pattern occurring on its inputs during test of the CUD. The generation process of instance tables is sketched in Fig. 9. First, CA test patterns are applied to the CUD. These test patterns are obtained from a commercial CA test pattern generation tool that targets intra-cell defects. Next, a datalog containing information on the failing test patterns and corresponding failing primary outputs is obtained. From this datalog and the circuit netlist, a logic diagnosis is carried out (still using a commercial tool) and gives the list of suspected cells. From this list and the datalog information, one can finally generate an instance table for each suspected cell. Note that in case several test sequences, e.g., one static and one dynamic, are used for diagnosis of the CUD, the generation process is repeated so as to produce static and dynamic instance tables for all suspected cells. This is the case in [21].

Fig. 10 Example of static
and dynamic instance tables

```
------------------------------------------------------
                NOR Cell - NR2NHVTX1
------------------------------------------------------
        Z   Output   L412/C1381A
        A   Input    U59/Z
        B   Input    U28/Z
------------------------------------------------------
    Pattern 1   PASSING   Z: stuck-at-0
        Z   000011111111111 – 1
        A   111100000000000 – 0
        B   000000000000000 – 0
    Pattern 2   FAILING   Z: stuck-at-1
        Z   011100000000000 – 0
        A   000011111111111 – 1
        B   100011111111111 – 1
------------------------------------------------------
```

The format of a static instance table is illustrated in Fig. 10 for a given two-input NOR cell and two static cell-patterns. In this example, the first part of the file gives information on how the cell is linked to other cells in the circuit, while the second part represents, respectively, the pattern number, the pattern status (failing, passing), and the cell output Z with the associated fault model for which exercising conditions are reported. These conditions shown right below each cell-pattern in Fig. 10 represent the stimulus arriving at the cell inputs during the shift phase (before '-') and applied during the launch cycle (after '-'). For example, cell-pattern 2 consists in applying a 1 on input A and B, and failing in detecting a stuck-at 1 on Z.

4.2.3 Generation of New Data

New data are generated after post-processing of instance tables. They are composed of various instances, each of them being associated to one suspected cell in the CUD and represent a feature vector that characterizes the real behavior of the cell during test application. From each new data instance, one can extract one or more defect candidates that must be classified as good or bad candidate with a corresponding probability to be the root cause of failure. This classification is done by comparing the new data instance with the training data of the corresponding suspected cell and identify those training data instances that match (or not) with the new data instance.

The formats of a new data instance as used in [21–25] and [26] are illustrated in Figs. 11 and 12 respectively. This format is quite close to the format of a training data instance but has a different meaning. In each instance, the value '1' (resp. '0') is associated to a failing (resp. passing) cell-pattern P_i for a given defect candidate, meaning that the candidate is **actually** detectable (resp. undetectable) by the cell-pattern P_i at the output of the cell during test of the CUD and hence can (cannot) be the real defect. In such instance, the value '0.5' is associated to a cell-pattern for a given defect candidate when this pattern cannot appear at the inputs of a suspected

P1	P2	P3	P4	P5	P6	P7	P8	P9	P10	P11	P12	P13	P14	P15	P16	
0.5	0.5	0.5	0.5	0	0	0	0.5	0	1	0	0	0	0	0	0	D_i

Fig. 11 Format of a new data instance for a two-input cell as used in [C–H]

P1'	P2'	P3'	P4'	P5'	P6'	P7'	P8'	P9'	P10'	P11'	P12'	
0	0	0.5	1	0	1	0	0	0.5	0	0	0	Di

Fig. 12 Format of a new data instance for a two-input cell as used in [G]

cell during real test application with an ATE. The median value '0.5' was chosen to avoid missing information in new data instances while not biasing the features of these data.

4.2.4 Diagnosis of Defects in Sequential Cells

All the work carried out in [21–26] was about diagnosis of defects occurring in combinational standard cells of a customer returns. However, defects in SoCs may also occur in sequential standard cells of logic blocks. This section shows how the previous diagnosis flow can handle sequential cells and related defects by adding new information to the training dataset [27].

The two main differences between a combinational cell and a sequential cell are that (i) the latter has a clock input pin and (ii) the fact that the previous logic value of a sequential cell output can affect the current output value of the cell. To take this difference into account, each cell-pattern for a sequential cell is considered as a tuple in which the first value represents the input clock signal (pulsing or not), the second value is associated to the main input of the cell (e.g., D), and the third value is associated to a virtual input pin representing the previous value of the output pin of the cell (e.g., Q). *Note that in case of sequential cells with multiple real inputs (e.g., D flip-flop with a D, Scan-In, Scan-Enable and Clock input signals), the cell-pattern representation is expanded accordingly.* In each tuple, the first value is either U (i.e., pUlse) or 0, depending on whether or not there is an active clock signal. The second value can be 0, 1, R or F. The third value can only be static (i.e., 0 or 1). An example of training dataset for all defect types (static and dynamic) that may occur in a sequential cell is shown in Fig. 13. Note that the CA views used during the generation of training data do not contain information about cell-patterns with non-pulsing clock signals (i.e., none of the cell internal defects can be detected at the cell output without clock pulse). Consequently, the training data do not include such cell-patterns as can be observed in the example of Fig. 13. Note also that instance tables of sequential cells may contain cell-patterns with no transition on the main inputs of the cell. To allow the ML algorithm understanding this information, the

P1	P2	P3	P4	P5	P6	P7	P8	Pattern
U00	U01	U10	U11	UR0	UR1	UF0	UF1	Defect
0	0	1	0	0	0	1	0	D1
1	1	0	0	0	0	1	1	D2
0.5	0.5	0.5	0.5	1	1	0	0	D11
0.5	0.5	0.5	0.5	1	0	0	1	D12

Fig. 13 Example of training dataset for all defect types (static and dynamic) in a sequential cell. The pin order is clock-data-previous output

solution consists in including static cell-patterns (e.g., P1 to P4 in Fig. 13) in the training data of sequential cells.

With the above representation of training data for sequential cells, one can see that the diagnosis flow in Fig. 6 can be used in a straightforward manner without any change. The two main steps (model training by using a training dataset, implementation of the NB classifier to make inference) remain the same irrespective of the type of manipulated standard cells.

5 Applications on Industrial Cases

This section shows the experimental results of ML usage for CA model generation as well as CA diagnosis.

5.1 CA Model Generation Results

The CA model generation is performed with a python program. The ML algorithms were taken from the publicly available python module called scikit-learn [28]. The dataset was composed of 1712 standard cells coming from standard cell libraries developed using three technologies (C40 (446 cells), 28SOI (825 cells) and C28 (441 cells)). All these cells already had a CA model generated by a commercial tool. The CA-matrix is generated for each cell. The method was experimented in two different ways. First, the ML model was trained and evaluated using cells belonging to one technology. Second, the model was trained on one technology and evaluated it on another one.

5.1.1 Predicting Defect Behavior on the Same Technology

The ML model is first trained on cells of 28SOI standard cell libraries. As mentioned earlier, cells were grouped according to their number of transistors and inputs. For *m* cells available in each group, the ML model is trained over *m-1* cells and evaluate its **prediction accuracy** on the *m*-th cell. A loop ensured that each cell is used as the *m*-th cell. On average, a group contains 8.6 cells. In the following, all possible open and short defects (static and dynamic) in each cell are considered. Results presented below report the prediction for open defects. Results achieved for short defects are similar.

Table 5 presents the prediction accuracy achieved for open defects. Non-empty boxes report the **average** prediction accuracy obtained **for a group of cells**. Empty boxes mean that there is zero or one cell available and that the group cannot be evaluated. A grey background indicates that the maximum prediction accuracy in this group is 100%, i.e., the ML model can perfectly predict the defective behavior of at least one cell. In contrast, white background indicates that no cell was perfectly predicted in that group (all prediction accuracies are less than 100%). For example, let us consider the bold value in Table 5. One has 24 cells having 4 inputs and 24 transistors: (i) 15 cells are perfectly predicted (100% accuracy), which leads to a green background, (ii) the prediction accuracy for the 9 remaining cells ranges from 99.82% to 99.99% and (iii) the average prediction accuracy over all 24 cells is 99.97%.

These results show that the ML model can accurately predict the behavior of a cell affected by a given defect and that our method could be used to generate CA models. The goal of the next subsection is to leverage on existing CA models to generate CA models for a new technology.

5.1.2 Predicting Defect Behavior on Another Technology

Experiments are also conducted on cells belonging to two different technologies. Evaluation was slightly different compared to the previous one. Here, the ML model was trained over all available cells of a given technology and the evaluation was done on one cell of another technology. A loop was used to allow all cells of the second technology to be evaluated. Cells were grouped according to their number of inputs and transistors.

Table 6 shows the prediction accuracy achieved on open defects of the C28 cells after training on the 28SOI cells. Results are averaged over all cells in each group (same number of inputs and number of transistors). The average prediction accuracies are globally lower compared to those of Table 1. After investigating on this point, one can noticed that the behavior of most of the cells (68% of cells) is accurately predicted (accuracy > 97%), while accuracy for few cells is quite low. This phenomenon is discussed later in this section.

To verify the efficiency of the method when different transistor sizes are considered, the ML model was trained over the 28SOI standard cells and used to

Table 5 Prediction accuracy for cells in the same technology

Prediction accuracy (%)		Number of inputs				
		2	3	4	5	6
Number of transistors	6	99.98	99.99			
	8	99.91	99.96	99.91		
	9		100.0			
	10	99.98	99.81	99.96		
	12	99.72	99.73	100.0	99.91	99.93
	14	99.70	99.56	99.83	99.92	99.96
	16	99.99	100.0	99.94		99.98
	18	99.99	99.94			
	20	100.0	99.98	100.0	99.73	
	22		99.84	99.98	99.62	
	24	100.0	99.84	**99.97**		99.85
	26	100.0	99.70	100.0		99.89
	28	99.49	99.98	100.0	99.88	99.81
	30	99.75	100.0	100.0		
	32	100.0	100.0			99.98
	42		100.0			
	44		100.0			
	46		99.81			
	47		99.98	99.95		

predict the behavior of C40 cells. Table 7 shows the prediction accuracy achieved on open defects of the C40 cells after training on the 28SOI cells. Results are averaged over all cells in each group (same number of inputs and transistors). This time, 80% of cells are accurately predicted (accuracy > 97%), proving that our ML-based characterization methodology could be used to generate CA models for a (large) part of cells of a new technology.

5.1.3 Analysis and Discussion

The cells for which the defect characterization methodology gives excellent prediction accuracy as well as those for which the prediction accuracy was quite low are firstly analyzed. Then, the limitation of the detailed method for CA model generation is investigated. After running several experiments on different configurations using one fault model at a time, one noticed the following behaviors:

- Accuracy for most of the cells is excellent, i.e., more than 97% prediction accuracy for 70% of cells. In this case, **the CA model generated by ML fit the real behavior achieved with electrical simulation**.
- Accuracy for few cells (30%) is quite low and the ML prediction is not accurate.

Table 6 Average prediction accuracy for cells in different technologies

Prediction accuracy (%)		Number of inputs				
		2	3	4	5	6
Number of transistors	6	98.21	99.47			
	8	94.56	96.86	99.00		
	9					
	10	94.69	96.01	99.27		
	12	87.73	98.05	99.10		99.76
	14	85.69	97.35	98.75		
	16	91.74		99.20		
	18	88.18	96.28			
	20	90.29	94.37			
	22	78.73		98.37		
	24	87.91	96.88	99.37		99.79
	26	87.24	98.92			
	28	88.18	98.68			
	30			97.52		
	32	88.73	95.6			
	42					
	44					
	46					
	47					

For the first cell category with good prediction score, cells have been analyzed manually to identify why they led to good results. The analysis showed that all these cells had at least one cell in the training dataset with the same transistor structure or a similar one.

For the second cell category – cells leading to poor prediction accuracy – the manual analysis showed that they have: (i) new logic functions that do not appear in the cells of the training dataset, or (ii) a transistor configuration which is completely new when compared to cells in the training dataset.

5.1.4   Hybrid Flow for CA Model Generation

Considering the above analysis, it appears that the ML-based CA model generation flow cannot be used for all cells in a standard cell library to be characterized. A mixed solution, which consists in combining ML-based CA model generation and conventional (simulation-based) CA model generation, should be preferably used. This is illustrated in the following.

The flow sketched in Fig. 14 is proposed for accelerating the CA model generation. Typically, when the CA model for a new cell is needed, one first check if the ML-based generation will lead to high-quality CA models. This is done by analyzing the structure of the new cell and check whether the training dataset

Table 7 Average prediction accuracy for cells with different transistor size

Prediction accuracy (%)		Number of inputs				
		2	3	4	5	6
Number of transistors	6	100.0	99.80			
	8	87.39	99.14	99.03		
	9		97.19			
	10	92.07	95.49	99.32	98.46	
	12	91.71	98.07	99.24	98.47	99.46
	14	90.1	95.84	98.63	98.79	99.52
	16	91.17	93.59	99.23		99.59
	18	88.5	97.15	97.14	97.74	
	20	83.87	97.73	97.15	98.94	
	22	87.26		98.98	98.44	
	24	93.96	99.34	98.58	98.84	99.63
	26	87.52	97.55	99.04	99.02	99.92
	28	98.19		98.79	99.31	99.44
	30			99.13	99.37	99.58
	32	92.91			98.92	99.78
	42					
	44	92.03	98.82			
	46		99.23			
	47		98.29	99.76		

contains a cell with identical or similar structure (as presented in V.B). If the ML algorithm is expected to give good results, the new cell is prepared (representation in a CA-matrix) and submitted to the trained ML algorithm.

The output information is then parsed to the desired file format. Conversely, if the ML algorithm is expected to give poor prediction results, the standard generation flow presented in Fig. 14 is used to obtain the CA model. A feedback loop uses this new simulated CA model to supplement the training datasets and improve the ML algorithm for further prediction.

To estimate the improvement in CA model generation time achieved with the flow in Fig. 2, the following experiments are performed. The Random Forest model is trained on 28SOI standard cells and generated CA models for a subgroup of the C40 standard cell libraries. A subgroup is composed of cells representing all the cell functions available in C40 libraries. In our experiments, this subgroup contained 409 cells: 118 (29%) have a cell with an identical structure in the training dataset, 87 (21%) have a cell with an equivalent structure (as explained in Section V.B) in the training dataset, and 204 (50%) have no identical or equivalent structure in the training dataset (a simulation-based generation is thus needed). For these 204 cells, the generation time was calculated and found to be equal to ~172 days (~ 5.7 months) considering a single SPICE license. Using the ML-based CA model generation for the 118 + 87 = 205 (50%) remaining cells requires 21,947 s (~ 6 h), again considering a single SPICE license. Considering that a

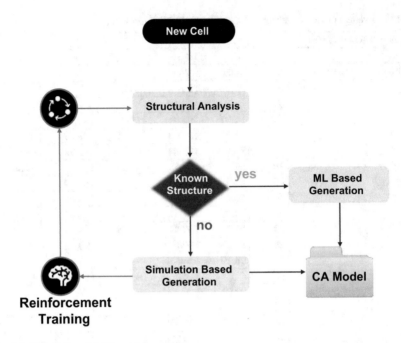

Fig. 14 Hybrid flow for CA model generation

simulation-based generation for these 205 cells would require ~78 days, one can estimate the reduction in generation time to **99.7%**. Now, if one considers the whole C40 subgroup composed of 409 cells, the hybrid generation flow would require ~172 days + ~6 hours, to be compared with ~172 days + ~78 days = ~250 days by using only the simulation-based generation. This represents a reduction in generation time of about **38%**. After investigating results of these experiments, one can observed that the ML-based CA model generation works well for about 80% of cells of the C40 subgroup. Surprisingly, the structural analysis revealed that only 50% (205 cells) could be evaluated using the ML-based generation part of the flow. This shows that there is still room for further improvement of the structural analysis in this flow, and hence get better performance of the ML-based CA model generation process.

To conclude, experiments have been carried out on a reasonable size (1712) of standard cell population. Considering that more than 10,000 cells have usually to be characterized for a given technology, the hybrid flow in Fig. 14 is expected to provide even better results, especially owing to the reinforcement training that uses simulation generated models for supplementing the training datasets, and hence reduce the number of electrical simulations.

5.2   CA Diagnosis Results

The CA diagnosis flow described in Sect. 4.2 and targeting defects in both combinational and sequential cells of CUD has been implemented in a Python program. For validation purpose, authors in [21–27] have experimented the proposed flow in three different ways:

- First, they conducted experiments on ITC'99 benchmark circuits with defect injection campaigns targeting **combinational cells** in each circuit. Various results are reported in [21–27] to show the superiority of the framework when compared to commercial diagnosis solutions.
- Next, they considered a test chip developed by STMicroelectronics and designed using a 28 nm FDSOI technology, and they conducted two defect injection campaigns targeting **sequential cells** [27]. Results are reported in Sect. 5.2.1 and demonstrate the effectiveness of the diagnosis framework.
- Finally, they considered **a customer return** from STMicroelectronics and performed a silicon case study with a real defect subsequently analyzed and identified during PFA. Results are reported in Sect. 5.2.2.

5.2.1   Simulated Test Case Studies

Authors in [27] conducted experiments on a silicon test chip developed by STMicroelectronics and designed with a 28 nm FDSOI technology. The test chip is only composed of digital and memory blocks, and one PLL. The digital blocks are made of 3.8 million cells. Other features (number of primary inputs, primary outputs, and scan flip-flops) are given in Table 8.

A first simulated case study was done with a **static** defect injection campaign. All possible static defects were successively injected into three scan flip-flops (SFF) of a single full-scan digital block. This block was tested with a static CA test sequence achieving a stuck-at + static CA fault coverage of 100%. The average numbers of passing and failing test patterns are given in Table 9. Results obtained after executing the CA diagnosis flow and averaged over all defect injections have shown an accuracy of **100%** (the injected defect was always reported in the list of suspects) and a resolution of **1.25.** The resolution ranges between 1 and 3, and Fig. 15 shows the distribution of this resolution with respect to the total number of simulated cases. As can be seen, in most of the cases, the number of suspects is equal to 1 (perfect resolution).

A second simulated case study with another defect injection campaign was performed on the same test chip. All possible dynamic defects were successively

Table 8 Main features of the silicon test chip

#cells	#PIs	#POs	#SFF
3.8M	97	32	17.5 k

Table 9 Average pattern count in instance tables of the first simulated case study

#passing patterns	#unique passing patterns	#failing patterns	#unique failing patterns
43.4	24.0	15.5	8.6

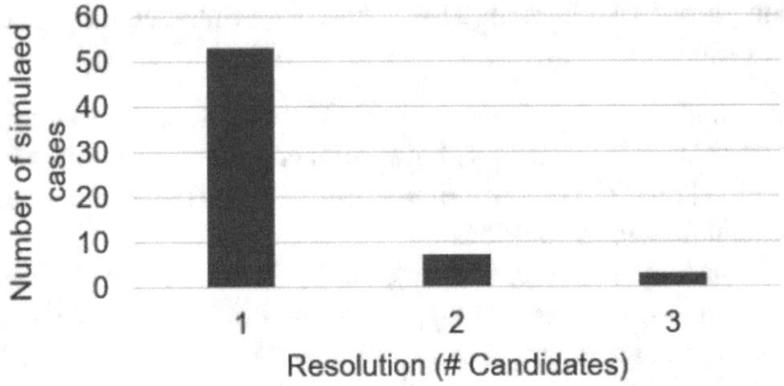

Fig. 15 Distribution of the resolution with respect to the simulated cases

injected into three scan flip-flops of a single full-scan digital block. This time, a dynamic CA test sequence was applied and achieved a transition + dynamic CA fault coverage of 89.8%. The average number of failing test patterns was 7.9. Again, the results obtained after executing the CA diagnosis flow and averaged over all defect injections have shown an accuracy of 100%. The average resolution obtained for dynamic defect injection experiments was 1.37. Again, the resolution ranged between 1 and 3, and in most of the case, the number of suspects was equal to 1.

5.2.2 Silicon Test Case Studies

Next, a silicon case study was performed on a customer return designed with a 28 nm FDSOI technology from STMicroelectronics [27]. The test conditions used to run the experiments were as follows: a nominal supply voltage of 0.83 V, a scan test frequency of 10 MHz, a launch-to-capture clock speed (for the dynamic CA test sequence application) adjusted with respect to the nominal clock frequency of the circuit, and a temperature of 25 °C. The process was considered as typical. The CA diagnosis flow was experimented and the following results were obtained. Initially, the circuit failed on the tester after application of the static CA test sequence when applied at the nominal voltage. This information was stored in a "static" datalog. Then, a logic diagnosis gave a short list of suspected cells among which a six-input SFF cell made of 56 transistors and having a Reset, an Enable, a Test-Input and Test-Enable input pins. The cell contains 758 potential short or open defects. A static instance table was then generated for this suspected cell and contained 5 failing and 75 passing cell-level test patterns. From the new data generated after post-processing of this instance table, the NB classifier identified four suspected

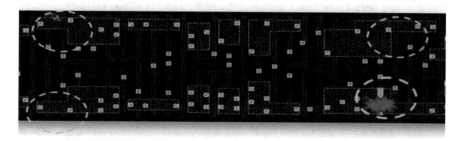

Fig. 16 Layout view of the suspected cell and the incriminated transistor. Yellow circles indicate defect candidates and red mark indicates actual observed defect

defects among which defect D62 (a short between the gate and source of NMOS 19).

The above diagnosis results were provided to the Failure Analysis team of STMicroelectronics, who made a PFA in the past on this customer return based on the results found by their in-house intra-cell diagnosis tool. The result obtained with the CA diagnosis flow was validated as defect D62 was found to be the real defect. This was done after performing a polysilicon level inspection on the layout of the cell (c.f. Fig. 16) and observing the failure analysis cross-sectional view.

6   Conclusion and Discussion

This chapter has provided an overview of the various machine learning approaches and techniques proposed to support cell-aware generation, test, and diagnosis from leading-edge research in this domain. In a comprehensive form, it proposes a compendium of solutions existing in this field. More in detail, the chapter has presented, after some backgrounds on conventional approaches to generate and diagnose cell-aware defects, learning-based solutions to generate CA models and CA diagnosis. Experiments on silicon test cases have been done to validate those solutions and demonstrate their efficacy in terms of accuracy and resolution.

Results of this chapter prove the appropriateness of learning-based methods to solve the problem of CA models generation and CA diagnosis. The learning-based solutions to generate CA models was evaluated using two scenarios: i) the model was trained and evaluated using cells belonging to one technology, and ii) the model was trained on one technology and evaluated it on another one. Those evaluations have shown that the ML-based CA model generation flow cannot be used for all cells in a standard cell library to be characterized. A mixed solution, *Hybrid flow for CA model generation*, which consists in combining ML-based CA model generation and conventional (simulation-based) CA model generation, should be preferably used. The learning-based solution to generate CA diagnosis has been experimented in three different ways: (i) on ITC'99 benchmark circuits with defect injection campaigns targeting combinational cells, (ii) using a test chip developed by STMicroelectronics and designed using a 28 nm FDSOI technology with defect injection campaigns targeting sequential cells, and (iii) using a customer return from STMicroelectronics with a real defect subsequently analyzed and identified during PFA. In all cases, the learning-based solution to generate CA diagnosis succeeds in identifying the good defect candidate compared to commercial diagnosis solutions.

References

1. Ladhar, A., Masmoudi, M., Bouzaida, L.: Efficient and Accurate Method for Intra-Gate Defect Diagnoses in Nanometer Technology. In: Proceedings of IEEE/ACM Design Automation and Test in Europe (2009)
2. Sun, Z., Bosio, A., Dilillo, L., Girard, P., Virazel, A., Auvray, E.: Effect-Cause Intra-cell Diagnosis at Transistor Level. In: Proceedings of IEEE International Symposium on Quality Electronic Design (2013)
3. Hapke, F., et al.: Cell-aware test. IEEE Trans. Comput-Aid. Des. 33(9), 1396–1409 (2014)
4. Maxwell, P., Hapke, F., Tang, H.: Cell-Aware Diagnosis: Defective Inmates Exposed in their Cells. In: IEEE European Test Symposium (2016)
5. Liu, F., Nikolov, P.K., Ozev, S.: Parametric Fault Diagnosis for Analog Circuits Using a Bayesian Framework. In: Proceedings of IEEE VLSI Test Symposium (2006)
6. Wang, S., Wei, W.: Machine Learning-Based Volume Diagnosis. In: Proceedings of IEEE/ACM Design Automation and Test in Europe (2009)
7. Mukherjee, N. et al.: Digital Testing of ICs for Automotive Applications. In: Proceedings of IEEE International Conference on VLSI Design
8. Hapke, F., Krenz-Baath, R., Glowatz, A., Schloeffel, J., Hashempour, H., Eichenberger, S., et al.: Defect-Oriented Cell-Aware ATPG and Fault Simulation for Industrial Cell Libraries and Designs. In: Proceedings of IEEE International Test Conference (2009, November)
9. Goncalves, F.M., Teixeira, I.C., Teixeira, J.P.: Integrated Approach for Circuit and Fault Extraction of VLSI Circuits. In: Proceedings of IEEE International Symposium on Defect and Fault Tolerance in VLSI Systems (1996, November)
10. Goncalves, F.M., Teixeira, I.C., Teixeira, J.P.: Realistic Fault Extraction for High-Quality Design and Test of VLSI Systems. In: Proceedings of IEEE International Symposium on Defect and Fault Tolerance in VLSI Systems (1997, October)
11. Stanojevic, Z., Walker, D.M.: Fed–x - A Fast Bridging Fault Extractor. In: Proceedings of IEEE International Test Conference (2001, November)
12. Lorenzelli, F., Gao, Z., Swenton, J., Magali, S., Marinissen, E.J.: Speeding up Cell-Aware Library Characterization by Preceding Simulation with Structural Analysis. In: Proceedings of IEEE European Test Symposium (2021)
13. Venkataraman, S., Drummonds, S.D.: A Technique for Logic Fault Diagnosis of Interconnect Open Defect. In: IEEE VLSI Test Symposium (2000)
14. Li, C.-M., McCluskey, E.J.: Diagnosis of Resistive-Open and Struck-Open Defects in Digital CMOS ICs. IEEE Trans. CAD Integr. Circuits Syst. 24(11), 1748–1759 (2005)
15. d'Hondt, P., Ladhar, A., Girard, P., Virazel, A.: A Learning-Based Methodology for Accelerating Cell-Aware Model Generation. In: IEEE /ACM Design Automation and Test in Europe (2021)
16. Waicukauski, J.A., Lindbloom, E.: Failure diagnosis of structured VLSI. IEEE Des. Test Comput. 6(4), 49–60 (1989)

17. Abramovici, M., Breuer, M.A.: Fault Diagnosis Based on Effect-Cause Analysis: An Introduction. In: Proceedings of ACM Design Automation Conference, pp. 69–76 (1980)
18. Fan, X., Moore, W., Hora, C., Gronthood, G.: A Novel Stuck-At Based Method for Transistor Stuck-Open Fault Diagnosis. In: Proceedings of IEEE International Test Conference, pp. 386–395 (2005)
19. Hapke, F., Reese, M., Rivers, J., Over, A., Ravikumar, V., Redemund, W., Glowatz, A., Schloeffel, J., Rajski, J.: Cell-Aware Production Test Results from a 32-nm Notebook Processor. In: Proceedings of IEEE International Test Conference (2012). https://doi.org/10.1109/TEST.2012.6401533
20. Sun, A., Bosio, A., Dillilo, L., Girard, P., Virazel, A., Pravossoudovitch, S., Auvray, E.: Intracell defects diagnosis. J. Electron. Test. **30**(5), 541–555 (2014)
21. Mhamdi, S., Girard, P., Virazel, A., Bosio, A., Ladhar, A.: A Learning-Based Cell-Aware Diagnosis Flow for Industrial Customer Returns. In: Proceeding of IEEE International Test Conference (2020). https://doi.org/10.1109/ITC44778.2020.9325246
22. Mhandi, S., Virazel, A., Girard, P., Bosio, A., Auvray, E., Faehn, E., Ladhar, A.: Towards Improvement of Mission Mode Failure Diagnosis for System-on-Chip. In: Proceedings of IEEE International On-Line Testing Symposium (2019)
23. Mhandi, S., Girard, P., Virazel, A., Bosio, A., Ladhar, A.: Cell-Aware Diagnosis of Automotive Customer Returns Based on Supervised Learning, presented at IEEE Automotive Reliability and Test Workshop (2019)
24. Mhamdi, S., Girard, P., Virazel, A., Bosio, A., Faehn, E., Ladhar, A.: Cell-aware defect diagnosis of customer returns based on supervised learning. IEEE Trans. Device Mater. Reliab. **20**(2), 329–340 (2020)
25. Mhandi, S., Girard, P., Virazel, A., Bosio, A., Ladhar, A.: Learning-Based Cell-Aware Defect Diagnosis of Customer Returns. In: Proceedings of IEEE European Test Symposium (2020)
26. Mhandi, S., Girard, P., Virazel, A., Bosio, A., Ladhar, A.: Cell-Aware Diagnosis of Customer Returns Using Bayesian Inference. In: Proceedings of IEEE International Symposium on Quality Electronic Design (2021)
27. d'Hondt, P., Mhamdi, S., Girard, P., Virazel, A., Ladhar, A.: A Comprehensive Framework for Cell-Aware Diagnosis of Customer Returns, to appear in Microelectron. Reliab. J. (2021, October)
28. Pedregosa, F., et al.: Scikit-learn: machine learning in Python. J. Mach. Learn. Res. **12**(Oct), 2825–2830 (2011)

Machine Learning Support for Diagnosis of Analog Circuits

Haralampos-G. Stratigopoulos

1 Introduction

The fabrication process of Integrated Circuits (ICs) is imperfect and is likely that a fabricated IC will not meet the intent specifications. Main sources of failure include silicon defects and variations in process parameters. For this reason, before deployment in an application all fabricated ICs need to go through post-manufacturing testing which aims at screening out non-functional instances. Post-manufacturing testing procedures include wafer-level testing aiming at identifying gross instabilities in the fabrication process, final testing of packaged dies aiming at verifying that the actual design specifications of the IC are met, and burn-in testing, where the chip is exercised sufficiently long in stress conditions, in order to avoid early in-use system failures. The demanded outgoing quality level is defined by the criticality of the application. Safety-critical and mission-critical applications, i.e., automotive, biomedical, defense, aerospace, etc., require zero defective parts per million (DPPM), which poses a great challenge. Even more, an IC that passes post-manufacturing testing may fail during mission-mode due to latent defects, i.e., defects that manifest themselves later during the application, silicon aging, or environmental factors. For critical applications it is often demanded that the IC is self-tested so as to detect reliability hazards and take preventive actions, i.e., halt the application before a fatal consequence occurs. Self-testing requires on-chip resources and is performed on-line in idle times or concurrently with the operation.

Diagnosis of ICs refers to the analysis performed to identify the root cause of a hardware-level failure that has occurred. Diagnosis can be launched on an IC that fails post-manufacturing testing before deployment in an application or on an IC that

H.-G. Stratigopoulos (✉)
Sorbonne Université, CNRS, LIP6, Paris, France
e-mail: haralampos.stratigopoulos@lip6.fr

© The Author(s), under exclusive license to Springer Nature Switzerland AG 2023
P. Girard et al. (eds.), *Machine Learning Support for Fault Diagnosis of
System-on-Chip*, https://doi.org/10.1007/978-3-031-19639-3_7

fails in the field of application and is returned by the customer. In this latter scenario, failures are not easy to reproduce in the laboratory as the real mission conditions and executed workload are unknown and cannot be exhaustively modeled. The diagnosis output is the isolation of the fault responsible for the failure, its localization at transistor-level, as well as its quantification, e.g., the deviation of a circuit parameter from its nominal expected value.

Diagnosis is a crucial step in a product life-cycle. In the first phases of production, it can reveal important statistics, such as defect distribution and yield detractors. It can assist the designers in gathering valuable information regarding the underlying failure mechanisms. The objective here is to make use of the diagnosis results to enhance yield for future products through improvement of the fabrication environment and development of design techniques that minimize the failure rate, and also to evaluate and improve the quality of post-manufacturing tests. In the case of failures in the field of application, diagnosis of customer returns is important to identify the root-cause of failure so as to repair the system if possible, gain insight about environmental conditions that can jeopardize the system's health, and apply corrective actions that will prevent failure re-occurrence and, thereby, expand the safety features.

Diagnosis is a multi-step procedure. In a first step, diagnosis generates a set of candidate hardware-level faults (or defects) based on diagnostic measurements or observed symptoms. In a second step, the IC is physically examined, for example using a thermal camera or laser probing, to highlight anomalies in the operation and narrow down further the set of candidate defects. In the third and last step, the IC is submitted to Physical Failure Analysis (PFA) where de-layering and cross-sectioning of the die is performed to confirm the defect using imaging. Since PFA is destructive and irreversible, ideally the first step should pinpoint the actual defect. However, very often the first step results in ambiguity groups of candidate defects. According to industrial experience, the size of ambiguity groups should be less than 5–10 to increase the PFA success rate. Last but not least, an additional constraint is that the diagnosis should be completed in reasonable time, i.e., typically maximum in a few days, given that the number of diagnoses performed per week per design can be in the order of thousands. Achieving a short diagnosis cycle is challenging considering the status of diagnosis tools that are available today.

In summary, diagnosis metrics include: (a) resolution, i.e., the size of an ambiguity group; (b) accuracy, i.e., whether the reported diagnosed defect corresponds to the actual defect that has occurred; and (c) diagnosis cycle time.

An IC comprises numerous interconnected blocks that are heterogeneous, i.e., processor, memory, analog, mixed-signal, RF, etc. Diagnosis starts by first identifying the source of the failure at the system-level, that is, the failed sub-block. This can be achieved by post-manufacturing testing using the test infrastructure already present on-chip to support post-manufacturing testing, i.e., on-chip test instruments, test access and control mechanisms, etc. [86, 114]. The rest of the diagnosis procedure aiming at identify the root cause of observed failure at transistor-level is specific to the failed block type.

While for digital blocks there exists several in-house frameworks and commercial Electronic Design Automation (EDA) tools for diagnosis [17, 57, 63, 78, 80, 116], for analog blocks there exists neither a commercial tool nor a standardized diagnosis approach. Analog diagnosis is still an ad-hoc, manual, tedious, and time-consuming process very often resulting in large ambiguity groups and no actionable diagnosis information. In fact, there is a vast literature on analog diagnosis [12, 19, 45], yet none of the proposed solutions has matured enough to meet industry standards.

In this Chapter, we discuss diagnosis for analog circuits, where the term "analog" is used in a broad sense and includes also mixed-signal and RF circuits. The focus will be on the use of machine learning support for diagnosis. Machine learning finds numerous applications in several other test-related tasks [103], i.e., test cost reduction, yield learning, adaptive testing, post-manufacturing circuit tuning, outlier detection, test metrics estimation, etc. Essentially, diagnosis can be viewed as a pattern recognition task. Each candidate fault has its own pattern, where in the diagnosis context the pattern is composed of the values of a set of diagnostic measurements, and the goal is to distinguish the patterns of any two different faults.

We will start by discussing fault modeling and fault simulation for analog circuits in Sects. 2 and 3, respectively. Fault modeling results in the fault hypotheses, i.e., the list of potential faults that may be responsible for a circuit failure. Fault simulation is used extensively in a diagnosis flow either in a pre-diagnosis phase to build the diagnosis tools or iteratively during the actual diagnosis procedure. We will provide a concise overview of the prior art on analog diagnosis approaches in Sect. 4, and in Sect. 5 we will present a diagnosis flow based on an ensemble of different machine learning models. In Sect. 6, we will demonstrate a machine learning-based diagnosis flow on an industrial case study. Section 7 concludes the Chapter and provides perspectives.

2   Fault Modeling

A prerequisite for launching diagnosis is the *a priori* generation of fault hypotheses. Fault hypotheses compose a fault model, i.e., a set of hardware-level faults (or defects) that may be responsible for the failure, hoping that the true occurring fault is part of this set. For analog ICs there is no widely accepted fault model, unlike digital ICs for which there are well established fault models, i.e., stuck-at, bridging, and delay faults, that have driven for years the development of Automatic Test Pattern Generation (ATPG) algorithms and diagnosis procedures. The reason is that for analog ICs the fault universe is immeasurable. In general, faults are categorized into parametric (or soft) faults and catastrophic (or hard) faults. A parametric fault refers to the deviation of a component value, i.e., transistor geometry, resistance value, etc. A catastrophic fault refers to a structural change in the circuit topology and is typically modelled as an open-circuit or short-circuit. Component deviation can be in any range and incomplete open-circuits and short-circuits can take any resistance value, thus the fault model size easily explodes.

Inevitably a reduced-size fault model needs to be considered, which may be sufficient for assessing the quality of a test, i.e., based on the resultant fault coverage, but it is problematic for diagnosis since the actual fault occurring may have not been included in the fault model. At best, the diagnosis procedure should respond that the fault is not found. However, due to incomplete fault modeling the diagnosis result may be misleading erroneously predicting another fault in the set.

To this end, a probable list of catastrophic faults can be defined based on historical silicon defective data and Inductive Fault Analysis (IFA) which is used to predict faults that are likely to occur [46]. To reduce the catastrophic fault size, often ideal short-circuits with $10\,\Omega$ resistance and ideal open-circuits with $1\,M\Omega$ resistance are assumed. To account for the fact that an ideal open-circuit does not exist and cannot be handled by a SPICE simulator, a weak pull-up or pull-down is assigned to each open-circuit [43]. The modeling approach relies on controlling the gate-to-source voltage by the drain-to-source voltage multiplied with a coefficient that depends on transistor parameters. Another simplification is to reduce the number of faults in the transistor terminals. In particular, for MOS transistors traditionally six faults are considered, i.e., shorts across gate-to-source, gate-to-drain, and drain-to-source, and opens in each terminal. However, all shorts have a similar effect on the transistor being stuck-on and all opens have a similar effect on the transistor being stuck-off. Thus, for MOS transistors one can use only gate open and drain-to-source short faults [115].

Parametric faults are more difficult to model. Often fixed percentage variations on component nominal values are considered, i.e., transistor length, transistor threshold voltage, resistance value, etc. A parametric fault can be due to variations in fabrication process-related parameters, i.e., doping concentration variations, but the origin may be as well a silicon defect. In the first case, all process parameters vary simultaneously following a distribution, thus the failure may be due to a combination of component variations. In this case, a parametric fault model can be defined based on Monte Carlo simulations using the process parameter distributions defined in the Process Design Kit (PDK) of the technology [109]. More specifically, successive Monte Carlo simulations are performed to approximate the boundary in the process parameter space that encloses all combinations of process parameters that result in circuit performances compliant to the specifications. The fault model can include circuit instances with process parameter samples distributed around the outer side of the boundary. However, this parametric fault model is not useful for diagnosis since it implies distributed faults, i.e., there is no specific fault location.

So far, we discussed faults occurring during manufacturing. In addition, faults can be induced in the field due to ageing phenomena, including Negative Bias Temperature Instability (NTBI) [94], Hot Carrier Injection (HCI) [117], Time-Dependent Dielectric Breakdown (TDDB) [117], and electromigration [118]. To simplify the analysis, often ageing effects are abstracted and modelled with parametric and catastrophic faults.

Another type of failure mechanism are latent defects that are created during manufacturing but are manifested after some period of operation. The most common latent defect is the rupture of the gate oxide of MOS transistors known as pinhole which accelerates the TDDB [51]. It has been recently shown that a pinhole can be modeled as a decrease in the effective value of the oxide thickness [51].

3 Fault Simulation

An indispensable step in diagnosis procedures is fault simulation. A main challenge here is that analog simulation can be very long, thus simulating a large number of faults entails a computational burden. For many types of analog ICs, i.e., data converters and Phase-Locked Loops (PLLs), simulation time may be in the order of days. Performing fault simulations at behavioral-level [16] may not provide the necessary resolution to carry out the diagnosis task. Besides, a behavioral-level model is not detailed enough to perform fault injection at this level. A solution to this problem is to consider a divide-and-conquer approach where the circuit is described at transistor-level only in the vicinity of the fault location. More specifically, an analog IC can be decomposed into sub-blocks. Only the sub-block wherein the fault is located is described at transistor-level, while the rest of the blocks are described at behavioral-level, i.e, in a hardware description language (HDL) such as Verilog. A mixed-mode simulation will complete faster compared to simulating the full transistor-level netlist.

Another requirement for efficient diagnosis procedures is the automation of fault simulation. Performing fault simulations manually by injecting one fault at a time is very tedious given the large fault model size. To this end, the recently proposed analog defect simulators [48, 115, 132] can help automating, speeding up, and tackling the complexity of fault simulation and will prove to be very useful for assisting diagnosis.

The main input to a defect simulator tool includes the circuit netlist and sub-block behavioral-level descriptions, a fault model, and test benches. In its basic operation, the tool outputs the list of all possible faults based on the fault model, it injects one fault into the netlist at a time, and simulates each faulty netlist to compute the different test values, i.e., diagnostic measurements.

4 Overview of Diagnosis Approaches for Analog Circuits

4.1 Rule-Based Diagnosis

Rule-based diagnosis approaches represent the experience of skilled diagnosticians in the form of rules which generally take the form "IF symptom(s) THEN fault(s)" [42, 45], as illustrated in Fig. 1. This approach can only locate the faulty block in a larger system or an assembly fault, i.e., broken interconnect, but it cannot diagnose faulty components down to the transistor-level. The main challenge with

Fig. 1 Rule-based diagnosis

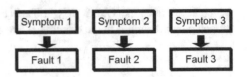

this approach is to acquire the knowledge to build the rules. The problem arises from the fact that different faults often have the same influence on the IC behavior and, thereby, result in the same symptoms. Moreover, typically limited diagnostic information is available. Only few faulty IC samples are available to build the rules, thus not permitting case-based reasoning.

4.2 Model-Based Diagnosis

Model-based diagnosis approaches target parametric faults. The idea is to first build a model linking diagnostic measurements to circuit parameters. Then, given the diagnostic measurements from the real failing device, the model is used to identify the faulty circuit parameter, as well as its deviation from the nominal value. The model can be constructed using nonlinear circuit equations, sensitivity analysis, and behavioral modeling.

4.2.1 Explicit Nonlinear Diagnosis Equations

Explicit nonlinear diagnosis equations take the form $F(\mathbf{p}, \mathbf{m}) = 0$, where F is a matrix with elements that are nonlinear functions of the circuit parameters $\mathbf{p}$ and diagnostic measurements $\mathbf{m}$. Diagnosis equations can be derived analytically using a combination of component connection models, component transfer functions, and composite circuit transfer functions [90, 95]. Given the diagnostic measurement pattern of the failed chip, denoted by $\mathbf{m}_{chip}$, the solution $\mathbf{p}^*$ to these equations can be reached using a Newton-Raphson iteration scheme, namely, $J_F\left(\mathbf{p}^k\right)\left(\mathbf{p}^{k+1} - \mathbf{p}^k\right) = -F\left(\mathbf{p}^k, \mathbf{m}_{chip}\right)$, where $\mathbf{p}^k$ is the k-th estimate of the solution and J_F is the Jacobian. This formulation goes along with diagnosability tests, i.e., ambiguity tests, to examine whether $\mathbf{p}$ can be uniquely determined given $\mathbf{m}_{chip}$ [47]; however, no automated method exists to select diagnostic measurements that satisfy the diagnosability criterion. Moreover, it is not always guaranteed that the Newton-Raphson scheme will converge to a solution.

4.2.2 Sensitivity Analysis

This approach employs linear error models [36, 98] of the form $\Delta \mathbf{m} = S\left(\Delta \mathbf{p}/\mathbf{p}\right)$, where $\Delta \mathbf{p}/\mathbf{p}$ is the normalized vector of parameter deviations, $\Delta \mathbf{m}$ is the vector

Fig. 2 Model-based diagnosis based on behavioral modeling

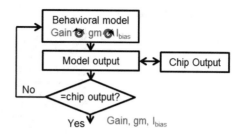

of diagnostic measurement deviations from the nominal expected values obtained on the fault-free device, and S is a sensitivity matrix evaluated at the nominal $\mathbf{p}$. Thus, we can write $\Delta\mathbf{p}/\mathbf{p} = \left(S^T S\right)^{-1} S^T \Delta\mathbf{m}_{chip}$, provided that $\left(S^T S\right)^{-1}$ exists. However, in the presence of fault ambiguity, $S^T S$ is not full rank. Secondly, even with numerically full rank, $S^T S$ may still be nearly singular, in which case the solution will be unstable. Several algorithms have been proposed to determine fault ambiguity in this formulation resulting in a column-reduced sensitivity matrix with full rank [36, 56]. Clearly, linear error models are inadequate for substantial deviations of $\mathbf{p}$. To address this issue, an iterative procedure is implemented that requires to update the sensitivity matrix at each step [36, 98]; however, there is no formal proof that guarantees convergence.

4.2.3 Behavioral Modeling

Behavioral model-based techniques rely on generating an approximate behavioral model of the circuit [31, 68, 75]. During fault diagnosis, this reference behavioral model is perturbed by varying behavioral-level parameters until its response matches the response of the faulty chip, as illustrated in Fig. 2. A match is found if $|\mathbf{m}_{chip} - \mathbf{m}_{model}| < \epsilon$, where $\mathbf{m}_{model}$ is the diagnostic measurements provided by the behavioral model and ϵ is a small error quantity. When a match is found, the behavioral parameters that have deviated from their nominal values are retained. Given the mapping between circuit components and behavioral parameters, a candidate set of components which may have caused the failure is retained. The main disadvantage of this approach is the limited diagnosis resolution. It generates the faulty behavioral parameters; however, a behavioral parameter can depend on several components and each component can be assigned several different faults, thus the number of fault candidates can be prohibitively high. Another challenge is to build a behavioral model that faithfully reproduces the analog circuit behavior.

Fig. 3 Fault dictionary-based diagnosis

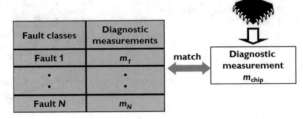

4.3 Fault Dictionary-Based Diagnosis

Perhaps the most common diagnosis approach is based on the use of a fault dictionary [8, 23, 62, 76, 99, 100], as illustrated in Fig. 3. A fault dictionary contains fault hypothesis/diagnostic measurement pattern pairs. More specifically, given a list of N faults $F_i, i = 1, \cdots, N$, one fault is injected at a time in the netlist, and the circuit is simulated to obtain the diagnostic measurement pattern $\mathbf{m}(F_i), i = 1, \cdots, N$, where $\mathbf{m}$ is composed of d test values. During diagnosis, the same diagnostic measurement pattern is obtained on the failed chip and is compared to the logged diagnostic measurement patterns in the faulty dictionary using a similarity measure $d(\mathbf{m}(F_i), \mathbf{m}_{chip})$. The diagnosed fault is the one that presents the most similar diagnostic measurement pattern, i.e., $\min_i d(\mathbf{m}(F_i), \mathbf{m}_{chip})$. This is in essence a pattern recognition, e.g., classification, approach. As such, it is mostly suitable for catastrophic faults whose diagnostic measurement patterns are more separable. Diagnostic measurement patterns of parametric faults tend to overlap and, thereby, are less distinguishable.

4.4 DfT-Assisted Diagnosis

Design-for-test (DfT) consists in embedding test structures on-chip with the aim to facilitate testing, i.e., improve fault coverage and reduce costs. The cost is dictated by the test application time, as well as the Automatic Test Equipment (ATE) requirements. Built-in Self-test (BIST) is a special form of DfT where the test procedure takes place entirely on-chip without needing to interface the chip to external ATE. DfT and BIST are traditionally used for post-manufacturing test, while BIST can be reused in the field of operation to perform on-line test in idle times or concurrent error detection.

In general, the DfT circuitry can comprise one or more of the following test structures: test access points, digitally-controlled re-configuration schemes, and test instruments, i.e., test stimulus generators, actuators, sensors, checkers, and test response analyzers.

For analog circuits, DfT can be functional, i.e., targeting measuring the performances promised in the datasheet, or fault-oriented, i.e., targeting detecting faults. DfT can be generic and applicable to many circuit classes [32, 65, 85] or can be specific to the circuit class, i.e., linear time-invariant circuits [24, 64, 82, 106], PLLs [66, 73, 113, 123], data converters [11, 15, 26, 39, 49, 71, 77, 88, 91] and RF transceivers [1, 2, 29, 41, 59, 119]. Circuit class specific DfT approaches can also be specific to the architecture within a circuit class. For example, for the class of Analog-to-Digital Converters (ADCs), different DfT approaches exist for Successive Approximation Register (SAR) ADCs, pipeline ADCs, $\Sigma\Delta$ ADCs, etc.

DfT can assist the diagnosis task in multiple ways. First, for a large circuit, observing primary outputs can result in large ambiguity groups and limited diagnosis accuracy. DfT can help accessing internal sub-blocks and obtaining information-rich diagnostic measurements at sub-block level so as to break the ambiguity groups and improve the overall diagnosis accuracy [127]. For example, on-chip sensors, i.e., current sensors [29, 128], amplitude detectors [59], and temperature sensors [2], can be used in this context. Second, DfT can assist in forcing specially crafted test signals at internal nodes to more effectively sensitize faults and generate distinct diagnostic measurement patterns per fault. Third, DfT can help reconfiguring the circuit to better expose the faults and create distinct diagnostic measurements. An example here is topology modification by using 1-bit controlled Pull-Down (PD) and Pull-Up (PU) transistors to tie a node to power supply or ground, respectively [32]. Using K PD/PU transistors, we can reconfigure the circuit into 2^K different topologies.

As an example, let us consider in more detail the BIST-assisted diagnosis methodology recently proposed in [84]. The underlying BIST technique is Symmetry-based BIST (SymBIST), which is a generic fault-oriented BIST approach. SymBIST is based on constructing and monitoring invariant properties into the design, as illustrated in Fig. 4. Invariant properties refer to invariant signals, i.e., signals that by construction are confined within a tolerance window for any circuit input. Invariant signals can be built by exploiting symmetries into the design. For example, for a differential sub-block, the sum of the voltages at two symmetrical nodes of the two differential paths is constant [82, 102]. Other invariant signals can be built by summing up complementary signals or subtracting identical signals from replicated sub-blocks. Each invariant signal is then monitored by a checker which flags an error whenever the invariant property is violated, i.e., when the invariant signal slides outside the tolerance window [67, 105]. The checker outputs a 1-bit decision, where by convention 0 refers to invariant property violation. The checker is clocked such that its output observed during a time window corresponds to a bit-string of size ℓ. Invariant properties are identified and built to cover the entire design. The premise is that if a fault occurs, then one or more invariant properties will be violated. For K checkers, we have a diagnostic measurement pattern **m** of size $K \times \ell$ that is the concatenation of all checkers' output bit-strings. In fault-free operation, **m** is a vector of ones, while in the presence of a fault some bits will be flipped to zero. The problem then is to vary the test setup, i.e., test stimulus, internal re-configuration, nominal width of tolerance window, etc., so as to obtain a

Fig. 4 SymBIST principle of
operation

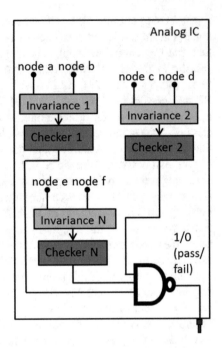

different diagnostic measurement pattern for any two different faults F_i and F_j, i.e.,
$\mathbf{m}(F_i) \neq \mathbf{m}(F_j)$. A fault dictionary-based diagnosis approach is used to identify
the fault that has occurred given the diagnostic measurement pattern $\mathbf{m}_{chip}$, i.e.,
checkers' response, of the failed chip.

An example is shown in Fig. 5 considering 1 checker with a 5-bit output, 2
defects, and 2 test setups. The invariant signal in the fault-free scenario is depicted
in the middle and is shown to be confined within the tolerance window. For setup
1, the invariant signal violates the lower limit of the tolerance window for defect
1 and the upper limit of the tolerance window for defect 2. Thus, the diagnostic
measurement pattern has all ones for both defects and the two defects cannot be
distinguished using test setup 1. For test setup 2, we observe that the invariant signal
in the case of defect 2 slides within the tolerance window for some time. As a result,
the diagnostic measurement becomes 11011 and is now distinguishable from the
diagnostic measurement of defect 1 which remains 11111. In this way, using test
setup 2 we can diagnose which of the two defects has occurred.

4.5 Diagnosis with Defect Simulator in-the-Loop

An iterative diagnosis flow invoking a defect simulator, such as the one in [115], was
recently demonstrated in [79]. First the failed sub-block of the chip is identified. Let
us assume that provisions have been made to facilitate diagnosis. For example, some

Fig. 5 SymBIST example
for diagnosis

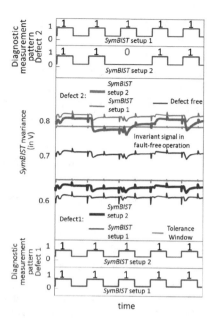

test structures have been added on-chip to extract diagnostic measurements at the
sub-block level of the chip. Additional diagnostic measurements can be obtained
with physical examination of the failed chip, i.e., using imaging or probing, and can
be reproduced at simulation level. The next step is to perform fault injection into
the netlist of the sub-block using a defect simulator. The complete fault model is
employed and all the diagnostic measurements are simulated for each fault. Then,
comparing the simulated and measured diagnostic measurements, we can narrow
down the list of candidate faults, that is, we can exclude faults for which there is a
mismatch between simulated and measured diagnostic measurements. If only one
candidate is found, then we can proceed to the next step which is PFA. Otherwise, if
there are ambiguity groups, then additional measurements are defined, for example
by adding probes into the chip, and the defect simulator is invoked again but this
time starting with the reduce fault model containing only the candidate faults found
in the first iteration. The procedure is repeated, where in each iteration the ambiguity
groups are progressively resolved and a progressively smaller fault model is used,
until only one fault candidate is left.

4.6 Machine Learning-Based Diagnosis

The machine learning concept is well-suited for performing diagnosis. Applications
of machine learning in the context of diagnosis include: (a) defect filter development
for the automatic identification of the fault type that has occurred, i.e., parametric

or catastrophic, so as to launch the appropriate fault diagnosis procedure; (b) parametric fault diagnosis based on regression modeling; (c) using classifiers to establish the mapping between diagnostic measurement patterns and fault classes in fault dictionary-based diagnosis. These applications will be discussed in detail in Sect. 5. We will show that a complete diagnosis flow can be designed based on an ensemble of different machine learning models.

5 Machine Learning Support

5.1 A Unified Diagnosis Flow Based on Machine Learning

A unified diagnosis flow based on machine learning is illustrated in Fig. 6 [61]. Devices subjected to diagnosis can be devices failing during post-manufacturing testing and customer returns, i.e., devices that fail in the field of application.

The diagnosis starts by obtaining the diagnostic measurements on the failed device. The diagnostic measurements are typically pre-selected and have been also simulated on the circuit netlist beforehand under different conditions, i.e., process variations and fault injection. Examples of diagnostic measurements include stan-

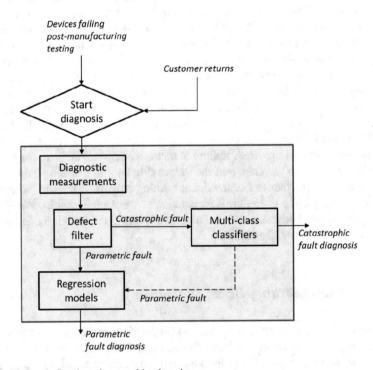

Fig. 6 Diagnosis flow based on machine learning

dard specification-based tests used in post-manufacturing testing, circuit responses to specially crafted test stimuli, DfT-assisted measurements at the sub-block level, and measurement of voltages or currents on internal nodes, which are performed on the actual device using probing. The available set of diagnostic measurements may show some redundancy, i.e., the same diagnostic resolution and accuracy can be achieved with a shorter diagnosis cycle time using a compacted set of diagnostic measurements. Diagnostic measurement extraction and selection will be discussed in more detail in Sect. 5.2.

The diagnostic measurement pattern is first processed through a defect filter whose function is to automatically identify the fault type, i.e., parametric or catastrophic. The design of a defect filter using non-parametric kernel density estimation (KDE) will be discussed in detail in Sect. 5.3.

Thereafter, the appropriate diagnosis procedure is applied. For parametric faults, the flow uses regression functions that map the diagnostic measurements to low-level process parameter values, circuit macro-model parameter values, or circuit component values. Based on the predicted values, the parametric fault is quantified and can be localized or attributed to process variations. Parametric fault diagnosis based on regression modeling will be discussed in more detail in Sect. 5.4. For catastrophic faults, the flow uses the fault dictionary approach where the mapping between diagnostic measurements and fault classes is established based on multi-class classifiers. This approach will be discussed in more detail in Sect. 5.5.

The central components of the diagnosis flow, i.e. defect filter, regression models, and multi-class classifiers, are learning machines which are tuned in a pre-diagnosis learning phase. Diagnostic measurement extraction and selection can also make use of machine learning algorithms.

5.2 Diagnostic Measurement Extraction and Selection

The diagnosis success indicators, i.e., resolution, accuracy, and diagnosis cycle time, are largely dependent on the selection of diagnostic measurements. This makes the selection of diagnostic measurements a critical step. The primary goal is to identify diagnostic measurements that make the effect of the different faults separable so as to reduce the ambiguity groups. The standard specification-based tests typically cannot provide the required resolution. Therefore, it is needed to identify additional and alternative diagnostic measurements obtained at the circuit outputs or at sub-block level by monitoring internal nodes.

In machine learning terminology, diagnostic measurements serve as features or inputs to the machine learning model, and diagnostic measurement identification is called feature extraction. The extraction of appropriate features is key for machine learning model construction as it determines the learning capacity and the lower bound of the generalization error. In regression modelling used for parametric fault diagnosis, the goal is to select features that correlate well with the parameters that are being predicted. In classification used for catastrophic fault diagnosis, the goal is to select features such that devices with different catastrophic faults are separable in the feature space, i.e., we can allocate hyper-dimensional boundaries in the feature space that clearly separate the clusters of different catastrophic faults such that they do not overlap.

5.2.1 Extraction

Feature extraction involves test stimuli generation, defining test responses on the circuit output or internal nodes, and then extracting features on the measured test responses. For analog circuits feature extraction is a circuit-specific problem and, although there exists generic feature extraction approaches applicable to many circuit classes, better diagnosis results can be achieved if we incorporate a priori domain knowledge, i.e., features tailored to the particularities of the circuit under diagnosis.

For analog circuits, standard test stimuli include sinusoidals, multi-tone sinusoidals, ramp signals, etc., and standard feature extraction involves performing DC probing, output sampling, and post-processing using Fast Fourier Transform (FFT).

The key is that the diagnosis practitioner thinks out of the box as non-conventional features can prove to be very relevant. In this regard analog circuits offer more possibilities compared to their digital counterparts. On the test stimulus side, non-conventional signal waveforms, such as white noise [100], chirp signals [93], piece-wise linear signals [121], applied also on nodes other than the circuit's inputs, i.e., bias nodes, power supply node, etc., identification curves [35], and power supply ramping [99] have shown to be very efficient as they are rich in frequencies and can excite the circuit across its bandwidth. On the test response side, we can consider various post-processing algorithms, i.e., Principal Component analysis (PCA) [21, 129], Wavelet transform [8, 18], etc., statistical features [122, 130], i.e. variance, entropy, kurtosis, etc., or simple feature construction by associating tests [44, 116], i.e. subtracting tests. One can also rely on dedicated on-chip test structures for extracting useful features. For example, one can rely on the SymBIST principle [84], Process Control Monitors (PCMs) at die-level [1, 3, 28, 92, 133], amplitude detectors [120], current sensors [29], digital signatures of analog waveforms [14], etc. Another approach is to optimize the features towards a low generalization error for the machine learning models [6, 121]. For example, we can consider a parameterizable input test stimulus and craft it using a gradient descent approach where evaluating the fitness in each step involves learning and evaluation of the diagnosis resolution, as shown in Fig. 7.

5.2.2 Selection

Very often there is no clear rationale on how to choose features. The reason is the large number of process parameters and their intricate interactions which makes impossible to foresee the correlation between features on one hand and circuit

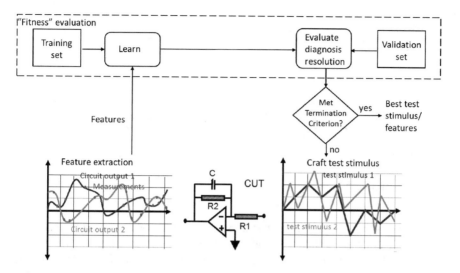

Fig. 7 Feature optimization by involving the machine learning model

parameters or fault classes on the other hand. The brute-force approach is to extract as many features as possible in an *ad hoc* fashion and then, in a subsequent step, select the most relevant and eliminate redundancy. In machine learning terminology, this procedure is called feature selection.

Reducing the set of features is desired for two reasons. First, a smaller set of features means smaller on-chip resources and faster diagnosis cycle time. Second, the volume of the training data required to train the defect filter, regression models, and classifiers, increases with the dimensionality of the feature space. The higher the dimensionality of the feature space is, the more training data is needed to span the feature space and improve learning. By using many features we may run into the problem of the curse of dimensionality. In particular, for a given training set size, by increasing the dimensionality of the feature space the error on the training set decreases. The generalization error initially decreases too but after some dimensionality it starts increasing. The underlying reason is that the training data tend to have a larger amount of surface area when projected in higher dimensional spaces, meaning they become more sparse. Since the density of training samples in a given neighborhood decreases, it implies that learning in this neighborhood becomes less accurate or even random. Typically, the required training set size increases exponentially with the feature dimensionality. For example, consider a two-class classification problem. By increasing the feature dimensionality, the training data become sparse leaving large empty spaces between them. In this case, many classification boundaries can be allocated separating perfectly the two classes, but any new sample lying in the space empty of training data will be somewhat randomly classified.

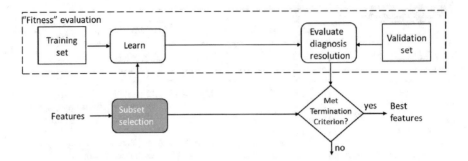

Fig. 8 Wrapper method for feature selection

Feature selection algorithms search in the power-set of features and select best subsets. If there are d features available, then there are 2^d possible subsets and searching for the best subset is an NP-complete problem. There exist several heuristic feature selection algorithms that a diagnosis practitioner can choose from [7, 13, 52, 53, 72, 81, 87, 101, 110]. They can be categorized into filter and wrapper methods. Filter methods select features regardless of the machine learning model, i.e., based on feature correlation [83]. In wrapper methods the fitness of each feature subset is evaluated using the machine learning model guiding the search, as shown in Fig. 8. The fitness is defined based on the diagnosis accuracy predicted by the machine learning model, and can also incorporate the cost for obtaining the features, i.e., on-chip resources needed, test time, etc. Popular wrapper feature selection algorithms include Genetic Algorithms (GAs) [37, 50, 96] and floating search algorithms [89].

GAs maintain a population of chromosomes of fixed size. In our case, chromosomes are bit-strings of length equal to the number of diagnostic measurements, where the j-th bit is set to 1 if the j-th diagnostic measurement is present in the subset and 0 otherwise. As shown in Fig. 9, starting with a base population, new chromosomes are generated using the mutation and crossover operators. In crossover, parts of two different parent chromosomes are mixed to produce an offspring. In mutation, bits of a single parent chromosome are randomly perturbed to create a child. At the end of each generation, each chromosome is evaluated to determine its fitness criteria. Only the fittest chromosomes are likely to survive and breed into the next generation. GAs evolve with the juxtaposition of bit templates, quickly optimizing the target fitness criteria.

5.3 Defect Filter

The role of the defect filter is to identify the fault type, i.e., parametric or catastrophic, based on the diagnostic measurement pattern and, thereafter, forward the device to the appropriate diagnosis tier. The defect filter is required since

Fig. 9 Feature selection
using genetic algorithms

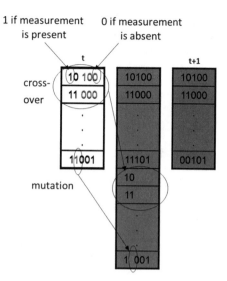

different diagnoses flows are used for parametric and catastrophic faults and applying the wrong flow may inadvertently lead to poor diagnosis. Thus, the defect filter enables a unified catastrophic/parametric fault diagnosis approach without needing to specify in advance the fault type.

The defect filter can be built based on an one-class classifier in the space of diagnostic measurements. The classifier allocates a boundary enclosing data from devices with process variations such that a device with a catastrophic fault has a footprint outside the boundary.

In its simplest form, a defect filter can be a hyper-rectangle [4]. However, a hyper-rectangle is a crude approximation of the area surrounding the distribution of devices with process variations. Another approach is using nonlinear guard-bands in the space of diagnostic measurements [70]. However, the correct positioning of guard-bands requires detailed information on defective devices which is not readily available. In addition, the positioning will be to some degree random due to the non-statistical nature of defective devices.

Herein, we describe in more detail a defect filter based on the joint probability density function of the diagnostic measurements [111], denoted by $f(\mathbf{m})$, where $\mathbf{m} = [m_1, \ldots, m_d]$ is the d-dimensional diagnostic measurement pattern.

The form of $f(\mathbf{m})$ is unknown, thus we will estimate it using a training set, in particular using the d-dimensional diagnostic measurement patterns of n circuit instances, denoted by $\mathbf{M}^1, \cdots, \mathbf{M}^n$. The estimate, denoted by $\tilde{f}(\mathbf{m})$, is derived using a non-parametric kernel density estimator defined by [97]

$$\tilde{f}(\mathbf{m}, \alpha) = \frac{1}{n} \sum_{j=1}^{n} \frac{1}{(\lambda_j(\alpha) \cdot h)^d} K_e \left(\frac{\mathbf{m} - \mathbf{M}^j}{\lambda_j(\alpha) \cdot h} \right), \tag{1}$$

Fig. 10 Non-parametric
kernel density estimation

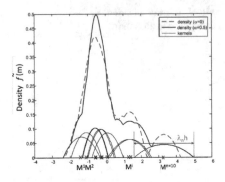

where

$$h = \left\{ 8c_d^{-1}(d+4)(2\sqrt{\pi})^d \right\}^{1/(d+4)} n^{-1/(d+4)} \tag{2}$$

is a smoothing parameter called bandwidth,

$$c_d = 2\pi^{d/2}/(d \cdot \Gamma(d/2)) \tag{3}$$

is the volume of the unit d-dimensional sphere,

$$K_e(\mathbf{t}) = \begin{cases} \frac{1}{2}c_d^{-1}(d+2)\left(1 - \mathbf{t}^T\mathbf{t}\right) & \text{if } \mathbf{t}^T\mathbf{t} < 1 \\ 0 & \text{otherwise} \end{cases} \tag{4}$$

is the Epanechnikov kernel, λ_j are local bandwidth factors defined by

$$\lambda_j(\alpha) = \left\{ \tilde{f}(\mathbf{M}^j, 0)/g \right\}^{-\alpha}, \tag{5}$$

and g is the geometric mean

$$\log g = n^{-1} \sum_{j=1}^{n} \log \tilde{f}(\mathbf{M}^j, 0). \tag{6}$$

The density estimate in (1) is a weighted sum of kernels centered on the n observations, as shown with the one-dimensional example of Fig. 10. The bandwidth h defines the half-width of the kernels. The parameter $\lambda_j(\alpha)$ multiplies the bandwidth of the kernel of the j-th observation. The default value of α is $\alpha = 0$, resulting in $\lambda_j(0) = 1$ for all n observations. By increasing α, the tails of the density estimate become smoother and longer, but less heavier [97].

Non-parametric means that no assumption is made regarding the parametric form of $f(\mathbf{m})$, thus any diagnostic measurement pattern can be handled. The density

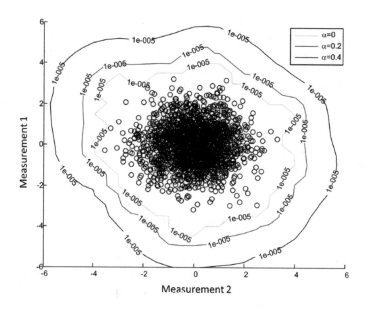

Fig. 11 Defect filters in a 2-dimensional diagnostic measurement space. The black circles correspond to devices with process variations which are used to construct the defect filters

function is estimated using only devices with process variations. Such a training set can be generated by a Monte Carlo simulation. The fact that no devices with catastrophic faults are required to estimate the density function makes the defect filter independent of fault dictionaries, which is an appealing property.

Noticing that the density estimate vanishes at some point, we can choose naturally to filter a device as catastrophic if its measurement pattern $\mathbf{m}_{chip}$ lies in an area that has zero probability density, i.e., it satisfies

$$\tilde{f}(\mathbf{m}_{chip}, \alpha) = 0, \tag{7}$$

since in this case it is inconsistent with the statistical nature of the bulk of the data from devices with process variations that was used to estimate the density. The solutions to the above equation compose the frontier of the defect filter. The parameter α controls the extent of the filtering. The larger α is, the more "lenient" the defect filter is. This is depicted in two dimensions in Fig. 11, where the frontiers of three progressively "lenient" defect filters are displayed with contours of low density.

The devices with catastrophic faults that are filtered out are forwarded to a multi-class classifier that is trained in the pre-diagnosis phase to map the diagnostic measurement pattern to the underlying catastrophic fault, as it will described in more detail in Sect. 5.5. On the other hand, if $\tilde{f}(\mathbf{m}_{chip}, \alpha) > 0$, the device is considered to contain process variations, that is, a parametric fault has occurred. For parametric fault diagnosis, we use nonlinear inverse regression functions that are trained in the pre-diagnosis phase to map the diagnostic measurement pattern to the values of circuit parameters of interest, as it will be described in more detail in Sect. 5.4. The defect filter is always tuned to filter out devices with catastrophic faults. However, this could inadvertently result in some devices with parametric faults being also screened out and forwarded to the classifier. To correct this leakage, the multi-class classifier is trained during the pre-diagnosis phase to include detection of devices with process variations as well, i.e., the class of process variations is added. Thus, in the unlikely case where a device with a parametric fault is presented to a classifier, the classifier kicks it back to the regression tier.

5.4 Parametric Fault Diagnosis Based on Regression Modeling

Regression can be used to develop an alternative model-based diagnosis approach for parametric faults. In particular, regression can be used to approximate the functions relating the diagnostic measurement pattern to the values of circuit parameters of interest. These functions are intricate and explicit formulations using circuit equations or linear error models are difficult to derive and are valid for small parameter deviations. Regression can be used to learn these functions for the entire circuit parameter ranges using training data. Formally, for each circuit parameter $\{p_j\}_{j=1,\cdots,n}$, we train a regression function $f_j : \mathbf{m} \mapsto p_j, j = 1, \ldots, n$, as shown in Fig. 12 [23, 25, 38, 61]. On the failed device, we obtain the diagnostic measurement pattern $\mathbf{m}_{chip}$ and we use the regression functions to predict the values of circuit parameters. This way we can diagnose off-target circuit parameter values.

This approach is similar to the alternate test principle where regression modeling is used to learn the mapping between low-cost measurements and the performances promised in the datasheet [5, 10, 14, 107, 108, 121, 125]. The goal here is to circumvent the explicit measurement of performances which is very costly as it entails long test times and sophisticated ATE. By employing a measurement pattern obtained on a single test configuration using low-cost ATE we can achieve significant test cost reduction.

Fig. 12 Parametric fault diagnosis using regression functions

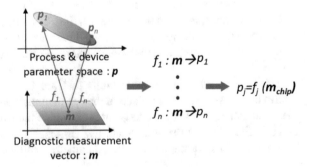

Regression-based parametric fault diagnosis gives flexibility as to what circuit parameters are diagnosed. Circuit parameters can vary from low-level process parameters, i.e., threshold voltage, oxide thickness, and junction capacitances, to component values, i.e., transistor geometries, resistor values, and capacitor values, to design parameters, i.e., transistor transconductance values, gains of stages, and biasing values.

The learning phase employs a training set of devices with typical and extreme process variations on which both the circuit parameters and the diagnostic measurements are obtained. The device instances can be generated by Monte Carlo simulation and corner analysis using the statistical PDK of the technology. Real training data can also be generated on fabricated chips from various lots, wafers, and wafer sites as long as the circuit parameters can be measured. This typically requires some on-chip test infrastructure. For example, low-level process parameters can be measured using PCMs that are typically placed in the scribe lines of the wafer.

The diagnosis practitioner can use one of the several available regression tools, i.e., polynomial regression, neural networks, Multivariate Adaptive Regression Splines (MARS), Support Vector Machine (SVM) regression, etc. [21, 27, 54]. There are many regression modeling approaches to choose from and each comes with a suite of training algorithms. The main issue is choosing the tool with the right complexity. By using a simplistic regression model we may under-fit the training data, resulting also in high generalization error, while by using a more sophisticated regression modeling approach that can draw highly nonlinear approximations we may over-fit the training data, resulting inadvertently in poor generalization. To give a simple example, let us consider a set of training observations (x, y) generated according to the function $y = \alpha x$. If the observations are noisy, then using a high-order polynomial model to derive an exact fit through all observations will result in zero error on the training observations, but this will result in a poor generalization on new observations, which is the main issue at stake in machine learning. In this case, a first-order polynomial would be the best choice. Most advance regression tools offer some inherent criterion and form of regularization to control the complexity of the learned model.

5.5 Fault Dictionary-Based Diagnosis Using Classifiers

Classifiers offer an elegant solution for fault dictionary-based diagnosis. The first step in fault dictionary-based diagnosis is to create a list of the N most probable catastrophic short- or open-circuit fault locations, denoted by $\{F_i\}, i = 1, \cdots, N$, for example based on IFA and historical defect data. The catastrophic faults are injected sequentially in the netlist of the device and we perform Monte Carlo simulation, where in each pass a different short or open resistance R is used. These values are sampled from the resistance distributions for short- and open-circuits, denoted by $p(R|F_i)$, as shown in Fig. 13 [60]. In this way, a set of n samples is

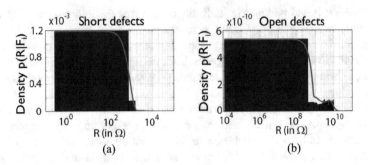

Fig. 13 Estimated probability density function of resistance (in Ω) for (**a**) open defects and (**b**) short defects, plotted in logarithmic scale

created for each catastrophic fault i, denoted by $\{F_i^j\}$, $j = 1, \cdots, n$. It is also possible to enhance each set with more points that represent process spread. This is recommended if we can afford the extra simulation effort. For example for each fault value F_i^j we can run a Monte Carlo simulation of the circuit using the PDK of the technology. For each value $\{F_i^j\}$ we obtain the d-dimensional diagnostic measurement pattern denoted by $\mathbf{m}_i^j = [m_{i,1}^j, m_{i,2}^j, \cdots, m_{i,d}^j]$. The fault cluster for fault F_i, $i = 1, \cdots, N$ is $FC_i = \{\mathbf{m}_i^j\}$, $j = 1, \cdots, n$. It can be further enhanced by more samples by estimating the density function of the samples $\{\mathbf{m}_i^j\}$, $j = 1, \cdots, n$, for example using non-parametric KDE [97], and, thereafter, sample it to generate an arbitrarily large number of synthetic samples [112]. In summary, the i-th fault cluster FC_i of the i-th fault, $i = 1, \cdots, N$, consists of n samples $\{\mathbf{m}_i^j\}$, $j = 1, \cdots, n$ allocated in the space of diagnostic measurements. Additionally, we can define the process variation cluster FC_{PV} corresponding to circuit instances with process variations, which can be generated with Monte Carlo simulation and corner analysis where in this case there is no fault injected. The set comprising the fault clusters $\{FC_i\}$, $i = 1, \cdots, N$ and the process variation cluster FC_{PV} compose the fault dictionary.

Thereafter, the fault dictionary is used in the pre-diagnosis phase to train a multi-class classifier that allocates a boundary in the space of diagnostic measurements to separate one fault cluster from another. The boundary divides the diagnostic measurement space into subspaces each corresponding to a single fault, while there is also one subspace corresponding to process variations. On the failed device, we obtain the d-dimensional diagnostic measurement pattern $\mathbf{m}_{chip} = [m_{chip}^1, m_{chip}^2, \cdots, m_{chip}^d]$ and present it to the classifier. The classifier examines where the footprint of $\mathbf{m}_{chip}$ lies with the respect to the boundary, i.e., in which subspace it lies, and the fault corresponding to the subspace is diagnosed. If $\mathbf{m}_{chip}$ falls into the process variations cluster, then parametric fault diagnosis is launched.

This approach is illustrated in Fig. 14 for a 2-dimensional diagnostic measurement pattern and 5 faults. As it can be seen, while faults 1–3 are clearly separable, faults 4–5 overlap in this diagnostic measurement space resulting in fault ambiguity

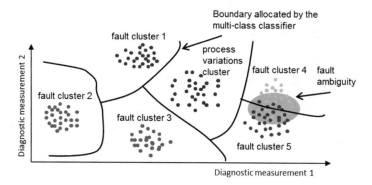

Fig. 14 Fault dictionary-based diagnosis using a multi-class classifier

around the boundary. The goal is that the diagnostic measurement pattern is selected so as to result in clear separation of the fault clusters. The accuracy of the fault classification defines the diagnosis resolution and accuracy.

Note that if we define a single fault cluster with circuit instances that violate one or more performance specifications, either because of catastrophic faults or process variations, and we use an additional nominal cluster with circuit instances that satisfy all performance specifications, then a fault detection classification-based system can be developed [104].

Furthermore, as suggested by practitioners in the field of pattern recognition [69, 124], the overall classification accuracy can be improved by combining the response of different classifiers [62]. Thus, instead of using a single classifier one can use an ensemble of c classifiers $\{C_j\}$, $j = 1, \cdots, c$, each trained separately, also possibly in a different diagnostic measurement space. Each classifier j assigns a score to each fault i, denoted by $s_j(F_i)$, instead of just making a deterministic judgment about which fault is present in the faulty device. Thereafter, the individual scores of the classifiers are combined to assign a single score to each fault i, denoted by $s(F_i)$. Various combination methods have been proposed in the literature, including averaging, weighted averaging, majority vote, etc. [69, 124]. The simplest is the averaging method which consists of computing the average value of scores obtained by different classifiers, i.e., $s(F_i) = \frac{1}{c} \sum_{j=1}^{c} s_j(F_i)$. The weighted averaging method requires a validation set to assign weights to the classifiers, yet this validation set is typically not available at the time when the diagnosis tools are built. The majority vote method renders a deterministic diagnosis rather than a ranking of catastrophic faults that are likely to have occurred. Using an ensemble of classifiers, the output of the catastrophic diagnosis flow is the ranking of the catastrophic faults according to their likelihood of occurrence in the faulty device.

Similar to regression tools for parametric fault diagnosis, there is a multitude of classifier tools one can use, i.e., k-nearest neighbors, discrimination analysis, neural networks, SVMs, decision trees, etc. [21, 27, 54]. The selection criteria can be based on the dataset size, feature dimensionality, presence of outliers, training runtime, etc. Most importantly, the practitioner needs to understand how the classifier works and tune its user-defined parameters which may require trial and error. A good practice is to put to the test different classifiers, including advance ones that have internal regularization, and then choose the one(s) that offer the highest classification and decision confidence.

The discussed catastrophic fault diagnosis flow using classification has been explored extensively in the literature where different combinations of diagnostic measurements (including test stimuli and test response post-processing) and classifiers have been studied on different case studies [8, 9, 20, 22, 23, 30, 34, 40, 55, 62, 76, 99, 100, 126, 129, 131].

6 Industrial Case Study

6.1 Device Under Test

We will show the application of the diagnosis flow presented in Sect. 5 to an industrial case study [62]. The Device Under Test (DUT) is a Controller Area Network (CAN) transceiver designed by NXP Semiconductors in a BiCMOS-DMOS process. The circuit netlist has 1032 elements of which 613 are transistors. A high-level block diagram of the circuit is shown in Fig. 15. This device is produced in high-volume and constitutes an essential part in the electronic system of automobiles. It is deployed in a safety-critical application, thus it has to meet stringent specifications and demands practically zero test escapes. Therefore, it is of vital importance to diagnose the sources of failure, in order to achieve better quality control and, when possible, improve the design such that similar failures do not emerge in the field during the lifetime of the operation.

6.2 Real Dataset

We have at hand a set of 29 devices from different lots that failed at least one of the specifications during production test. Time-consuming failure analysis was carried out for all these devices and it was observed in all cases that the cause of failure is a short-circuit defect. For example, Fig. 16 shows a Focused Ion Beam (FIB) image of the short-circuit defect observed in DUT 18 and Fig. 17 shows a Scanning Electron Microscope (SEM) image of the short-circuit defect observed in DUT 26. For the purpose of the experiment, we assume that the actual defects that have occurred in each of these devices are unknown and we set out to diagnose them by applying the diagnosis flow presented in Sect. 5.

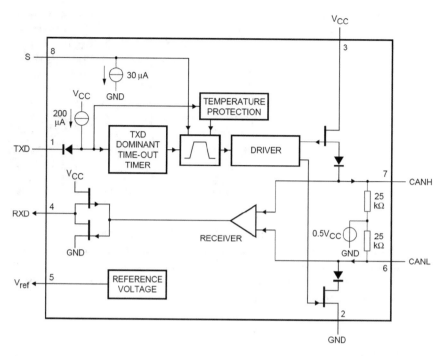

Fig. 15 High-level block diagram of the CAN transceiver

Fig. 16 FIB image of the short-circuit defect diagnosed in DUT 18

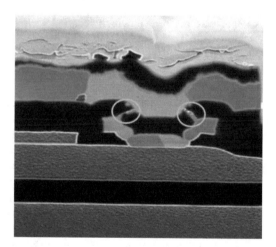

6.3 Fault Modeling

For this particular device produced in high volume under a mature technology where process variation is well understood and controlled, device failures due to parametric deviation of process and device parameters are very unlikely to occur. Furthermore,

Fig. 17 SEM image of the short-circuit defect diagnosed in DUT 26

for this particular technology, open-circuit defects are less likely to occur than short-circuit defects. In fact, in analog designs, typically one has space to do via doubling, which makes open-circuit defects even less likely. As a result, more than 90% of the observed defects in production are short-circuits. Thus, only catastrophic short-circuit defects are considered for fault modeling.

6.4 Fault Dictionary

An IFA was performed which resulted in a list of $N = 923$ probable short-circuit defects. Each short-circuit is modeled with $n = 3$ different bridge resistance values, i.e., $5\,\Omega$, $50\,\Omega$, $200\,\Omega$. These values are chosen according to defect data characterization analysis for this particular technology. Subsequently, a total of $3 \times 923 = 2769$ fault simulations were carried out to generate the 923 fault clusters. In this large-scale industrial case study, we cannot afford extra simulation effort to consider process variation in fault simulation. Thus, each simulation consists of inserting a short-circuit defect in the netlist with a specific bridge resistance value, while the circuit parameters are fixed at their nominal design values. In each fault simulation we collect the same diagnostic measurements.

The standard production tests for this device include digital, analog, and IDDQ tests. We consider as diagnostic measurements $d = 97$ non-digital tests (i.e., voltage, current, timing and hysteresis measurements) which dominate the test time. Each measurement is scaled in the range $[-1, 1]$.

Fault simulation took approximately 12 h. Notice that fault simulation is a one time effort. Building the diagnosis tools and performing the diagnosis of a faulty device takes only a few minutes.

6.5   Missing Values Problem

In this real-world case study, the injection of a defect in the device netlist might render the system of equations during circuit simulation unsolvable. Therefore, it is highly likely that there exist diagnostic measurements that are unattainable for specific defects and specific resistance values. The problem of missing values also concerns the real diagnostic measurement pattern. Indeed, a diagnostic measurement might hit the instrument limit, in which case its value is artificially "forced" to equal the instrument limit. In this case, we can only use the pass/fail information provided by the diagnostic measurement and we should consider the absolute value as missing. To account for missing values, the recommendations in [74] are followed. In short, missing values force us to exclude either diagnostic measurements or defects from the analysis. In the former case, we remove information that may be useful for performing diagnosis. In the latter case, we are bound to obtain misleading diagnosis results if the defect that is present in the faulty device has been inadvertently excluded from the analysis.

6.6   Classifiers

As mentioned in Sect. 5.5, numerous classifiers, ranging from simple to more elaborate ones, can be employed to diagnose catastrophic faults. In general, the efficiency of a classifier depends on the distribution of fault clusters and the extent to which they overlap. In the context of diagnosis, there is no solution for choosing the optimal classifier among an array of different classifiers. The reason is that classifiers can be compared only on the basis of a real validation set, but such a set is not available at the time we build the diagnosis tools. In this section, we describe in detail five different classifiers that were considered for this experiment, based on pass/fail verification, Euclidean distance, Mahalanobis distance, non-parametric KDE, and SVM. Our intention is to list a number of popular classifiers. In theory, any other classifier can be used in the same context. In addition, we show how these classifiers assign to each defect a normalized score between [0, 1], where the highest score is given to the most probable defect. The normalized scores are combined to consider the ensemble of classifiers as opposed to using a single classifier, so as to improve the diagnosis accuracy.

6.6.1   Pass/Fail Verification

This classifier simply examines the similarity of the patterns $\mathbf{m}_{chip}$ and $\mathbf{m}_i^j$ by verifying the pass/fail information for each diagnostic measurement. Formally, we consider the specification indicator $I_{j,k}^i$, such that (a) $I_{j,k}^i = 1$ if both $\mathbf{m}_{chip}$ and $\mathbf{m}_i^j$ comply with the specification of the k-th diagnostic measurement or if both $\mathbf{m}_{chip}$

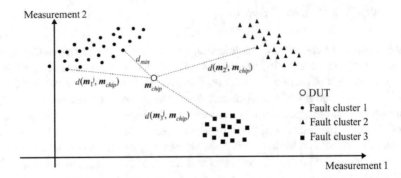

Fig. 18 Classifier based on Euclidean distance

and $\mathbf{m}_i^j$ fail the specification of the k-th diagnostic measurement and (b) $I_{j,k}^i = 0$ if only one of $\mathbf{m}_{chip}$ and $\mathbf{m}_i^j$ complies with the specification of the k-th diagnostic measurement. The normalized score between $[0,1]$ for defect F_i is defined as

$$s_1(F_i) = \frac{1}{n} \sum_{j=1}^{n} \frac{1}{d} \sum_{k=1}^{d} I_{j,k}^i. \tag{8}$$

6.6.2 Euclidean Distance

This classifier relies on the distances between the patterns $\mathbf{m}_{chip}$ and $\mathbf{m}_i^j$, $i = 1, \cdots, N$, $j = 1, \cdots, n$, as illustrated in Fig. 18. The Euclidean distance is used to determine pattern proximity

$$d(\mathbf{m}_i^j, \mathbf{m}_{chip}) = \sqrt{(m_{i,1}^j - m_{chip}^1)^2 + \cdots + (m_{i,d}^j - m_{chip}^d)^2}. \tag{9}$$

We define the minimum distance as

$$d_{\min} = \min_{i,j} d(\mathbf{m}_i^j, \mathbf{m}_{chip}) \tag{10}$$

which allows us to scale the distances between $[0,1]$

$$d'(\mathbf{m}_i^j, \mathbf{m}_{chip}) = d_{\min}/d(\mathbf{m}_i^j, \mathbf{m}_{chip}). \tag{11}$$

The pattern $\mathbf{m}_i^j$ with the shortest distance from the pattern $\mathbf{m}_{chip}$ is mapped to 1. We assign a score to each defect F_i by computing the average normalized distance over all resistance values $j = 1, \cdots, n$

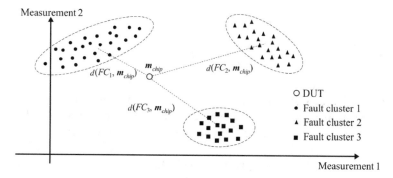

Fig. 19 Classifier based on Mahalanobis distance

$$s_2(F_i) = \frac{1}{n} \sum_{j=1}^{n} d'(\mathbf{m}_i^j, \mathbf{m}_{chip}). \tag{12}$$

6.6.3 Mahalanobis Distance

This classifier considers the Mahalanobis distance between the pattern $\mathbf{m}_{chip}$ and each fault cluster $FC_i, i = 1, \cdots, N$. As shown in Fig. 19, this form of distance represents the difference between the pattern $\mathbf{m}_{chip}$ and the mean of the fault cluster FC_i, normalized by the within-cluster covariance which is a measure of the spread of the cluster around the center of its mass

$$d(FC_i, \mathbf{m}_{chip}) = \sqrt{(\mathbf{m}_{chip} - \mathbf{u}_i)^T \times S_i^{-1} \times (\mathbf{m}_{chip} - \mathbf{u}_i)}, \tag{13}$$

where

$$\mathbf{u}_i = [u_{i,1}, \cdots, u_{i,d}] \tag{14}$$

is the mean vector with

$$u_{i,k} = \frac{1}{n} \sum_{j=1}^{n} m_{i,k}^j, \tag{15}$$

S_i is the covariance matrix

$$S_i = \begin{bmatrix} s_{11}^i & \cdots & s_{1d}^i \\ \cdots & \ddots & \cdots \\ s_{d1}^i & \cdots & s_{dd}^i \end{bmatrix} \tag{16}$$

with

$$s_{k,\ell}^i = \frac{1}{n} \sum_{j=1}^{n} (m_{i,k}^j - u_{j,k})(m_{i,\ell}^j - u_{j,\ell}). \tag{17}$$

This method favors fault clusters for which the distance between their center of mass and the pattern $\mathbf{m}_{chip}$ is small and penalizes fault clusters for which this distance is large compared to their spread. By defining the minimum distance as

$$d_{min} = \min_i d(FC_i, \mathbf{m}_{chip}), \tag{18}$$

we assign a score to each defect F_i between $[0,1]$

$$s_3(F_i) = d_{min}/d(FC_i, \mathbf{m}_{chip}). \tag{19}$$

6.6.4 Non-Parametric Kernel Density Estimation

According to the Bayes' theorem, the *posterior* probability that a faulty DUT with pattern $\mathbf{m}_{chip}$ contains fault F_i is expressed as

$$P(F_i|\mathbf{m}_{chip}) = \frac{f_i(\mathbf{m}_{chip}|F_i)P(F_i)}{p(\mathbf{m}_{chip})}, \tag{20}$$

where $P(F_i)$ is the *prior* probability of defect F_i, $f_i(\mathbf{m}_{chip}|F_i)$ is the conditional joint probability density function of $\mathbf{m}_{chip}$ given the presence of defect F_i (also called the likelihood), and $p(\mathbf{m}_{chip})$ is the probability density function of $\mathbf{m}_{chip}$. A faulty DUT will most likely contain defect F_m if

$$P(F_m|\mathbf{m}_{chip}) > P(F_i|\mathbf{m}_{chip}), \quad \forall i \neq m. \tag{21}$$

Combining (20) and (21), the above inequality becomes

$$f_m(\mathbf{m}_{chip}|F_m)P(F_m) > f_i(\mathbf{m}_{chip}|F_i)P(F_i), \quad \forall i \neq m. \tag{22}$$

The *prior* probabilities of faults can be extracted from IFA. Here, for the purpose of simplicity and without loss of generality, we assume that they are equal. Under this scenario, a faulty DUT will most likely contain defect F_m if

$$f_m(\mathbf{m}_{chip}|F_m) > f_i(\mathbf{m}_{chip}|F_i), \quad \forall i \neq m. \tag{23}$$

This method relies on the estimation of the densities $f_i(\mathbf{m}|F_i)$, $i = 1, \cdots, N$ using the available observations $\mathbf{m}_i^j$, $j = 1, \cdots, n$, contained in the i-th fault cluster FC_i. To estimate $f_i(\mathbf{m}|F_i)$ we use non-parametric KDE, as explained in Sect. 5.3,

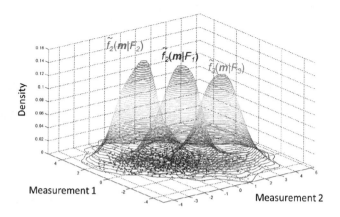

Fig. 20 Classifier based on non-prametric kernel density estimation

so as not to make any assumption regarding its parametric form, e.g., Gaussian, and allow the observations to speak for themselves [97]. Figure 20 shows the estimated densities $\tilde{f}_i(\mathbf{m}|F_i)$ for three defects in a 2-dimensional diagnostic measurement space.

Given a DUT with pattern $\mathbf{m}_{chip}$, we assign a normalized score between $[0,1]$ to each defect

$$s_4(F_i) = \frac{\tilde{f}_i(\mathbf{m}_{chip}|F_i) - \tilde{f}_{\min}}{\tilde{f}_{\max} - \tilde{f}_{\min}}, \tag{24}$$

where

$$\tilde{f}_{\min} = \min_i \tilde{f}_i(\mathbf{m}_{chip}|F_i) \tag{25}$$

$$\tilde{f}_{\max} = \max_i \tilde{f}_i(\mathbf{m}_{chip}|F_i). \tag{26}$$

As before, the defect that achieves the highest density $\tilde{f}_i(\mathbf{m}_{chip}|F_i)$ is mapped to 1. Furthermore, if $s_4(F_i)$ is zero for every defect i, then the pattern $\mathbf{m}_{chip}$ is considered to be "foreign" to all fault clusters. In this case, we can conclude that the fault that has occurred had not been modeled in the fault dictionary. Thus, unlike the other methods that always assign a score to each fault, the non-parametric KDE method is the only one that in theory can identify an "unexpected" fault. This is a very appealing attribute of this classifier.

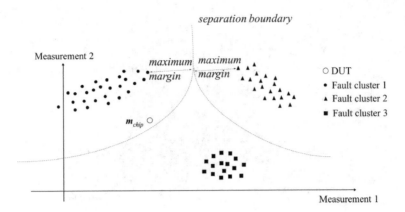

Fig. 21 Classifier based on SVM

6.6.5 Support Vector Machine

This classifier aims to allocate nonlinear boundaries in the space of diagnostic measurements to separate the N fault clusters. In particular, we use SVMs [33] to learn the boundaries that traverse the middle of the Euclidean distance between the N fault clusters. This is shown in Fig. 21 for a 2-dimensional diagnostic measurement space.

The SVM classifier was originally used to solve binary classification problems. For multi-class classification with N fault clusters ($N > 2$), we can reduce the problem into either $\binom{N}{2}$ or N distinct binary classification problems and apply either the "one-against-one" or the "one-against-all" strategies. Experiments on large problems show that the "one-against-one" strategy is more suitable for practical use [58]. In this approach, the classification is carried out by a max-wins voting strategy, where each binary classifier assigns the DUT to one of two fault clusters, then the vote for the assigned fault cluster is increased by one vote, and finally the fault cluster with the largest number of votes determines the fault cluster to which the DUT belongs to.

This method assigns normalized scores between [0,1] to each defect according to

$$s_5(F_i) = N_i/N_{\max}, \tag{27}$$

where N_i denotes the number of classifiers that assign the pattern $\mathbf{m}_{chip}$ to defect F_i and

$$N_{\max} = \max_i N_i. \tag{28}$$

6.7 Diagnosis Results

Table 1 shows the 5 most highly ranked defects according to their scores for each of the 29 failed devices. The first column shows the DUT number, the second column shows the actual defect that is present, the third column shows the ranking of defects, and the fourth column shows the corresponding (rounded) final scores. As it can be observed from Table 1, the proposed method diagnoses correctly 17 out of the 29 failed devices with the true defect matching with the first choice and for 4 failed devices the true defect appears in the first three choices. In some cases the ranking indicates with high confidence the location of the defect. For example, for DUT 2, the five defects that come first in the ranking (e.g. 320, 341, 126, 374, 111) are

Table 1 Diagnosis results

DUT	True defect	Defect ranking	Normalized scores
1	107	**107** 90 920 114 347	0.924 0.923 0.923 0.923 0.923
2	320	**320** 341 126 374 111	0.948 0.867 0.833 0.827 0.822
3	125	47 616 **125** 681 360	0.914 0.839 0.838 0.837 0.837
4	101	**101** 117 459 50 388	0.831 0.829 0.826 0.817 0.817
5	216	**216** 666 192 516 120	0.831 0.795 0.792 0.788 0.785
6	300	524 608 744 294 789	0.900 0.890 0.862 0.855 0.850
7	20	**20** 126 24 27 111	0.889 0.866 0.862 0.850 0.849
8	27	**27** 111 126 446 341	0.891 0.856 0.837 0.834 0.834
9	104	111 **104** 465 721 126	0.848 0.844 0.839 0.823 0.822
10	21	310 682 524 789 608	0.867 0.858 0.855 0.855 0.851
11	101	**101** 117 459 50 388	0.831 0.829 0.826 0.818 0.817
12	19	**19** 541 106 562 595	0.810 0.794 0.780 0.780 0.780
13	19	**19** 541 562 595 106	0.799 0.791 0.788 0.771 0.771
14	140	401 **140** 457 40 919	0.936 0.912 0.911 0.910 0.910
15	20	**20** 24 126 27 111	0.887 0.865 0.862 0.853 0.849
16	101	**101** 117 459 50 388	0.831 0.829 0.826 0.817 0.817
17	107	**107** 90 920 114 347	0.924 0.923 0.923 0.923 0.923
18	31	117 **31** 50 388 622	0.901 0.888 0.882 0.881 0.880
19	101	252 305 366 363 31	0.883 0.857 0.846 0.844 0.843
20	19	**19** 541 106 562 595	0.821 0.794 0.793 0.780 0.780
21	156	524 608 744 789 682	0.903 0.893 0.872 0.872 0.866
22	20	**20** 126 24 27 111	0.882 0.870 0.867 0.864 0.853
23	107	**107** 90 920 114 347	0.924 0.923 0.923 0.923 0.923
24	22	**22** 19 541 338 106	0.826 0.808 0.808 0.795 0.795
25	107	**107** 90 920 114 347	0.924 0.923 0.923 0.923 0.923
26	380	666 192 516 676 457	0.910 0.906 0.905 0.904 0.903
27	376	383 456 112 34 196	0.924 0.920 0.830 0.826 0.824
28	28	666 192 516 355 676	0.910 0.907 0.898 0.896 0.896
29	300	524 608 744 475 215	0.896 0.896 0.866 0.864 0.862

short-circuits across nodes of a transistor pair. The ranking of these defects can be subsequently used to speed up a classical FA method by placing the emphasis on the locations of the chip where the defect has probably occurred.

By comparing the diagnosis predictions to the true defect existing in each DUT, we identify the defects that we are unable to diagnose. We were unable to diagnose correctly defects 21, 28, 156, 300, 376, 380, and in one case defect 101. Furthermore, in some cases the true defects are not ranked as the first priority, such as the cases of DUT 3, 9, 14, and 18. The reason for the above fault ambiguities is that there are different defects whose patterns tend to overlap in the diagnostic measurement space. In other words, the impact of these defects on the diagnostic measurements is very similar. Fault ambiguity can be observed as early as in the fault simulation phase. To resolve fault ambiguity additional diagnostic measurements need to be considered.

7   Conclusion and Discussion

Diagnosis for analog circuits is still a manual, tedious, and time-consuming procedure in industry, unlike digital circuits for which a large degree of automation has been achieved and there are available EDA tools. Apart from modern analog defect simulators, which can be used to speed up and automate iterative fault simulations during the diagnosis procedure, all other tasks in a diagnosis flow remain manual. Identification of diagnostic measurements for analog circuits, including test stimulus generation and feature extraction on test responses, is a circuit-specific problem and there is no standardized approach. However, the practitioner can borrow ideas from the literature for literally every analog circuit class. To improve diagnosis resolution, diagnostic measurements need also to be derived at sub-block level requiring DfT support, yet analog DfT infrastructures are still *ad-hoc*. Another challenge is the definition of the analog fault model. The analog fault universe is infinite increasing prohibitively fault simulation time and entailing a risk that the actual fault that has occurred is unmodeled. On the algorithmic side, machine learning can provide support to develop all the required diagnosis tools, i.e., for fault type identification and for fault localization and quantification. There is a vast literature on diagnosis algorithms using different machine learning models. Machine learning has shown to be successful in a number of tasks and diagnosis is not an exemption. Machine learning can be used to train a system that predicts the occurring fault from a set of features, i.e., diagnostic measurements, on the faulty device. Thus, the problem is twofold: identify effective diagnostic measurements in terms of fault prediction capability and cost, i.e., area overhead of required on-chip test resources and test time, and train the system, which comprises density estimation, regression, and classification, all standard learning problems for which numerous tools are readily available. Machine learning algorithms can help circumvent to a large extent the manual diagnosis procedures and have shown to be effective in mining information into the fault dictionaries to resolve ambiguity groups and improve the success rate of the subsequent PFA. It is up to the EDA industry now to adopt these algorithms, automate their execution, and showcase them on real case studies.

Acknowledgments This work was supported by the ANR EDITSoC project under Grant ANR-17-CE24-0014-02.

References

1. Abdallah, L., Stratigopoulos, H.G., Mir, S., Kelma, C.: Experiences with non-intrusive sensors for RF built-in test. In: Proceedings of the IEEE International Test Conference, Paper 17.1 (2012).
2. Abdallah, L., Stratigopoulos, H.G., Mir, S., Altet, J.: Defect-oriented non-intrusive RF test using on-chip temperature sensors. In: Proceedings of the IEEE VLSI Test Symposium (2013)
3. Ahmadi, A., Stratigopoulos, H.G., Huang, K., Nahar, A., Orr, B., Pas, M., Carulli, J.M., Makris, Y.: Yield forecasting across semiconductor fabrication plants and design generations. IEEE Trans. Comput. Aided Des. Integr. Circuits Syst. **36**(12), 2120–2133 (2017)
4. Akbay, S.S., Chatterjee, A.: Fault-based alternate test of RF components. In: IEEE International Conference on Computer Design, pp. 518–525 (2007)
5. Akbay, S.S., Halder, A., Chatterjee, A., Keezer, D.: Low-cost test of embedded RF/Analog/Mixed-signal circuits in SOPs. IEEE Trans. Adv. Packag. **27**(2), 352–363 (2004)
6. Akbay, S.S., Torres, J.L., Rumer, J.M., Chatterjee, A., Amtsfield, J.: Alternate test of RF front ends with IP constraints: frequency domain test generation and validation. In: Proceedings of the IEEE International Test Conference, pp. 4.4.1–4.4.10 (2006)
7. Alippi, C., Catelani, M., Fort, A., Mugnaini, M.: Automated selection of test frequencies for fault diagnosis in analog electronic circuits. IEEE Trans. Instrum. Meas. **54**(3), 1033–1044 (2005)
8. Aminian, M., Aminian, F.: A modular fault-diagnosis system for analog electronic circuits using neural networks with wavelet transform as a preprocessor. IEEE Trans. Instrum. Meas. **56**(5), 1546–1554 (2007)
9. Aminian, F., Aminian, M., Collins, H.W. Jr.: Analog fault diagnosis of actual circuits using neural networks. IEEE Trans. Instrum. Meas. **51**(3), 544–550 (2002)
10. Ayari, H., Azais, F., Bernard, S., Compte, M., Renovell, M., Kerzerho, V., Potin, O., Kelma, C.: Smart selection of indirect parameters for dc-based alternate RF IC testing. In: Proceedings of the IEEE VLSI Test Symposium, pp. 19–24 (2012)
11. Azais, F., Bernard, S., Bertrand, Y., Renovell, M.: Optimizing sinusoidal histogram test for low cost ADC BIST. J. Electron. Test. Theory Appl. **17**(3–4), 255–266 (2001)
12. Bandler, J.W., Salama, A.E.: Fault diagnosis of analog circuits. Proc. IEEE **73**(8), 1279–1325 (1985)
13. Barragan, M.J., Léger, G.: A procedure for alternate test feature design and selection. IEEE Des. Test **32**(1), 18–25 (2015)
14. Barragán, M.J., Fiorelli, R., Léger, G., Rueda, A., Huertas, J.L.: Alternate test of LNAs through ensemble learning of on-chip digital envelope signals. J. Electron. Test. Theory Appl. **27**(3), 277–288 (2011)
15. Barragan, M., Alhakim, R., Stratigopoulos, H.G., Dubois, M., Mir, S., Gall, H.L., Bhargava, N., Bal, A.: A fully-digital BIST wrapper based on ternary test stimuli for the dynamic test of a 40 nm CMOS 18-bit stereo audio $\Sigma\Delta$ ADC. IEEE Trans. Circuits Syst. Regul. Pap. **63**(11), 1876–1888 (2016)
16. Barragan, M.J., Stratigopoulos, H.G., Mir, S., Gall, H.L., Bhargava, N., Bal, A.: Practical simulation flow for evaluating analog/mixed-signal test techniques. IEEE Des. Test **33**(6), 46–54 (2016)

17. Benware, B., Schuermyer, C., Sharma, M., Herrmann, T.: Determining a failure root cause distribution from a population of layout-aware scan diagnosis results. IEEE Des. Test Comput. **29**(1), 8–18 (2012)
18. Bhunia, S., Roy, K.: A novel wavelet transform-based transient current analysis for fault detection and localization. IEEE Trans. Very Large Scale Integr. VLSI Syst. **13**(4), 503–507 (2005)
19. Binu, D., Kariyappa, B.: A survey on fault diagnosis of analog circuits: Taxonomy and state of the art. AEU Int. J. Electron. Commun. **73**, 68–83 (2017)
20. Binu, D., Kariyappa, B.S.: RideNN: A new rider optimization algorithm-based neural network for fault diagnosis in analog circuits. IEEE Trans. Instrum. Meas. **68**(1), 2–26 (2019)
21. Bishop, C.M.: Neural Networks for Pattern Recognition. Oxford University Press, Oxford (1995)
22. Catelani, M., Fort, A.: Fault diagnosis of electronic analog circuits using a radial basis function network classifier. Measurement **28**(3), 147–158 (2000)
23. Chakrabarti, S., Cherubal, S., Chatterjee, A.: Fault diagnosis for mixed-signal electronic systems. In: Proceedings of the IEEE Aerospace Conference, pp. 169–179 (1999)
24. Chatterjee, A.: Concurrent error detection and fault-tolerance in linear analog circuits using continuous checksums. IEEE Trans. Very Large Scale Integr. VLSI Syst. **1**(2), 138–150 (1993)
25. Chatterjee, S., Chatterjee, A.: Test generation based diagnosis of device parameters for analog circuits. In: Design, Automation & Test in Europe Conference and Exhibition, pp. 596–602 (2001)
26. Chen, T., Jin, X., Geiger, R.L., Chen, D.: USER-SMILE: ultrafast stimulus error removal and segmented model identification of linearity errors for ADC built-in self-test. IEEE Trans. Circuits Syst. Regul. Pap. **65**(7), 2059–2069 (2018)
27. Cherkassky, V., Mulier, F.: Learning from Data. John Wiley & Sons, New York (1998)
28. Cilici, F., Barragan, M.J., Mir, S., Lauga-Larroze, E., Bourdel, S.: Assisted test design for non-intrusive machine learning indirect test of millimeter-wave circuits. In: IEEE European Test Symposium (2018)
29. Cimino, M., Lapuyade, H., Deval, Y., Taris, T., Bégueret, J.B.: Design of a 0.9V 2.45 GHz self-testable and reliability-enhanced CMOS LNA. IEEE J. Solid-State Circuits **43**(5), 1187–1194 (2008)
30. Collins, P., Yu, S., Eckersall, K.R., Jervis, B.W., Bell, I.M., Taylor, G.E.: Application of Kohonen and supervised forced organization maps to fault diagnosis in CMOS opamps. Electron. Lett. **30**(22), 1846–1847 (1994)
31. Cota, E.F., Negreiros, M., Carro, L., Lubaszewski, M.: A new adaptive analog test and diagnosis system. IEEE Trans. Instrum. Meas. **49**(2), 223–227 (2000)
32. Coyette, A., Esen, B., Dobbelaere, W., Vanhooren, R., Gielen, G.: Automatic generation of test infrastructures for analog integrated circuits by controllability and observability co-optimization. Integr. VLSI J. **55**, 393–400 (2016)
33. Cristianini, N., Shawe-Taylor, J.: Support Vector Machines and Other Kernel-Based Learning Methods. Cambridge Univercity Press, Cambridge (2000)
34. Cui, J., Wang, Y.: A novel approach of analog circuit fault diagnosis using support vector machines classifier. Measurement **44**(1), 281–289 (2011)
35. Czaja, Z.: Using a square-wave signal for fault diagnosis of analog parts of mixed-signal electronic embedded systems. IEEE Trans. Instrum. Meas. **57**(8), 1589–1595 (2008)
36. Dai, H., Souders, M.: Time-domain testing strategies and fault diagnosis for analog systems. IEEE Trans. Instrum. Meas. **39**(1), 157–162 (1990)
37. Deb, K., Pratap, A., Agarwal, A., Meyarivan, T.: A fast and elitist multiobjective genetic algorithm: NSGA-II. IEEE Trans. Evol. Comput. **6**(2), 182–197 (2002)
38. Devarakond, S.K., Sen, S., Bhattacharya, S., Chatterjee, A.: Concurrent device/specification cause–effect monitoring for yield diagnosis using alternate diagnostic signatures. IEEE Des. Test Compu. **29**(1), 48–58 (2012)

39. Dufort, B., Roberts, G.W.: On-chip analog signal generation for mixed-signal built-in self-test. IEEE J. Solid-State Circuits **34**(3), 318–30 (1999)
40. Epstein, B.R., Czigler, M., Miller, S.R.: Fault detection and classification in linear integrated circuits: An application of discrimination analysis and hypothesis testing. IEEE Trans. Comput. Aided Des. Integr. Circuits Syst. and Systems **12**(1), 102–113 (1993)
41. Erdogan, E.S., Ozev, S.: Detailed characterization of transceiver parameters through loop-back-based BiST. IEEE Trans. Very Large Scale Integr. VLSI Syst. **18**(6), 901–911 (2010)
42. Erdogan, E.S., Ozev, S., Cauvet, P.: Diagnosis of assemply failures for system-in-package RF tuners. In: IEEE International Symposium on Circuits and Systems, pp. 2286–2289 (2008)
43. Esen, B., Coyette, A., Gielen, G., Dobbelaere, W., Vanhooren, R.: Effective DC fault models and testing approach for open defects in analog circuits. In: Proceedings of the IEEE International Test Conference, Paper 3.2 (2016)
44. Fang, L., Lemnawar, M., Xing, Y.: Cost effective outliers screening with moving limits and correlation testing for analogue ICs. In: Proceedings of the IEEE International Test Conference, Paper 31.2 (2006)
45. Fenton, W., McGinnity, T.M., Maguire, L.P.: Fault diagnosis of electronic systems using intelligent techniques: a review. IEEE Trans. Syst. Man Cybern. Part C Appl. Rev. **31**(3), 269–281 (2001)
46. Ferguson, F.J., Shen, J.P.: A CMOS fault extractor for inductive fault analysis. IEEE Trans. Comput. Aided Des. Integr. Circuits Syst. **7**(11), 1181–1194 (1988)
47. Fontana, G., Luchetta, A., Manetti, S., Piccirilli, M.C.: A fast algorithm for testability analysis of large linear time-invariant networks. IEEE Trans. Circuits Syst. Regul. Pap. **64**(6), 1564–1575 (2017)
48. Gil, V.G., Arteaga, A.J.G., Léger, G.: Assessing AMS-RF test quality by defect simulation. IEEE Trans. Device Mater. Reliab. **19**(1), 55–63 (2019)
49. Gines, A., Léger, G.: Sigma-delta testability for pipeline A/D converters. In: Proceedings of the Design, Automation and Test in Europe Conference (2014)
50. Goldberg, D.E.: Genetic Algorithms in Search, Optimization, and Machine Learing. Addison-Wesley, Boston (1989)
51. Gomez, J., Kama, N., Coyette, A., Vanhooren, R., Dobbelaere, W., Gielen, G.: Pinhole latent defect modeling and simulation for defect-oriented analog/mixed-signal testing. In: Proceedings of the IEEE VLSI Test Symposium (2020)
52. Grasso, F., Luchetta, A., Manetti, S., Piccirilli, M.C.: A method for the automatic selection of test frequencies in analog fault diagnosis. IEEE Trans. Instrum. Meas. **56**(6), 2322–2329 (2007)
53. Guyon, I., Elisseeff, A.: An introduction to variable and feature selection. J. Mach. Learn. Res. **3** (2003)
54. Hastie, T., Tibshirani, R., Friedman, J.: The Elements of Statistical Learning: Data Mining, Inference, and Prediction. Springer, New York (2001)
55. He, W., He, Y., Li, B.: Generative adversarial networks with comprehensive wavelet feature for fault diagnosis of analog circuits. IEEE Trans. Instrum. Meas. **69**(9), 6640–6650 (2020)
56. Hemink, G.J., Meijer, B.W., Kerkhoff, H.G.: Testability analysis of analog systems. IEEE Trans. Comput. Aided Design **9**(6), 573–583 (1990)
57. Holst, S., Wunderlich, H.: Adaptive debug and diagnosis without fault dictionaries. In: IEEE European Test Symposium, pp. 7–12 (2007)
58. Hsu, C.W., Lin, C.J.: A comparison of methods for multi-class support vector machines. IEEE Trans. Neural Netw. **13**(2), 415–425 (2002)
59. Huang, Y.C., Hsieh, H.H., Lu, L.H.: A built-in self-test technique for RF low-noise amplifiers. IEEE Trans. Microwave Theory Tech. **56**(2), 1035–1042 (2008)
60. Huang, K., Stratigopoulos, H.G., Mir, S.: Bayesian fault diagnosis of RF circuits using nonparametric density estimation. In: Proceedings of the IEEE Asian Test Symposium, pp. 295–298 (2010)

61. Huang, K., Stratigopoulos, H.G., Mir, S.: Fault diagnosis of analog circuits based on machine learning. In: Proceedings of the Design, Automation & Test in Europe Conference, pp. 1761–1766 (2010)
62. Huang, K., Stratigopoulos, H.G., Mir, S., Hora, C., Xing, Y., Kruseman, B.: Diagnosis of local spot defects in analog circuits. IEEE Trans. Instrum. Meas. **61**(10), 2701–2712 (2012)
63. Huang, Q., Fang, C., Mittal, S., Blanton, R.D.: Towards smarter diagnosis: A learning-based diagnostic outcome previewer. ACM Trans. Des. Autom. Electron. Syst. **25**(5) (2020)
64. Huertas, J.L., Rueda, A., Vasquez, D.: Testable switched-capacitor filters. IEEE J. Solid-State Circuits **28**(7), 719–724 (1993)
65. Huertas, G., Vázquez, D., Peralías, E.J., Rueda, A., Huertas, J.L.: Testing mixed-signal cores: a practical oscillation-based test in an analog macrocell. IEEE Des. Test Comput. **19**(6), 73–82 (2002)
66. Ince, M., Yilmaz, E., Fu, W., Park, J., Nagaraj, K., Winemberg, L., Ozev, S.: Digital built-in self-test for phased locked loops to enable fault detection. In: IEEE European Test Symposium (2019)
67. Kolarik, V., Mir, S., Lubaszewski, M., Courtois, B.: Analog checkers with absolute and relative tolerances. IEEE Trans. Comput. Aided Des. Integr. Circuits Syst. **14**(5), 607–612 (1995)
68. Kook, S., Banerjee, A., Chatterjee, A.: Dynamic specification testing and diagnosis of high-precision sigma-delta ADCs. IEEE Des. Test Comput. **30**(4), 36–48 (2013)
69. Kuncheva, L.: "Fuzzy" versus "nonfuzzy" in combining classifiers designed by Boosting. IEEE Trans. Fuzzy Syst. **11**(6), 729–741 (2003)
70. Kupp, N., Drineas, P., Slamani, M., Makris, Y.: On boosting the accuracy of non-RF to RF correlation-based specification test compaction. J. Electron. Test. Theory Appl. **25**(6), 309–321 (2009)
71. Laraba, A., Stratigopoulos, H.G., Mir, S., Naudet, H.: Exploiting pipeline ADC properties for a reduced-code linearity test technique. IEEE Trans. Circuits Syst. Regul. Pap. **62**(10), 2391–2400 (2015)
72. Larguech, S., Azais, F., Bernard, S., Comte, M., Kerzérho, V., Renovell, M.: Efficiency evaluation of analog/RF alternate test: comparative study of indirect measurement selection strategies. Microelectron. J. **46**(11), 1091–1102 (2015)
73. Le-Gall, H., Alhakim, R., Valka, M., Mir, S., Stratigopoulos, H., Simeu, E.: High frequency jitter estimator for SoCs. In: IEEE European Test Symposium (2015)
74. Little, R., Rubin, D.: Statistical Analysis with Missing data, 2nd edn. John Wiley & Sons, New York (2002)
75. Liu, F., Ozev, S., Brooke, M.: Identifying the source of BW failures in high-frequency linear analog circuits based on S-parameters measurements. IEEE Trans. Comput. Aided Des. Integr. Circuits Syst. **25**(11), 2594–2605 (2006)
76. Maidon, Y., Jervis, B.W., Dutton, N., Lesage, S.: Diagnosis of multifaults in analogue circuits using multilayer perceptrons. IEE Proc. Circuits Dev. Syst. **144**(3), 149–154 (1997)
77. Malloug, H., Barragan, M.J., Mir, S.: Practical harmonic cancellation techniques for the on-chip implementation of sinusoidal signal generators for mixed-signal BIST applications. J. Electron. Test. Theory Appl. **34**(3), 263–279 (2018)
78. Maxwell, P., Hapke, F., Tang, H.: Cell-aware diagnosis: Defective inmates exposed in their cells. In: IEEE European Test Symposium (2016)
79. Melis, T., Simeu, E., Auvray, E.: Automatic fault simulators for diagnosis of analog systems. In: International Symposium on On-Line Testing and Robust System Design (2020)
80. Mhamdi, S., Girard, P., Virazel, A., Bosio, A., Faehn, E., Ladhar, A.: Cell-aware defect diagnosis of customer returns based on supervised learning. IEEE Trans. Device Mater. Reliab. **20**(2), 329–340 (2020)
81. Mir, S., Lubaszewski, M., Courtois, B.: Fault-based ATPG for linear analog circuits with minimal size multifrequency test sets. J. Electron. Test. Theory Appl. **9**(1-2), 43–57 (1996)
82. Mir, S., Lubaszewski, M., Kolarik, V., Courtois, B.: Fault-based testing and diagnosis of balanced filters. Analog Integr. Circ. Sig. Process **11**(1), 5–19 (1996)

83. Papakostas, D.K., Hatzopoulos, A.A.: Correlation-based comparison of analog signatures for identification and fault diagnosis. IEEE Trans. Instrum. Meas. **42**(4), 860–3 (1993)
84. Pavlidis, A., Faehn, E., Louërat, M.M., Stratigopoulos, H.G.: BIST-assisted analog fault diagnosis. In: IEEE European Test Symposium (2021)
85. Pavlidis, A., Louërat, M.M., Faehn, E., Kumar, A., Stratigopoulos, H.G.: *SymBIST*: symmetry-based analog and mixed-signal built-in self-test for functional safety. IEEE Trans. Circuits Syst. Regul. Pap. **68**(6), 2580–2593 (2021)
86. Portolan, M.: Automated testing flow: The present and the future. IEEE Trans. Comput. Aided Des. Integr. Circuits Syst. **39**(10), 2952–2963 (2020)
87. Prasad, V.C., Babu, N.S.C.: Selection of test nodes for analog fault diagnosis in dictionary approach. IEEE Trans. Instrum. Meas. **49**(6), 1289–1297 (2000)
88. Provost, B., Sánchez-Sinencio, E.: On-chip ramp generators for mixed-signal BIST and ADC self-test. IEEE J. Solid-State Circuits **38**(2), 263–273 (2003)
89. Pudil, P., Novovicova, J., Kittler, J.: Floating search methods in feature selection. Pattern Recogn. Lett. **15**, 1119–1125 (1994)
90. Rapisarda, L., Decarlo, R.A.: Analog multifrequency fault diagnosis. IEEE Trans. Circuits Syst. **CAS-30**(4), 223–234 (1983)
91. Renaud, G., Diallo, M., Barragan, M.J., Mir, S.: Fully differential 4-V output range 14.5-ENOB stepwise ramp stimulus generator for on-chip static linearity test of ADCs. IEEE Trans. Very Large Scale Integr. VLSI Syst. **27**(2), 281–293 (2019)
92. Şandru, E.D., David, E., Kovacs, I., Buzo, A., Burileanu, C., Pelz, G.: Modeling the dependency of analog circuit performance parameters on manufacturing process variations with applications in sensitivity analysis and yield prediction. IEEE Trans. Comput. Aided Des. Integr. Circuits Syst. **41**(1), 129–142 (2021)
93. Sarson, P.: Test time efficient group delay filter characterization technique using a discrete chirped excitation signal. In: Proceedings of the IEEE International Test Conference (2016)
94. Schroder, D., Babcock, J.: Negative bias temperature instability: Road to cross in deep submicron silicon semiconductor manufacturing. J. Appl. Phys. **94**(1), 1–18 (2003)
95. Sen, N., Saeks, R.: Fault diagnosis for linear systems via multifrequency measurements. IEEE Trans. Circuits Syst. **26**(7), 457–465 (1979)
96. Siedlecki, W., Sklansky, J.: A note on genetic algorithms for large-scale feature selection. Pattern Recogn. Lett. **10**, 335–347 (1989)
97. Silverman, B.W.: Density Estimation for Statistics and Data Analysis. Chapman & Hall/CRC, London (1986)
98. Slamani, M., Kaminska, B.: Analog circuit fault diagnosis based on sensitivity computation and functional testing. IEEE Des. Test Comput. **9**(1), 30–39 (1992)
99. Somayajula, S.S., Sanchez-Sinencio, E., de Gyvez, J.P.: Analog fault diagnosis based on ramping power supply current signature clusters. IEEE Trans. Circuits Syst. II, Analog Digit. Signal Process. **43**(10), 703–712 (1996)
100. Spina, R., Upadhyaya, S.: Linear circuit fault diagnosis using neuromorphic analyzers. IEEE Trans. Circuits Syst. II, Analog Digit. Signal Process. **44**(3), 188–196 (1997)
101. Starzyk, J.A., Liu, D., Liu, Z.H., Nelson, D.E., Rutkowski, J.O.: Entropy-based optimum test points selection for analog fault dictionary techniques. IEEE Trans. Instrum. Meas. **53**(3), 754–761 (2004)
102. Stessman, N.J., Vinnakota, B., Harjani, R.: System-level design for test of fully differential analog circuits. IEEE J. Solid-State Circuits **31**(10), 1526–1534 (1996)
103. Stratigopoulos, H.G.: Machine learning applications in IC testing. In: Proceedings of the IEEE European Test Symposium (2018)
104. Stratigopoulos, H.G.D., Makris, Y.: Non-linear decision boundaries for testing analog circuits. IEEE Trans. Comput. Aided Des. Integr. Circuits Syst. **24**(11), 1760–1773 (2005)
105. Stratigopoulos, H.G.D., Makris, Y.: An adaptive checker for the fully differential analog code. IEEE J. Solid-State Circuits **41**(6), 1421–1429 (2006)
106. Stratigopoulos, H.G.D., Makris, Y.: Concurrent detection of erroneous responses in linear analog circuits. IEEE Trans. Comput. Aided Des. Integr. Circuits Syst. **25**(5), 878–891 (2006)

107. Stratigopoulos, H.G., Makris, Y.: Error moderation in low-cost machine-learning-based Analog/RF testing. IEEE Trans. Comput. Aided Des. Integr. Circuits Syst. **27**(2), 339–351 (2008)

108. Stratigopoulos, H.G., Mir, S.: Adaptive alternate analog test. IEEE Des. Test Comput. **29**(4), 71–79 (2012)

109. Stratigopoulos, H.G., Sunter, S.: Fast Monte Carlo-based estimation of analog parametric test metrics. IEEE Trans. Comput. Aided Des. Integr. Circuits Syst. **33**(12), 1977–1990 (2014)

110. Stratigopoulos, H.G.D., Drineas, P., Slamani, M., Makris, Y.: Non-RF to RF test correlation using learning machines: a case study. In: IEEE VLSI Test Symposium, pp. 9–14 (2007)

111. Stratigopoulos, H.G., Mir, S., Acar, E., Ozev, S.: Defect filter for alternate RF test. In: Proceedings of the IEEE European Test Symposium, pp. 101–106 (2009)

112. Stratigopoulos, H.G., Mir, S., Makris, Y.: Enrichment of limited training sets in machine-learning-based analog/RF test. In: Proceedings of the Design, Automation & Test in Europe Conference, pp. 1668–1673 (2009)

113. Sunter, S., Roy, A.: On-chip digital jitter measurement, from megahertz to gigahertz. IEEE Des. Test Comput. **21**(4), 314–321 (2004)

114. Sunter, S., Côté, J.F., Rearick, J.: Streaming access to ADCs and DACs for mixed-signal ATPG. IEEE Des. Test **33**(6), 38–45 (2016)

115. Sunter, S., Jurga, K., Laidler, A.: Using mixed-signal defect simulation to close the loop between design and test. IEEE Trans. Circuits Syst. Regul. Pap. **63**(12), 2313–2322 (2016)

116. Tikkanen, J., Siatkowski, S., Sumikawa, N., Wang, L.C., Abadir, M.S.: Yield optimization using advanced statistical correlation methods. In: Proceedings of the IEEE International Test Conference (2014)

117. Tsividis, Y.: Operational and Modeling of the MOS Transistor, 2nd edn. McGraw-Hill, New York (1999)

118. Tu, K.: Recent advances on electromigration in very-large-scale-integration of interconnects. J. Appl. Phys. **94**(9), 5451–5473 (2003)

119. Valdes-Garcia, A., Silva-Martinez, J., Sanchez-Sinencio, E.: On-chip testing techniques for RF wireless transceivers. IEEE Des. Test Comput. **23**(4), 268–277 (2006)

120. Valdes-Garcia, A., Venkatasubramanian, R., Silva-Martinez, J., Sanchez-Sinencio, E.: A broadband CMOS amplitude detector for on-chip RF measurements. IEEE Trans. Instrum. Meas. **57**(7), 1470–1477 (2008)

121. Variyam, P.N., Cherubal, S., Chatterjee, A.: Prediction of analog performance parameters using fast transient testing. IEEE Trans. Comput. Aided Des. Integr. Circuits Syst. **21**(3), 349–361 (2002)

122. Vasan, A.S.S., Long, B., Pecht, M.: Diagnostics and prognostics method for analog electronic circuits. IEEE Trans. Ind. Electron. **60**(11), 5277–5291 (2013)

123. Veillette, B.R., Roberts, G.: On-chip measurement of the jitter transfer function of charge-pump phase-locked loops. IEEE J. Solid-State Circuits **33**(3), 483–491 (1998)

124. Verikas, A., Lipnickas, A., Malmqvist, K., Bacauskiene, M., Gelzinis, A.: Soft combination of neural classifiers: A comparative study. Pattern Recogn. Lett. **20**(4), 429–444 (1999)

125. Voorakaranam, R., Akbay, S.S., Bhattacharya, S., Cherubal, S., Chatterjee, A.: Signature testing of analog and RF circuits: algorithms and methodology. IEEE Trans. Circuits Syst. I **54**(5), 1018–1031 (2007)

126. Wang, Z., Gielen, G., Sansen, W.: Probabilistic fault detection and the selection of measurements for analog integrated circuits. IEEE Trans. Comput. Aided Des. Integr. Circuits Syst. **17**(9), 862–872 (1998)

127. Wey, C.L.: Built-in self-test (BIST) structure for analog circuit fault diagnosis. IEEE Trans. Instrum. Meas. **39**(3), 517–21 (1990)

128. Wey, C.L., Krishnan, S.: Built-in self-test (BIST) structures for analog circuit fault diagnosis with current test data. IEEE Trans. Instrum. Meas. **41**(4), 535–9 (1992)

129. Xiao, Y., He, Y.: A novel approach for analog fault diagnosis based on neural networks and improved kernel PCA. Neurocomputing **74**(7), 1102–1115 (2011)

130. Yuan, L., He, Y., Huang, J., Sun, Y.: A new neural-network-based fault diagnosis approach for analog circuits by using kurtosis and entropy as a preprocessor. IEEE Trans. Instrum. Meas. **59**(3), 586–595 (2010)
131. Zhang, C., He, Y., Yuan, L., Xiang, S.: Analog circuit incipient fault diagnosis method using DBN based features extraction. IEEE Access **6**, 23053–23064 (2018)
132. Zivkovic, V., Schaldenbrand, A.: Requirements for industrial analog fault-simulator. In: International Conference on Synthesis, Modeling, Analysis and Simulation Methods and Applications to Circuit Design, pp. 61–64 (2019)
133. Zjajo, A., Barragan, M.J., de Gyvez, J.P.: BIST method for die-level process parameter variation monitoring in analog/mixed-signal integrated circuits. In: Proceedings of the Design, Automation & Test in Europe Conference, pp. 1301–1306 (2007)

Machine Learning Support for Board-Level Functional Fault Diagnosis

Mengyun Liu, Xin Li, and Krishnendu Chakrabarty

1 Introduction

To realize complex functions, various chips are assembled into a printed circuit board (PCB). Even though all the chips assembled on a board have passed their own chip-level tests, testing of boards is still required. This is because the board-level working environment is different from the environment for chip-level standalone testing [1]. In many cases, a board may fail functional test even if all chips on it pass automated test equipment (ATE) tests [1].

1.1 Flow of Board-Level Manufacturing Test

A board typically integrates a large number of processor chips, ASICs, memory chips, passive components, as well as thousands of I/Os [2]. The objective of board-level manufacturing test is to verify the printed wiring and the physical contacts between wires and the pins of components. In addition, at-speed data transmission between different components are also tested. Figure 1 shows a typical manufacturing test flow of electronic systems.

First, automated optical or X-ray inspection (AOI or AXI) is applied to detect process flaws. It uses optics or X-rays to capture images of the board under test.

M. Liu (✉)
Hardware Team, Nvidia Corporation, Santa Clara, CA, USA
e-mail: mengyunl@nvidia.com

X. Li and K. Chakrabarty
Department of Electrical and Computer Engineering, Duke University, Durham, NC, USA
e-mail: xinli.ece@duke.edu; krish@duke.edu

© The Author(s), under exclusive license to Springer Nature Switzerland AG 2023
P. Girard et al. (eds.), *Machine Learning Support for Fault Diagnosis of System-on-Chip*, https://doi.org/10.1007/978-3-031-19639-3_8

These images are then processed using image processing software that detect unexpected components according to the position, the size, and the shape information.

Second, in-circuit test (ICT) performs a schematic verification by testing the individual components of a board one at a time. It uses small probes to make contact with test points on the board, and checks for shorts, opens, and other basic characteristics to determine whether the assembly is correct.

Third, functional test checks if circuitry functions meet the requirements of the specification. This test is carried out using at-speed clock to check the interactions between different components. When a board fails functional test in the manufacturing line, the failed board is sent to the diagnosis department for further analysis and repair (shown as stage 3 in Fig. 1). Typically, technicians run additional functional tests and measurements based on their professional experience. This process may take up to several days, and even several weeks if the board remains faulty after a suspect component is replaced [3]. The cost associated with the diagnosis and repair due to board-level failures is one of the highest contributors to board manufacturing cost [4].

Finally, burn-in test is carried out to detect reliability problems. As technology scales down, the critical width and spacing geometries become smaller. Therefore, tiny residues and particles play increasingly important roles as stimuli of reliability failures. Reliability failures indicate that some devices shipped to customers fail before an expected period of time. However, these defects cannot be easily screened out either by the chip-level or the board-level functional tests. Burn-in test is carried out based on the fact that the failure rate of semiconductor products shows a bathtub curve [5]. As shown in Fig. 2, the first region is the "infant mortality" region. In this region, the failure rate decreases dramatically as time increases. The failures in this region are mainly caused by design and fabrication defects. The second region is the useful life region. In this region, products show a stable failure rate, and these failures are caused by random defects. The third region is the wear-out region. In this region, the failure rate increases as time increases and the major root cause of

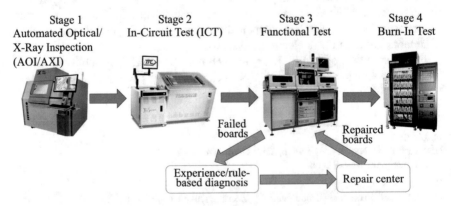

Fig. 1 The flow of board-level manufacturing test

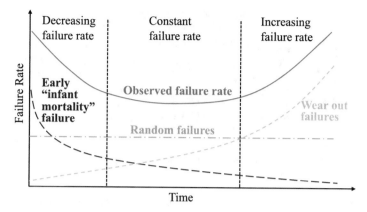

Fig. 2 The bath tub curve of the failure rate

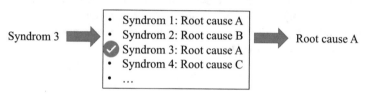

Fig. 3 Illustration of rule-based diagnosis methods

these failures is aging problems. Therefore, carrying out burn-in test can avoid infant failures in the first region, and hence reduce reliability problems.

1.2 Overview of Board-Level Fault Diagnosis

Fault diagnosis can determine the root cause of a malfunctioning board by collecting and analyzing information about its status using tests and other information sources. The cost associated with board-level fault diagnosis and repair is one of the highest contributors to board manufacturing cost. Therefore, a large number of board-level fault diagnosis techniques have been studied in the existing literature [6, 7, 10–12].

Rule-based diagnosis approaches take the form "IF syndrome(s), THEN fault(s)" to locate faults, based on the knowledge of experienced technicians [6] (shown in Fig. 3). In order to achieve a high accuracy of diagnosis, thousands of rules may be required to cover all the relevant information for the board under diagnosis. The benefits of rule-based diagnosis is its simplicity and fast diagnosis. However, these approaches cannot deal with unseen scenarios, and it is difficult to build a complete set of rules for a complex board.

Model-based diagnosis approaches construct approximate representations of the system under diagnosis [7]. The model is typically constructed in a hierarchical

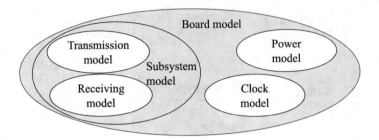

Fig. 4 Illustration of model-based diagnosis methods

manner; higher-level models are based on lower-level models, as shown in Fig. 4. This hierarchical model offers the potential to expand or upgrade the model by adding or changing lower-level models. However, the success of both rule-based and model-based diagnosis depends on the expertise of technicians who are familiar with the system. As boards become more complex, it is becoming very difficult to acquire the necessary knowledge.

2 Machine Learning-Based Functional Fault Diagnosis

The success of rule-based diagnosis and model-based diagnosis methods depends on the expertise of a technician who is familiar with the system. As boards become increasingly complex, it is becoming extremely difficult to acquire the necessary knowledge, and locate faults quickly and accurately based on expert rules or abstract models. To overcome this knowledge-acquisition bottleneck, machine-learning techniques have recently been advocated for board-level functional fault diagnosis.

Machine-learning techniques have been applied to various domains and reported to be effective [8, 9]. The application of machine-learning techniques to semiconductor testing has been studied in recent years. This is because semiconductor testing comprises multiple stages, from initial verification to connection test, functional test, and reliability test. At each stage, huge volumes of testing data are generated and collected. These testing data, especially *parametric results*, record a variety of information about the chip or board under test. With the help of powerful machine-learning techniques, effective information can be extracted from history testing data, and this information is used to facilitate the board-level functional fault diagnosis.

An adaptive diagnosis method based on the decision tree model has been proposed in [10]. To train the prediction model, fault syndromes that provide a complete description of the failures are extracted by parsing the log files with keywords. The actual repair action recorded in the database is used as outputs for the training. The diagnosis procedure is constructed as a binary tree, with the most discriminative syndrome as the root and final repair suggestions are available as the leaf nodes of the tree. In this way, the syndrome to be collected in the next step is determined based on the observations of syndromes collected thus far in the diagnosis procedure (i.e., the node in the decision tree). Hence the number of syndromes required for diagnosis can be significantly reduced.

More advanced machine-learning techniques have been adopted to train a machine-learning-based functional fault diagnosis model. An approach that combines the benefits of artificial neural networks (ANNs) and support vector machines (SVMs) through weighted majority voting has been proposed in [11]. Multiple diagnosis models are trained with different machine learning algorithms. The proposed system uses weights to combine the repair suggestions provided by each diagnosis model to identify a single set of recommended repair suggestions and this further improves the accuracy of board-level fault identification. Similarly, an approach that uses multi-kernel SVM has been studied in [3]. Classical SVMs are designed for linear classifications. However, in many practical scenarios, including fault diagnosis, classical SVMs fail to find an optimal linear classifier for separating different functional faults. One solution is to transform the problem to a higher-dimensional feature space through a non-linear mapping, also known as the kernel. This method leverages a linear combination of various kernels to achieve accurate faulty-component classification based on the errors observed.

However, the above techniques fail to provide appropriate repair suggestions when the diagnostic logs are fragmented and some error outcomes, or syndromes, are not available during diagnosis. To address this practical issue, a functional fault diagnosis system that handles missing data has been proposed in [12]. If no missing syndrome is detected, the diagnosis system continues with the standard diagnosis system. Otherwise, the data set is handled by the missing-value preprocessing. Multiple missing-value preprocessing methods are studied, including numerical imputation and zero imputation. For numerical imputation, a missing value is predicted on a numerical scale based on the mean value. For zero imputation, all missing syndromes are deemed as pass, which can be denoted as 0. Moreover, a syndrome-selection technique based on the minimum-redundancy-maximum relevance criteria is also incorporated to further improve the efficiency of the proposed methods.

The above methods leverage machine-learning algorithms by extracting knowledge from historical records of successfully repaired boards and construct effective board-level fault-diagnosis models.

3 Board-Level Functional Fault Identification Using Streaming Data

As discussed in Sect. 2, machine-learning algorithms have been shown to be effective for board-level fault identification [10–13]. These methods overcome the knowledge-acquisition bottleneck by collecting high-volume production data and leveraging machine-learning algorithms. Diagnosis systems can be trained using the data logs corresponding to both successful and unsuccessful repair cases.

A drawback of all the above methods is that they focus on static test environments and they require a complete dataset for learning. These methods assume that all the data are stored in fast-access memory that can be accessed any time with high bandwidth if needed. However, this scenario is not realistic as the volume of data in the manufacturing line increases rapidly with time. An incremental learning algorithm based on support-vector machines (SVMs) has been proposed recently to reduce training time [3]. However, this model selects training samples randomly from all the training data in each epoch without considering the fact that the statistical properties of functional errors, which the model is trying to predict, may change over time. This change in the target attributes (i.e., target concepts) over time is referred to as *concept drift* [14]. Such changes are reflected in incoming boards and they degrade the accuracy of classifiers learned from past training tuples.

The assumption of a static test is unrealistic because in the process of board-level functional fault identification, it is impractical to obtain all the test data and repair records in the early stages of high-volume manufacturing. In fact, as manufacturing proceeds, test data and repair records arrive in a streaming format characterized by large volumes of data instances. Compared to a static test and the diagnosis flow, processing data streams imposes two new requirements on diagnosis algorithms: (1) the ability to adapt to concept drift, and (2) the availability of a limited amount of memory (relative to the volume of streaming data).

To overcome these challenges, in this section, we present a diagnosis workflow that utilizes online learning to train classifiers incrementally with a small chunk of data at each step. These online learning algorithms adapt to concept drift quickly with carefully designed update rules. A hybrid algorithm is also proposed to handle the scenario that data for varying numbers of boards are collected at different times. This hybrid algorithm concurrently implements two basic models. For each data chunk, this algorithm chooses the better model with high probability [15, 16].

3.1 Problem Formulation

Functional fault identification is a key requirement in board manufacturing. The functional fault identification problem can be formulated as a binary classification problem, where the goal is to determine whether a board failure can be attributed to a given root cause. For a fault F_j, an instance $\mathbf{x}$ is a vector of functional test results. We label each board whose failure can be attributed to the functional fault F_j with "+1" and other boards with "−1". The learning algorithm constructs a classifier that outputs a class prediction for a given instance.

In the scenario of online learning for streaming data, we assume that learning samples from a stream S appear incrementally as a sequence of labeled samples $\{\mathbf{x}_t, y_t\}$ for $t = 1, 2, \ldots$, where t denotes the time stamp, and we denote the instance coming in round t by $\mathbf{x}_t \in \mathbb{R}^n$. Similarly, $\mathbf{x}_t$ is a vector of functional test results. Each instance $\mathbf{x}_t$ is associated with a label y_t that indicates whether this board failure can be attributed to a target fault. In real-life applications, instead of arriving one by one, samples may arrive in the form of data chunks. A data chunk containing k_t samples of the data stream is defined as $B_t = \{(\mathbf{x}_i, y_i) \in \mathbb{R}^n \times \{+1, -1\} \mid i = 1, 2, \cdots, k_t\}$. The evaluation or updating of classifiers is performed after processing all examples from a data chunk.

Alternatively, functional fault identification can be formulated as a multi-class classification problem, in which the goal is to identify the most likely root cause for the failure of a board. In this way, an instance $\mathbf{x}$ is a vector of features extracted from functional test results, which is the same as in a binary classifier. We label each board based on the functional faults that the board can be attributed to. For example, a board whose failure can be attributed to the functional fault F_1 (e.g., optical transceiver fault) is labeled with "1"; a board whose failure can be attributed to the functional fault F_2 (e.g., RF switch fault) is labeled with "2". The learning algorithm constructs a multi-class classifier that outputs a class prediction for a given instance.

In the scenario of online learning for streaming data, we assume that learning samples from a stream S appear incrementally as a sequence of data chunks $S = \{B_1, B_2, \cdots\}$. A data chunk containing k_t samples of the data stream is defined as $B_t = (X_t, Y_t) = \{(\mathbf{x}_i^{(t)}, y_i^{(t)}) \in \mathbb{R}^n \times \{1, 2, \cdots, M\} \mid i = 1, 2, \cdots, k_t\}$, where M is the number of fault types, t is the current time stamp. To evaluate the diagnosis system for streaming data, the model is alternately tested and trained for each incoming data chunk following the *Interleaved-Test-Train* method.

3.2 Online Learning for Binary Classifiers

Online incremental learning algorithms have been proposed for handling streaming data [17–20]. In order to deal with concept drift, classifiers implement *forgetting*, *adaptation* and *drift detection* mechanisms [14]. In order to overcome the challenge of limited memory, classifiers record only the key information extracted from the previous round of streaming data instead of all the past samples. Moreover, classifiers can learn the target concepts incrementally instead of training from scratch to save training time. In this manner, the process of model training is accelerated, and the memory space requirement is reduced. By executing online learning algorithms for streaming data, the trained model predicts more accurately when the data distributions shift.

The online learning algorithm, passive aggressive (PA) algorithm, was introduced in [21]; the main idea behind a PA model is similar to that of training an SVM. An SVM updates a weight function by minimizing the loss over all samples. Unlike an SVM, to adapt to concept drift, PA incrementally updates the weight vector $\mathbf{w}_t$ by minimizing the loss over all the new incoming samples. In order to retain the information that is learned in previous rounds, PA also requires that the new weight function ($\mathbf{w}_{t+1}$) remains as close as possible to the current one ($\mathbf{w}_t$) by minimizing the L2 norm between these two weight vectors ($\|\mathbf{w}_{t+1} - \mathbf{w}_t\|^2$).

However, we found that minimizing the L2 norm cannot ensure that the new weight function retains the most important information that is learned before. This is because weights in different dimensions have different importance. A small change in an important dimension may adversely affect the classification results, which can lead to overall accuracy drop.

Example 1 Suppose that we have previously trained a PA model to identify fault F_j, and the weight function of the trained model is given by $\mathbf{w}_t = (0, 0, 1)$. The L2 norm of the weight update ($\|\mathbf{w}_{t+1} - \mathbf{w}_t\|^2$) is the same when the new weight function is set to $\mathbf{w}_{t+1} = (0.5, 0, 1)$ or set to $\mathbf{w}_{t+1} = (0, 0.5, 1)$. Next, suppose the test results of a previous failing board that is not attributed to fault F_j are $\mathbf{x}_3 = (1, 100, -5)$, $y_3 = -1$. When the new weight function is set to $\mathbf{w}_{t+1} = (0.5, 0, 1)$, this new model can still correctly diagnose the instance $\mathbf{x}_3$: $\text{sign}((1, 100, -5) \cdot (0.5, 0, 1)) = \text{sign}(-4.5) = -1$. However, when the new weight function is set to $w_{t+1} = (0, 0.5, 1)$, this new model makes wrong diagnosis for the instance $\mathbf{x}_3$: $\text{sign}((1, 100, -5) \cdot (0, 0.5, 1)) = \text{sign}(45) = +1$.

The above example indicates that evaluating the differences between weights in terms of the L2 norm does not consider the different importance of the various features. In order to consider the importance of each feature dimension, we calculate the average absolute value of each dimension d among all previous instances $(\mathbf{x}_1, \mathbf{x}_2, \cdots \mathbf{x}_t)$ and record them in a vector $\tilde{\mathbf{x}}$

$$\tilde{\mathbf{x}} = \frac{1}{t} \sum_{i=1}^{t} \text{abs}(\mathbf{x}_i) \tag{1}$$

where abs is an operator that returns the element-wise absolute value of a vector ($\mathbf{x}_i$), and t is the number of previous instances.

In the improved PA algorithm, the update rule for the weight function follows the new constrained optimization problem shown below:

$$\mathbf{w}_{t+1} = \underset{\mathbf{w}^* \in \mathbb{R}^n}{\arg\min} \frac{1}{2} \|\mathbf{w}^* \odot \tilde{\mathbf{x}} - \mathbf{w}_t \odot \tilde{\mathbf{x}}\|^2, \text{ s.t. } \ell(\mathbf{w}^*; (\mathbf{x_t}, y_t)) = 0 \tag{2}$$

where we use $\odot$ to denote element-wise multiplication. The constraint of the optimization problem is the same as in the original PA, which ensures that $\mathbf{w}_{t+1}$ can classify the most recent samples correctly. The new objective function of the optimization problem forces $\mathbf{w}_{t+1}$ and $\mathbf{w}_t$ to make similar predictions for previous instances, thus retaining the information learned in previous rounds.

3.3   Fault Diagnosis Based on Multi-Class Classifiers

The Adaptive Random Forest (ARF) algorithm was introduced in [22]; it is an adaptation of the original Random Forest (RF) algorithm. The original random forest algorithm has been successfully applied to various machine learning tasks [23–25]. A random forest model consists of T decision trees. For each decision tree, the training data are randomly drawn out of all training samples. Based on these selected samples, the decision tree is generated by the following steps. First, we randomly select $\sqrt{M}$ out of M features where M denotes the total number of functional tests. Second, a binary splitting is repeatedly applied at each node by selecting a feature to maximize the Gini index [26]. We stop splitting a node if all samples associated with this node share the same label or the maximum depth of the decision tree is reached. Each trained decision tree can predict a likely reason for the failure of a board. Then, a final diagnosis is generated by taking the majority vote among all the decision trees.

However, the original random forest algorithm is not feasible for streaming data, because the training process requires multiple passes over the entire dataset and the trained RF model cannot adapt to *concept drift* [22]. In order to deal with streaming data, the Adaptive Random Forest (ARF) algorithm introduced three major adaptations in [22]. The flow of ARF is shown in Fig. 5, and the major adaptations are marked in red.

First, the bootstrap sampling is replaced by assigning a weight to each sample. For each tree in an RF model, we need access to all the previous samples and randomly draw N_T out of N samples. Since this is a sampling method with replacement, a sample could be selected K times ($K \in \{0, 1, \cdots, N_T\}$), and the probability of selecting a sample K times obeys a binomial distribution. Furthermore, when N is large, this binomial distribution can be approximated as a Poisson distribution. In our task, test data and repair records arrive in a streaming format, and hence we cannot randomly draw samples out of a complete dataset. To address this problem, an online sampling algorithm was proposed in [27], which approximates the original random sampling with replacement by weighting samples according to a Poisson distribution. For example, when the weight is set to 0, it indicates that the corresponding sample is not selected for training the decision tree; when the weight is set to 1, it is equivalent as selecting this sample once; when the weight is set to a value larger than 1, it indicates that the sample is selected for multiple times.

Second, the original RF algorithm is designed to cope with stationary datasets. In order to adapt to *concept drift* in streaming data, drift warning and drift detection methods are adopted in ARF. A drift warning and detection method named ADWIN [28] uses a sliding window of a variable size to maintain the most recent samples. A sliding window is split into two subwindows, and we calculate averages of these two subwindows. Whenever the difference between these two averages surpasses a pre-defined threshold, we can conclude that a drift is detected, and all samples before that time are discarded. A lower threshold is set for drift warning, while

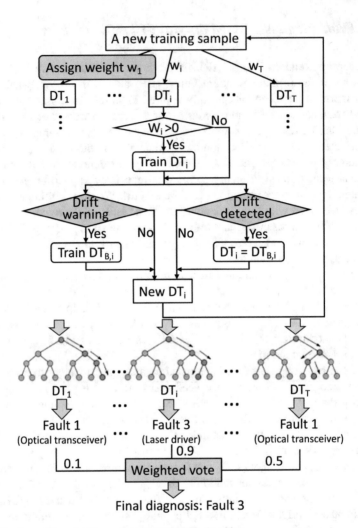

Fig. 5 The flow of the adaptive random forest (ARF) algorithm

a higher threshold is set for drift detection. In ARF, each tree is monitored by a drift detector. After a drift warning has been detected for one tree (DT_i), another "background" tree ($DT_{B,i}$) starts to grow in parallel and replaces the tree only if the warning escalates to a real drift.

Third, in the original RF model, each decision tree generates a prediction about each instance. Each tree's prediction is combined into a final classification through a "majority vote" mechanism. In ARF, each decision tree generates a prediction in the same way. However, different decision trees have different capabilities for handling concept drifts. Therefore, a weighted average is used to generate a final classification result. Assuming a tree DT_i has seen n_l instances since its last replacement and correctly classified c_l instances ($c_l \leq n_l$), then its weight is set to c_l/n_l. This weight reflects the performance of this tree on the current concept. For example, as shown in Fig. 5, two decision trees (DT_1 and DT_T) predict the reason for the failure as Fault 1, and only one decision tree (DT_i) predicts the reason for the failure as Fault 3. However, DT_i has better performance on the current concept, and hence it is assigned with a large weight of 0.9. Therefore, the ARF generates the final diagnosis as Fault 3.

3.4 A Hybrid Algorithm for Streaming Data with Different Chunk Sizes

In our application scenario, the number of failed boards and their corresponding repair records that arrive each day/week vary considerably. The online learning algorithms outperform traditional batch-based algorithms when the size of the data chunk is small. However, experimental results indicate that, when the size of the data chunk is large, batch-based algorithms using a sliding window provide better prediction performance.

Motivated by this observation, we aim at more accurate predictions by taking advantage of both online incremental learning and batch-based learning for different chunk sizes. We have developed a hybrid online learning algorithm to improve prediction accuracy when data is received in the form of chunks of different sizes [15]. This algorithm concurrently maintains an incremental-learning model and a batch-based model. Based on the chunk size and the performance of these two models, the proposed hybrid algorithm selects the more effective algorithm at each step of prediction.

3.5 Experimental Results

Experiments were performed on two complex boards that are currently in high-volume production. To evaluate the performance of fault diagnosis by training binary classifiers, we choose the sliding-window SVM as the batch-based model for the hybrid algorithm; we choose either the PA or the improved-PA as the online-learning model for the hybrid algorithm. Then, we use the *F1-score* of functional fault diagnosis as the criterion to compare these three basic models and the proposed hybrid algorithm. We use all the boards and simulate the real-world scenario that the data for varying numbers of failed boards are collected at different times.

The average *F1-score* for two boards using the complete streaming data is shown in Fig. 6. We observe that the proposed hybrid algorithm always achieves better performance than all the basic models. The hybrid model based on SVM and improved-PA achieves the best performance. On average, the *F1-score* equals 0.573

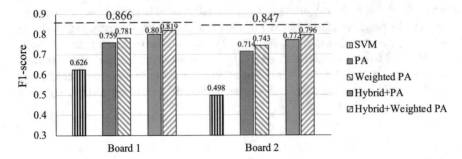

Fig. 6 Comparison of *F1-score* between different algorithms

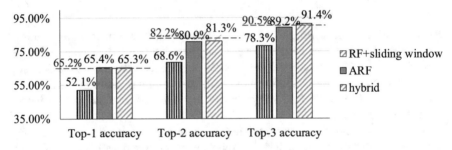

Fig. 7 Comparison of *top-N accuracy* between different algorithms

for sliding-window SVM, 0.740 for PA, and 0.765 for improved-PA. The proposed hybrid online learning algorithm improves the *F1-score* to 0.810. Moreover, instead of assuming the samples arrives in a data stream, we also build a fault diagnosis model by assuming that we have access to a static and complete dataset for the model training. The *F1-score* that can be achieved in this static scenario is denoted by red dashed lines in Fig. 6. The *F1-score* provided by the static method is 0.866 for Board 1 and 0.847 for Board 2, which can be thought of as "ceilings". We observe that, with the help of the online learning algorithm and the proposed hybrid algorithm, the performance gap between the "ceiling" and the actual online fault-diagnosis reduces from 0.285 to 0.048 in terms of the *F1-score*.

Next, we evaluate the effectiveness of the proposed hybrid online learning algorithm for multi-class classifiers in terms of *top-N accuracy*, where $N = 1, 2, 3$. We simulate the real-world scenario that the data for varying numbers of failed boards are collected at different times. The comparison of *top-N accuracy* between applying sliding-window RF, ARF and the hybrid algorithm is shown in Fig. 7. We observe that, compared with the sliding window RF, ARF greatly improve the *top-N accuracy*. Moreover, we also build a fault diagnosis model by assuming that we have access to a static and complete dataset for the model training. The *top-N accuracy* that can be achieved in this static scenario is denoted by red dashed lines in Fig. 7. We observe that, for the *top-1 accuracy*, both ARF and the hybrid algorithm achieve higher accuracy than the static setting. This is because the percentage of each type of fault may drift along with time. With a static assumption, the classifier can only learn the overall percentage of each class, while the online learning algorithms are able to learn the percentage distribution among the most recent samples. If there is no drift, the accuracy provided by the static method can be thought of as a "ceiling". In that sense, the proposed solution is able to reach this ceiling. On average, the *top*-1 *accuracy* equals 52.1% for sliding-window RF and 65.3% for the hybrid algorithm; the *top*-2 *accuracy* equals 68.6% for sliding-window RF and the hybrid algorithm improves it to 81.3%; the *top*-3 *accuracy* improves from 78.3 to 91.4%.

4 Knowledge Transfer in Board-Level Functional Fault Identification

As discussed in Sects. 2 and 3, the machine-learning-based diagnosis workflow can identify board-level functional faults with high accuracy. However, a drawback of all the above diagnosis methods is that they require a sufficient number of failed boards and the corresponding repair records to train a good model. In reality, it takes several months to accumulate an adequate database. Therefore, during the initial product ramp-up phase, it is not feasible to build a diagnosis system using traditional machine-learning algorithms due to lack of data. To save time and the effort of collecting labeled data, it is important to transfer the knowledge learned from mature products to the diagnosis model of a new product, such that the new diagnosis model can not only learn from its own failing boards, but also from other failing boards of the mature products. However, different test items are designed for different products, and their test results may obey different data distributions even if the same test item is considered.

The direct application of a trained model to a new product with different test items (i.e., features) and test results (i.e., data distribution) is likely to fail. Knowledge-transfer approaches have been proposed for alternate test [29] and yield forecasting [30]. However, for these applications, the test items are always the same for different domains. A knowledge-transfer approach was recently proposed for board-level functional fault diagnosis, which directly maps a trained model to a new product based on common keywords and board-structure similarities [31]. However, this approach requires the knowledge of board designs and test strategies to map the trained model. In this section, we present a functional fault diagnosis workflow that can effectively transfer the knowledge learned from mature boards (i.e., source domain) to a new board (i.e., target domain) [32, 33].

4.1 Problem Formulation

Functional fault identification is a key requirement in board manufacturing. The functional fault identification problem can be formulated as a binary classification problem, where the goal is to determine whether a board failure can be attributed to a given root cause. Alternatively, it can be formulated as a multi-class classification problem, in which goal is to identify the most likely root cause for the failure of a board.

In reality, a successful product typically experiences multiple updates and there are often similar products during a period of time. Our goal is to train a supervised learning model to identify board-level functional fault using a large number of *samples* from the mature product (i.e., source domain) and a limited number of *samples* from the new product (i.e., target domain). To differentiate between these two domains, we use subscripts S and T to represent the source domain and the target domain, respectively. For example, source-domain data is denoted as $\{X_S, y_S\} = \{(\mathbf{x}_i^S, y_i^S)\}_{i=1}^{n_S}$, where n_S is the number of learning samples in the source domain, and each instance $\mathbf{x}_S \in \mathbb{R}^{d_S}$, where d_S denotes the dimensionality of the source-domain feature space. Similarly, data pairs from the target domain are denoted by $\{X_T, y_T\} = \{(\mathbf{x}_i^T, y_i^T)\}_{i=1}^{n_T}$, where n_T is the number of samples in the target domain, and each instance $\mathbf{x}_T \in \mathbb{R}^{d_T}$, where d_T denotes the dimensionality of the target-domain feature space. Typically, we have much more data from the source domain than from the target domain ($n_S \gg n_T$). Note that d_S is not necessarily equal to d_T, because different test items may be applied to different products.

4.2 Knowledge Transfer for Binary Classifier

The proposed knowledge-transfer diagnosis workflow consists of three steps (shown in Fig. 8): (1) adapt differences between mature boards (i.e., source domain) and the new board (i.e., target domain); (2) utilize information from both the mature and the new boards to train a functional fault diagnosis model; (3) check the necessity of carrying out domain adaptation.

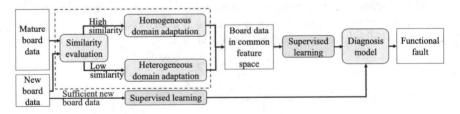

Fig. 8 The proposed knowledge-transfer method

For diagnosis models that are trained to determine whether a board failure can be attributed to a given root cause (i.e., binary classifiers), we analyze the similarity between domains, and explicitly map the source-domain data to the target domain. Second, after carrying out these domain-adaptation algorithms, we utilize information from both the mature and the new boards to train a functional fault diagnosis model. Moreover, in a real manufacturing process, domain adaptation is only required in the early stage when limited test and repair records are available for a new product. Therefore, in the third step, we concurrently maintain a domain-adaptation model and a model trained using target-only data. When we receive new repair records, we evaluate the performance of these two models and determine if the model trained using target-only data outperforms the domain-adaptation model. If the answer is affirmative in two consecutive evaluations, we can conclude that sufficient data has been collected for the new product, and we no longer need to transfer knowledge from mature boards.

4.2.1 Similarity Evaluation

The performance of knowledge transfer depends on the similarity between the two boards. Therefore, a similarity-evaluation metric is required.

To avoid the influence of trivial features, a feature-selection algorithm is first executed using the source-domain data. In this section, we adopt a feature-selection method based on information value (IV) [34]. For each test item (i.e., feature), a large IV implies a strong correlation between the test item and the target fault. Thus we keep the test items with large IVs and refer to them as key test items. Then, among these key test items, we calculate the proportion of test items that are also used for the new product. A large proportion indicates high similarity. Note that the feature-selection step is only carried out among the old boards. This is because only a few repair records are available for the new boards The use of such a small amount of repair data may easily lead to undesirable randomness in feature selection.

4.2.2 Homogeneous Domain Adaptation

Domain is defined as a combination of three components, i.e., an input space X, an output space Y and an associated probability distribution $p(x, y)$. For domain adaptation tasks, we have a source domain and a target domain. Two domains are considered as being different when any of these three components are different.

Most machine-learning algorithms are designed to address only a single-domain task. When the training data and testing data have different domains, these algorithms cannot directly recognize and apply relevant knowledge from previous learning experiences to a new domain, hence the prediction error increases. To help machine-learning algorithms adapt to different domains, we need to clearly identify the differences between domains.

Prior Shift

For prior shift, the prior probabilities are different in the source and the target domains ($p_S(y) \neq p_T(y)$), but the conditional distributions are equivalent ($p_S(y|x) = p_T(y|x)$). In functional fault diagnosis, the same conditional distribution ensures the following desirable outcome: When test results indicate the same pattern (e.g., failed in the same combination of test items for the same number of times), the same root cause is identified in both products. Different prior probabilities describe the scenario that the source and the target domains have different class proportions.

To adapt to a prior shift, we re-weight the source-domain classes with sampling techniques to achieve identical class proportions compared to the target domain. The class-based weighting technique has been extensively studied to handle datasets with imbalanced classes in traditional machine-learning settings [35].

Covariate Shift

For covariate shift, the data distributions are different ($p_S(x) \neq p_T(x)$), while the conditional distributions are equivalent ($p_S(y|x) = p_T(y|x)$). In functional fault diagnosis, different data distributions describe the scenario that the probabilities that a specific test-result pattern exists are different in different products.

The covariate shift indicates the different probabilities of seeing an instance. This motivates us to re-weight individual instances to correctly reflect the probability under the target distribution. To determine the appropriate weights for each instance, a re-weighting algorithm based on logistic regression was proposed in [36].

The logistic-regression-based re-weighting algorithm is composed of 3 steps: (1) label all source instances as 1, and label all target instances as -1; (2) train a logistic regression model, which predicts the probability of an instance belonging to the target domain. (3) reweight source instances with the predicted probability. This approach is effective because the source instances similar to the target instances are assigned larger weights, which alleviates the problem of biased data distributions between the source domain and the target domain.

4.2.3 Heterogeneous Domain Adaptation

In addition to the common test items shared by different products, some unique test items are designed to test board-specific functions. *Heterogeneous domain adaptation* tackles domain adaptation problems with different cross-domain feature spaces.

The major challenge here lies in bridging the disjoint cross-domain feature spaces. Cross-domain landmark selection (CDLS) method uses semi-supervised learning to transfer data from the source domain to the target domain [37]. Instead of viewing all cross-domain data to be equally important during domain adaptation,

CDLS identifies the adaptation capability of each instance by solving a constrained optimization problem.

Data from the source domain is denoted as $\{X_S, y_S\}$, while data in the target domain $\{X_T, y_T\}$ can be further partitioned into two parts, i.e., labeled data (with repair records) and unlabeled data (without repair records). This is because, in reality, technicians need to run additional functional tests and measurements to locate the root cause of the failure, and this process may take up to several days, or even several weeks if the board remains faulty after a suspect component is replaced [3]. Therefore, in the early stage, only a small volume of test data is collected, and even less repair records are available. Labeled data pairs from the target domain are denoted as $\{X_L, y_L\} = \{(x_i^L, y_i^L)\}_{i=1}^{n_L}$, where n_L is the number of failed boards with repair records from the new product. The remaining unlabeled instances are also considered during the adaptation process in CDLS; these are denoted as $X_U = \{x_j^U\}_{j=1}^{n_U}$, where n_U is the number of failed boards that have not been repaired from the new product.

Before projecting test results from the mature board (i.e., source domain) to the new board (i.e., target domain), we carry out feature extraction among all the new-board instances using principal component analysis (PCA) and convert the test results of the new board into an m-dimensional subspace, where m is set to a fixed value which defines the dimensionality after extraction. The goal of this dimensionality-reduction step is to prevent potential overfitting caused by the projection of data with high dimensionalities.

Next, we learn a feature transformation matrix $A \in \mathbb{R}^{m \times ds}$ to project the test results from the mature board to the PCA subspace of the new-board test results. The goal of this method is to associate cross-domain test results $X_S A^\top$ with X_T in the derived subspace. Such an association is achieved by eliminating the domain bias via two ways: (1) matching marginal distributions $P_T(X_T)$ and $P_S(X_S A^\top)$; (2) matching conditional distribution $P_T(y_T|X_T)$ and $P_S(y_S|X_S A^\top)$. In functional fault diagnosis, the process of matching the marginal distributions ensures that the probabilities that a specific test-result pattern exists are similar in different products. The process of matching the conditional distribution ensures that, when test results indicate the same pattern, the same root cause is identified in both products. In this way, the test results from the mature board can be mapped to the new board, and a good classifier can be built.

In order to eliminate these two biases, we utilize repair records and the corresponding test records (i.e., labeled data) from both boards to optimize the projection. In this way, the problem of cross-domain matching can be formulated as the minimization of the following objective function:

$$\min_{A} \quad E_{margin}(A, X_S, y_S, X_L, y_L)$$
$$+ E_{cond}(A, X_S, y_S, X_L, y_L) + \lambda \|A\|^2 \tag{3}$$

where E_{margin} and E_{cond} denote marginal distributions and conditional distributions, respectively.

4.3   Knowledge Transfer for Multi-Class Classifiers

Instead of training multiple binary classifiers, we can train a multi-class classifier, which tries to predict the most likely reason for the failure of boards.
Knowledge-transfer flows based on deep neural networks have been studied [38–
40]. These methods explore the domain-adaptation problem by learning the deep
representations which manifest invariant factors underlying different domains and
transfer these representations from the original tasks to similar new tasks. Moreover,
previous studies have shown that features in a neural network typically transition
from general to specific by layers [41]. In other words, the first few layers in a
network are used to extract general features that are transferable to similar tasks,
while the last few layers are tailored to specific tasks. Therefore, we propose a
knowledge-transfer flow based on domain adaptive networks. As shown in Fig. 9,
the proposed knowledge-transfer diagnosis workflow consists of three components:
(1) a feature extractor, (2) a domain-alignment component, and (3) a diagnosis
classifier. We first train a multi-layer perceptron (MLP) model.

4.3.1   Feature Extractor

In our application scenario, besides the common test items that are shared by different boards, some unique test items are designed to test board-specific functions.
Therefore, in order to learn a feature extractor using all source-domain data, we align
feature spaces among all source boards. For example, Fig. 10 lists the functional

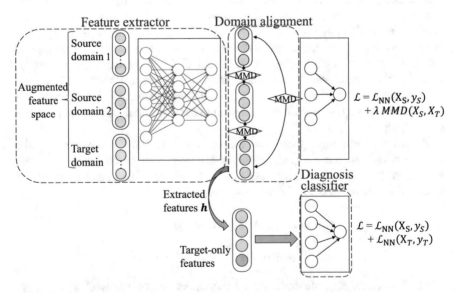

Fig. 9 The proposed domain-adaptation method

Source board 1	Tx. test for Channel 1	Tx. test for Channel 2	Rx. test for Channel 1	Rx. test for Channel 2	Output power test	0	0	0
Source board 2	Tx. test for Channel 1	Tx. test for Channel 2	Rx. test for Channel 1	Rx. test for Channel 2	0	LED test		0
Source board 3	Tx. test for Channel 1	0	Rx. test for Channel 1	0	Output power test	0	Eye pattern test	0
Target board	Tx. test for Channel 1	Tx. test for Channel 2	Rx. test for Channel 1	Rx. test for Channel 2	0	LED test	Eye pattern test	Voltage adjusting test

Augmented feature space   Target-only feature

Fig. 10 An example of feature alignment

tests that are carried out for each source board and the target board. We first align the common test items that are shared by all the boards. Then for any other test item, if it is not used in a source board, we add this test item as an extra feature and assign all the corresponding values in this source domain to zero.

Note that, target-only features are not added into the augmented feature space. This is because the feature extractor is trained using source-domain data, where all these target-only features are zero. In this way, the feature extractor cannot learn any knowledge about these features. Therefore, we leave the learning task for these unique features to the last stage, i.e., the diagnosis classifier, and only augment the feature space based on source boards.

After feature augmentation, as shown in Fig. 9, a MLP model is trained. The MLP model contains one input layer, one output layer, and two hidden layers. These two hidden layers contain 100 and 50 neurons, respectively. Previous studies have shown that the first several layers extract general features that are transferable among similar tasks. Therefore, we use the first two layers as a feature extractor, and the outputs from the second layer are used as the extracted domain-invariant features. We denote W_1 and W_2 as the connection weights of the first and the second layers, respectively. Then the extracted features q can be calculated as:

$$p = ReLU(W_1^T X + b_1) \tag{4}$$

$$h = W_2^T p + b_2 \tag{5}$$

$$q = ReLU(h) \tag{6}$$

where b_1 and b_2 are the biases of the first and the second layers, respectively. A non-linear activation function $ReLU(\cdot)$ is connected to the outputs of each layer [42].

4.3.2 Domain Alignment

To transfer knowledge from the source domain to the target domain, we need to align the features generated from the feature extractor. A domain adaptive neural network (DaNN) was proposed in [38], which utilizes a non-parametric probability distribution distance measure, i.e., the maximum mean discrepancy (*MMD*), as a regularization embedded in the supervised back-propagation training. By minimizing *MMD*, we can reduce the distribution mismatch between the extracted features induced by samples drawn from different domains [43]. Taking the *MMD* into consideration, the loss function can formulated as [38]:

$$\mathcal{L} = \mathcal{L}_{NN} + \lambda MMD(\boldsymbol{h}_S, \boldsymbol{h}_T) \tag{7}$$

where $\mathcal{L}_{NN}$ is the cross entropy loss over the source-domain data, which is generally used in NN computing. As defined in (5), we denote the features extracted from the source-domain data as $\boldsymbol{h}_S$, and the features extracted from the target-domain data as $\boldsymbol{h}_T$. The regularization constant λ controls the importance of *MMD* contribution to the loss function.

By adopting the *MMD* as a regularization in the loss function, we aim to train the network such that the supervised criterion is optimized and the hidden layer representations are encouraged to be invariant across different domains.

Compared with single-source domain adaptation, multi-source domain adaptation assumes that training data from multiple sources are available. In practical scenarios, there may be discrepancies among different source domains. Therefore, we extend the loss function defined in (7) by considering the *MMD* not only between the source and the target domains but also among multiple source domains.

Assume that we have access to N different source domains, and the features extracted from these source domains are denoted as $\boldsymbol{h}_S = \{\boldsymbol{h}^1, \boldsymbol{h}^2, \cdots, \boldsymbol{h}^N\}$. Similarly, the features extracted from the target domain is denoted as $\boldsymbol{h}_T$. Then the *MMD* across all domains is defined as:

$$MMD(\boldsymbol{h}_S, \boldsymbol{h}_T) = \frac{1}{N} \sum_{i=1}^{N} MMD(\boldsymbol{h}^i, \boldsymbol{h}_T)$$

$$+ \binom{N}{2}^{-1} \sum_{i=1}^{N-1} \sum_{j=i+1}^{N} MMD(\boldsymbol{h}^i, \boldsymbol{h}^j)$$

So the domain alignment component not only minimizes the distance of the features between the target domain and each source domain ($MMD(\boldsymbol{h}^i, \boldsymbol{h}_T)$), but also minimize the discrepancies between each source-domain pair ($MMD(\boldsymbol{h}^i, \boldsymbol{h}^j)$).

4.3.3 Diagnosis Classifier

Following this domain-adaptation flow, the feature extractor (i.e., the first two layers in the MLP model) maps the original test results X_S and X_T to a common latent feature space as h_S and h_T, respectively. The domain alignment component minimizes the discrepancies between different domains (i.e., $MMD(h_S, h_T)$) in the common feature space. Then, a diagnosis classifier is trained using both source-domain data and labeled target-domain data. As we discussed in Sect. 4.3.1, the target-only features are not considered in feature extractor. Therefore, we add these target-only features to the extracted general features to train the diagnosis model.

Note that, in functional fault diagnosis, the source and the target domains may have different root cause proportions. To ensure that the diagnosis classifier trained on all domains can achieve the best prediction performance on a target domain, we inform the classifier with the target domain's class portion. Similar to the method described in Sect. 4.2.2, we re-weight the source-domain classes with sampling techniques to achieve identical class proportions compared to the target domain.

In real diagnosis processes, testing results from the target board are forwarded through the feature generator and the classifier, whose classification results are used to predict the root cause of failures.

4.4 Experimental Results

Experiments were performed on four complex boards, three of which were in high-volume production and one of which was in the ramp-up phase. All the boards are optical network terminals from the same series, and hence share some similarities in design and test.

We compare two baseline methods, i.e., a target-only method and a common-feature method. The target-only method uses only a small portion of data (10 repair records from the target class) in the target board to train a diagnosis model. The common-feature method selects all the common test items between the source board and the target board as the features, and uses all the data from the source board to train a diagnosis model. Then we directly apply the trained model to the new board. We also show the results of using the homogeneous and heterogeneous domain-adaptation method, respectively. Experimental results show that domain-adaptation methods typically obtain much higher *F1-score*. The diagnosis results can provide debug technicians with a list of repair candidates. Repair engineers would run additional measurements to validate the prediction. By improving the *F-1* score from 33.2 to 54.6% in the very early stage, we provide debug engineers with a more accurate list of candidates, and hence reduce the time needed to locate the root cause by 40%.

We also evaluate the effectiveness of the proposed knowledge-transfer workflow. We plot the diagnosis results of two baseline methods (i.e., a target-only method and a common-feature method) and the selected domain-adaptation method with respect

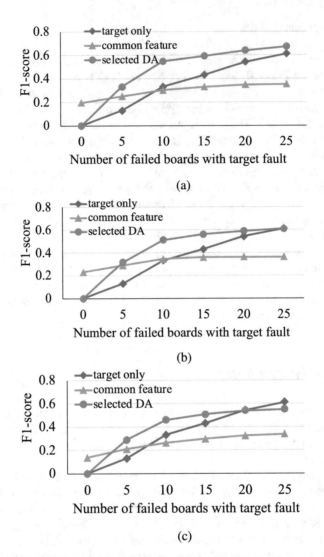

Fig. 11 Comparison of binary classification in terms of the *F1-score* between baseline methods and the selected domain-adaptation method with varying number of training samples in the target class. (**a**) Transfer board 1 to board 4. (**b**) Transfer board 2 to board 4. (**c**) Transfer board 3 to board 4

to the number of training samples in the target class (i.e., $m = 5, 10, 15, 20, 25$) in Fig. 11. The domain-adaptation method is selected based on the calculated similarity values. From the results, the performances of all methods increase for a larger value of m. We observe that, in the early stage (e.g., $m = 5, 10$ and 15), the selected domain-adaptation method always achieves the best performance. As the amount of repair records from the target board increases, we obtain a sufficient amount of data to directly train a functional fault identification model. As shown in Fig. 11c, the target-only method outperforms the selected domain-adaptation method for two consecutive times (i.e., $m = 20, 25$). This implies that sufficient data has been collected, and we no longer need to transfer knowledge from the old board.

As shown in Fig. 12, the domain-adaptation method significantly improves the performance of diagnosis for all settings. Moreover, instead of assuming that the target board is at the ramp-up stage and we have access to a limited number of target-domain data, we also build a fault diagnosis model by assuming that sufficient data has been collected for the target board. We use 70% of the samples from the target domain to train a diagnosis classifier, and the *top*-N *accuracy* that can be achieved in this setting (denoted by red dashed lines in Fig. 12) can be thought of as a "ceiling". We observe that, with the help of domain adaptation, for Board 1 and Board 3, the performance gap between the "ceiling" and the actual fault-diagnosis is greatly reduced. Note that, for Board 2 (Fig. 12b), the performance gap between the "ceiling" and the actual fault-diagnosis achieved by domain adaptation is relatively large. This is because one major root cause in Board Design 2 does not occur in any other board designs, and hence knowledge transfer cannot improve the performance for identifying this root cause. Moreover, we observe that, for Board 4 (Fig. 12d), the actual fault-diagnosis achieved by knowledge transfer is able to reach higher performance than the "ceiling". This is because Board 4 is still at the ramp-up phase, and we do not have sufficient data to train an accurate diagnosis model. In other words, the red dashed line is expected to be improved when more data is collected.

5   Chapter Summary

The goal of board-level functional fault diagnosis is to locate the root cause of failure and repair failed boards. The success of conventional rule-based diagnosis approaches and model-based diagnosis approaches depends on the expertise of a technician who is familiar with the system. As boards become increasingly complex, it is becoming extremely difficult to acquire the necessary knowledge. Machine-learning-based diagnosis model overcome the knowledge-acquisition bottleneck and obtain a high diagnosis accuracy.

This chapter presented a set of methods for machine-learning-based fault diagnosis. To address practical issues that arise in real testing data, two state-of-art machine-learning-based diagnosis models are discussed. Section 3 proposed a diagnosis model based on online incremental learning algorithms. It can quickly adapt to the time-dependent concept drift when new testing data arrives. With the help of online learning, the training time is saved while the diagnosis accuracy can be improved. Section 4 presented a workflow that can transfer the knowledge learned from mature boards to a new board. The proposed diagnosis model based on domain adaptation significantly reduces the requirement for the number of repair records from the new board, while achieving a relatively high diagnostic accuracy in the early stage of manufacturing a new product.

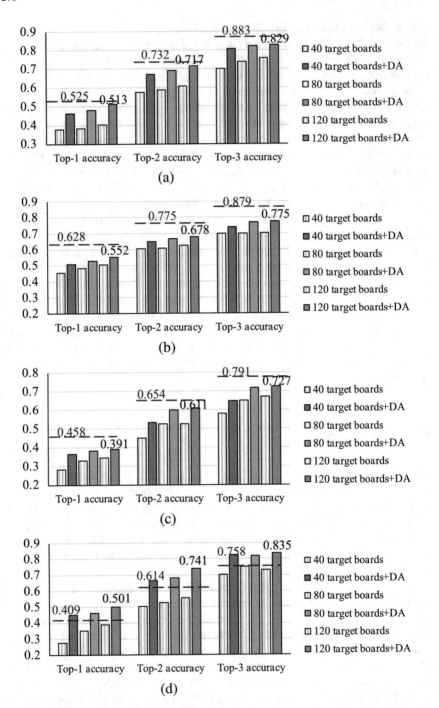

Fig. 12 The comparison of *top*-N *accuracy* between baseline methods and the proposed domain-adaptation method (DA) for different target boards. (**a**) Select board 1 as the target board. (**b**) Select board 2 as the target board. (**c**) Select board 3 as the target board. (**d**) Select board 4 as the target board

References

1. Conroy, Z., Richmond, G., Gu, X., Eklow, B.: A practical perspective on reducing ASIC NTFs. In: IEEE International Conference on Test, pp. 1–7. IEEE, Piscataway (2005)
2. Krishnamoorthy, A.V., Thacker, H.D., Torudbakken, O., Müller, S., Srinivasan, A., Decker, P.J., Opheim, H., Cunningham, J.E., Shubin, I., Zheng, X., Dignum, M.: From chip to cloud: optical interconnects in engineered systems. J. Lightw. Technol. **35**(15), 3103–3115 (2014)
3. Ye, F., Zhang, Z., Chakrabarty, K., Gu, X.: Board-level functional fault diagnosis using multikernel support vector machines and incremental learning. IEEE Trans. Comput.-Aided Des. Integr. Circuits Syst. **33**(2), 279–290 (2014)
4. Tourangeau, S., Eklow, B.: Test economics-what can a board/system test engineer do to influence supply operation metrics. In: IEEE International Test Conference, pp. 1–9. IEEE, Piscataway (2006)
5. Pantic, D.: Benefits of integrated-circuit burn-in to obtain high reliability parts. IEEE Trans. Reliabil. **35**(1), 3–6 (1986)
6. Fenton, W.G., McGinnity, T.M., Maguire, L.P.: Fault diagnosis of electronic systems using intelligent techniques: a review. IEEE Trans. Syst. Man Cybern. C **31**(3), 269–281 (2001)
7. Manley, D., Eklow, B.: A model based automated debug process. In: IEEE Board Test Workshop, pp. 1–7. IEEE, Piscataway (2002)
8. He, K., Zhang, X., Ren, S., Sun, J.: Deep residual learning for image recognition. In: Proceedings of the IEEE Conference on Computer Vision and Pattern Recognition, pp. 770–778. IEEE, Piscataway (2016)
9. Devlin, J., Chang, M.W., Lee, K., Toutanova, K.: Bert: pre-training of deep bidirectional transformers for language understanding. In: Proceedings of the Conference of the North American Chapter of the Association for Computational Linguistics: Human Language Technologies, vol. 1 (Long and Short Papers), pp. 4171–4186 (2019)
10. Ye, F., Zhang, Z., Chakrabarty, K., Gu, X.: Adaptive board-level functional fault diagnosis using decision trees. In: IEEE Asian Test Symposium, pp. 202–207. IEEE, Piscataway (2012)
11. Ye, F., Zhang, Z., Chakrabarty, K., Gu, X.: Board-level functional fault diagnosis using artificial neural networks, support-vector machines, and weighted-majority voting. IEEE Trans. Comput.-Aided Des. Integr. Circuits Syst. **32**(5), 723–736 (2013)
12. Jin, S., Ye, F., Zhang, Z., Chakrabarty, K., Gu, X.: Efficient board-level functional fault diagnosis with missing syndromes. IEEE Trans. Comput.-Aided Des. Integr. Circuits Syst. **35**(6), 985–998 (2015)
13. Jin, S., Chakrabarty, K.: Data-driven resiliency solutions for boards and systems. In: International Conference on VLSI Design and International Conference on Embedded Systems (VLSID), pp. 244–249. IEEE, Piscataway (2018)
14. Brzezinski, D., Stefanowski, J.: Combining block-based and online methods in learning ensembles from concept drifting data streams. Inf. Sci. **265**, 50–67 (2014)
15. Liu, M., Ye, F., Li, X., Chakrabarty, K., Gu, X.: Board-level functional fault identification using streaming data. In: IEEE VLSI Test Symposium (VTS), pp. 1–6. IEEE, Piscataway (2019)
16. Liu, M., Ye, F., Li, X., Chakrabarty, K., Gu, X.: Board-level functional fault identification using streaming data. IEEE Trans. Comput.-Aided Des. Integr. Circuits Syst. **40**(9), 1920–1933 (2020)
17. Street, W.N., Kim, Y.: A streaming ensemble algorithm (SEA) for large-scale classification. In: Proceedings of ACM SIGKDD International Conference on Knowledge Discovery and Data Mining, pp. 377–382. ACM, New York (2001)
18. Wang, H., Fan, W., Yu, P.S., Han, J.: Mining concept-drifting data streams using ensemble classifiers. In: Proceedings of ACM SIGKDD International Conference on Knowledge Discovery and Data Mining, pp. 226–235. ACM, New York (2003)
19. Hoens, T.R., Polikar, R., Chawla, N.V.: Learning from streaming data with concept drift and imbalance: an overview. Prog. Artif. Intell. **1**(1), 89–101 (2012)

20. Brzezinski, D., Stefanowski, J.: Reacting to different types of concept drift: the accuracy updated ensemble algorithm. IEEE Trans. Neural Netw. Learn. Syst. **25**(1), 81–94 (2013)
21. Crammer, K., Dekel, O., Keshet, J., Shalev-Shwartz, S., Singer, Y.: Online passive aggressive algorithms. J. Mach. Learn. Res. **7**, 551–585 (2006)
22. Gomes, H.M., Bifet, A., Read, J., Barddal, J.P., Enembreck, F., Pfharinger, B., Holmes, G., Abdessalem, T.: Adaptive random forests for evolving data stream classification. Mach. Learn. **106**(9), 1469–1495 (2017)
23. Liaw, A., Wiener, M.: Classification and regression by randomForest. R News **2**(3), 18–22 (2002)
24. Díaz-Uriarte, R., De Andres, S.A.: Gene selection and classification of microarray data using random forest. BMC Bioinf. **7**(1), 1–13 (2006)
25. Belgiu, M., Drăguţ, L.: Random forest in remote sensing: a review of applications and future directions. ISPRS J. Photogram. Remote Sens. **114**, 24–31 (2016)
26. Loh, W.Y.: Classification and regression trees. Wiley Interdiscipl. Rev. Data Min. Knowl. Discov. **1**(1), 14–23 (2011)
27. Bifet, A., Holmes, G., Pfahringer, B.: Leveraging bagging for evolving data streams. In: Joint European Conference on Machine Learning and Knowledge Discovery in Databases, pp. 135–150. Springer, Berlin (2010)
28. Bifet, A., Gavalda, R.: Learning from time-changing data with adaptive windowing. In: Proceedings of SIAM International Conference on Data Mining, pp. 443–448. Society for Industrial and Applied Mathematics, Philadelphia (2017)
29. Liaperdos, J., Stratigopoulos, H.G., Abdallah, L., Tsiatouhas, Y., Arapoyanni, A., Li, X.: Fast deployment of alternate analog test using Bayesian model fusion. In: Design, Automation & Test in Europe Conference & Exhibition (DATE), pp. 1030–1035. IEEE, Piscataway (2015)
30. Ahmadi, A., Stratigopoulos, H.G., Huang, K., Nahar, A., Orr, B., Pas, M., Carulli, J.M., Makris, Y.: Yield forecasting across semiconductor fabrication plants and design generations. IEEE Trans. Comput.-Aided Des. Integr. Circuits Syst. **36**(12), 2120–2133 (2017)
31. Ye, F., Zhang, Z., Chakrabarty, K., Gu, X.: Knowledge discovery and knowledge transfer in board-level functional fault diagnosis. In: International Test Conference, pp. 1–10. IEEE, Piscataway (2014)
32. Liu, M., Li, X., Chakrabarty, K., Gu, X.: Knowledge transfer in board-level functional fault identification using domain adaptation. In: IEEE International Test Conference (ITC), pp. 1–10. IEEE, Piscataway (2019)
33. Liu, M., Li, X., Chakrabarty, K., Gu, X.: Knowledge transfer in board-level functional fault diagnosis enabled by domain adaptation. IEEE Trans. Computer-Aided Des. Integr. Circuits Syst. **41**, 762–775 (2022)
34. Howard, R.A.: Information value theory. IEEE Trans. Syst. Sci. Cybern. **2**(1), 22–26 (1996)
35. Kotsiantis, S., Kanellopoulos, D., Pintelas, P.: Handling imbalanced datasets: a review. GESTS Int. Trans. Comput. Sci. Eng. **30**(1), 25–36 (2006)
36. Bickel, S., Brückner, M., Scheffer, T.: Discriminative learning under covariate shift. J. Mach. Learn. Res. **10**(9), 2137–2155 (2009)
37. Tsai, Y.H.H., Yeh, Y.R., Wang, Y.C.F.: Learning cross-domain landmarks for heterogeneous domain adaptation. In: Proceedings of the IEEE Conference on Computer Vision and Pattern Recognition, pp. 5081–5090. IEEE, Piscataway (2016)
38. Ghifary, M., Kleijn, W.B., Zhang, M.: Domain adaptive neural networks for object recognition. In: Pacific Rim International Conference on Artificial Intelligence, pp. 898–904. Springer, Cham (2014)
39. Long, M., Cao, Y., Wang, J., Jordan, M.: Learning transferable features with deep adaptation networks. In: International Conference on Machine Learning, pp. 97–105. PMLR (2015)
40. Yu, C., Wang, J., Chen, Y., Huang, M.: Transfer learning with dynamic adversarial adaptation network. In: IEEE International Conference on Data Mining (ICDM), pp. 778–786. IEEE, Piscataway (2019)
41. Yosinski, J., Clune, J., Bengio, Y., Lipson, H.: How transferable are features in deep neural networks? arXiv:1411.1792 (2014)

42. Nair, V., Hinton, G.E.: Rectified linear units improve restricted boltzmann machines. In: Proceedings of International Conference on International Conference on Machine Learning, pp. 807–814 (2010)
43. Gretton, A., Borgwardt, K.M., Rasch, M.J., Schölkopf, B., Smola, A.: A kernel two-sample test. J. Mach. Learn. Res. **13**(1), 723–773 (2012)

Machine Learning Support for Wafer-Level Failure Pattern Analytics

Li-C. Wang and Yueling (Jenny) Zeng

1 Introduction

Semiconductor fabrication is a complex process with many sources of variations from equipment to operation of the production line. These variations influence the yield. Due to their complex interactions, it is challenging to guarantee during fabrication that the yield always stays above a desired level. Usually, the yield is not observed until the *wafer probe* testing stage. When the yield of a wafer deviates significantly from an expected norm, we can call it a *yield excursion*.

In a production process, it might take some time to move a *wafer lot* through fabrication and bring the wafers to wafer probe. During the time, other lots are being fabricated. Hence, when a yield excursion is seen at wafer probe, the underlying cause might have already affected a large number of wafers. Therefore, it is desirable to detect yield excursions as early as possible.

While detecting yield excursions is important, not all yield excursions could be carefully investigated and corrected. Due to limited time and engineering resources, it is not uncommon that resources are prioritized for those excursions that have a more significant impact, i.e. those systematically affecting a large number of wafers. For those excursions that come and go, they might simply be recorded without further action.

Therefore in practice, early detection of a yield excursion can mean detecting yield excursions that are worth the investigation effort. Then, the problem becomes more challenging. For example, the amount of yield loss due to excursions can vary widely. Some excursions appear suddenly, and others develop over time. Excursions with significant and persistent yield loss might be easy to get noticed. Those less

L.-C. Wang (✉) · Y. (Jenny) Zeng
Department of ECE, University of California, Santa Barbara, CA, USA
e-mail: licwang@ucsb.edu; yuelingzeng@ucsb.edu

© The Author(s), under exclusive license to Springer Nature Switzerland AG 2023
P. Girard et al. (eds.), *Machine Learning Support for Fault Diagnosis of System-on-Chip*, https://doi.org/10.1007/978-3-031-19639-3_9

significant or less persistent ones might go unnoticed and yet, some of them might develop and become significant over time.

One common approach to detect a yield excursion is by checking to see if there is a formation of a *failure cluster* on the wafer. A failure cluster means there is a concentrated region of failing dies. Two failure clusters from two wafers may be considered as belonging to the same excursion, if the two clusters share some similar characteristics. For example, several wafers may have a cluster of failing dies all appearing along the wafer edge and at the 12 o'clock direction. This behavior can be seen as a signature for a yield excursion. However, detecting failure clusters and deciding which belong to the same excursion are not always as straightforward. A common way to make such a decision is to visualize failure clusters on *wafer maps*.

1.1 A Yield Excursion Example

Figure 1 recaps[1] the systematic yield event as reported in [46] on a single product. Wafers affected by this issue contain a failure cluster along the wafer edge at the 5 and/or 11 o'clock directions, as seen in Fig. 1a. The plot is based on stacking many wafer maps. On a single wafer map, the failure cluster may not be as pronounced as what is shown on this plot. In Fig. 1b, the numbers of wafers that contain either the 5 o'clock cluster or the 11 o'clock cluster or both, are shown. The x-axis shows the week number when those wafers were manufactured. Before week 19, this excursion occurred infrequently with weeks spanning between occurrences. As a result, the first two occurrences were seen but decided as one-off events due to

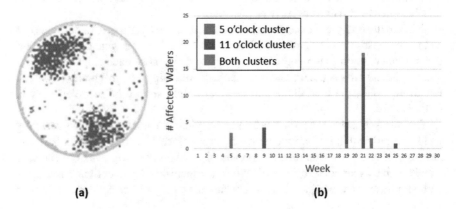

(a)   (b)

Fig. 1 The weekly trend and impact of a yield excursion for one product: (**a**) stacked wafer maps, (**b**) weekly occurrences of the problematic wafers [46]

---

[1] Figures 1, 2, and 3 in this section are figures from [46], ©2017 IEEE. Reprinted, with permission, from 2017 IEEE International Test Conference.

the small number of affected wafers and the varying locations of the clusters. The excursion was not elevated until week 19 when an entire wafer lot was affected.

There was a 14-week delay from the first occurrence to the elevation of the excursion for further investigation. During this time, the production line continued producing wafers. The yield on those wafers might have been impacted. Further, those wafers can have an unknown quality risk due to the underlying yield issue. Identifying the issue earlier could have led to a corrective action in place earlier, reducing the yield loss and the risk to degrade product quality.

1.2   Detecting Yield Excursion

Detecting abnormal yield occurrences is common in semiconductor manufacturing. Statistical Bin Limits (SBL) [19] and Below Minimum Yield (BMY) are two widely-adopted methods to identify wafers with abnormal yield. BMY is a coarse anomaly detection method. SBL considers yield from individual *test bins*. SBL provides additional resolution to detect yield abnormalities that may be missed by BMY.

There are also methods used to identify anomalies at the die level. Wafer spatial cluster detection algorithms such as Good Die in a Bad Neighborhood (GDBN) [29] and Unit Level Predictive Yield (ULPY) [28, 34], are methods commonly employed in post-processing of wafer level test data. GDBN determines the health of a die based on the failing statistics of its neighbors. ULPY also measures the health based on the neighbors where the influence of the neighbors takes the distance into account, i.e. a failing die in close proximity has more influence than those at distant.

Image processing techniques had also been proposed for wafer spatial pattern detection. In [55], the Hough Transform was used to detect scratch patterns. The authors in [20], proposed transforming wafer maps into so-called spatial correlograms to aid in detection and classification of failure patterns such as clusters, circles and others. A method was proposed for identifying similar patterns by comparing the correlograms of future wafers against a reference set of patterns.

Overall, we can see that historically there are three common perspectives to check for an abnormal yield occurrence: (1) checking some statistics of the yield, (2) checking the density of failures, and (3) checking failure patterns. The first two are more popular because their results are easier to be visualized and comprehended. Understanding the significance of a yield abnormality is important for making a decision to elevate the issue. Understanding and deciding on the significance of a failure pattern can be complicated. This is why after this introduction section, the rest of the chapter will be devoted to techniques that center on failure patterns.

1.3   Detecting a Systematic Failure Cluster

For the yield excursion shown in Fig. 1, the authors in [46] proposed a methodology
that can group wafers for exhibiting a failure cluster. The methodology involves two
main steps: (1) Use a density estimation method such as *kernel density estimation*
(KDE) [35] to convert a wafer map into a density map. From the density map, use a
threshold to identify a failure cluster region on the wafer. (2) Use some *features* to
describe the location and direction of the cluster and possibly some other attributes,
e.g. its size, total number of failures, and number of failures from each test bin, etc.
Then, use a *clustering* tool [31] to classify wafers into groups. *Clustering* is a very
common problem formulation in Machine Learning (ML) [31]. In this sense, the
overall methodology might be seen as a ML-supported methodology.

As reported in [46], the methodology was evaluated based on 30 weeks of
production data from 15 high-volume products manufactured in an analog circuit
technology. These products were designed for a variety of applications. In total,
around 40,000 wafers of various die sizes were analyzed.

One noticeable result from the methodology evaluation is summarized in Fig. 2,
which shows a plot similar to that shown in Fig. 1 before. The difference is that
Fig. 1 is based on one product and Fig. 2 is based on all 15 products.

Identifying failure clusters across multiple products increases the visibility of a
systematic issue that impacts an entire technology. When the analysis focused only
on one product as that shown in Fig. 1, during the first 11 weeks only 9 wafers had
the failure cluster patterns. After including the other 14 products in the analysis, this
number jumped to 55 as shown in Fig. 2.

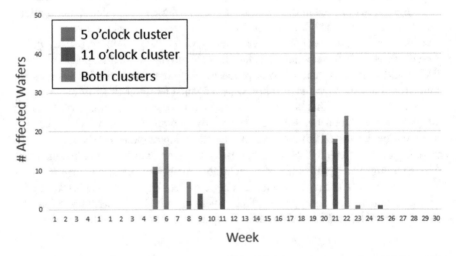

Fig. 2 The number of wafers from 15 products affected by the process excursion exhibiting with
a failure cluster either at 5 o'clock, 11 o'clock, or both directions, as that in Fig. 1 [46]

In Fig. 2, during the first 11 weeks, there were 5 wafers with clusters at the 5 o'clock direction (highlighted in blue) and 28 wafers with clusters at the 11 o'clock direction (highlighted in red). Additionally, there were also 22 wafers with clusters in both directions (highlighted in green). In contrast to 9 total wafers in Fig. 1, this total number of 55 wafers would have been sufficient to trigger an investigation with the in-house workflow [46].

Implementing the methodology as an automation tool facilitates analysis of failure clusters across multiple products and as a result, increases visibility of a process excursion. In the example shown, *Failure Analysis* (FA) could have been performed in week 11 or even earlier, instead of waiting until week 19, if the tool was in place.

1.3.1 Insight From Failure Analysis

For the yield excursion example discussed above, the FA report revealed what had happened (see [46]). During the period, process changes were slowly rolled out to improve the de-vail process prior to via metal deposition. The de-vail process removes extra material that overhangs a via hole. This small obstruction is called a *vail*. Vail can partially cover a via hole, hindering the metal deposition.

Changes to the de-vail process were thoroughly evaluated and approved. However, the changes resulted in significant yield loss whenever the process shifted to a particular corner. This yield loss occurred very infrequently at first, affecting roughly one lot a week across the entire technology. Additionally, the severity of the yield loss varied. After the yield excursion was elevated, failure analysis on failing parts found that the vails were not completely removed by the new de-vail process, resulting in partial-via defects. A cross section of one defective via is shown in Fig. 3.

Due to its infrequent occurrences and varying yield loss severity, this excursion was overlooked at first. This was discussed earlier with Fig. 1b. The location of the clusters were not consistent during their early occurrences and the issue only appeared infrequently. Even in hindsight, it would have not been obvious to group these failure modes together because the existing practice at the time did not consider cross-product analysis. With the methodology reported in [46] in place, cross-product analysis became feasible. Then, a more extensive analysis across 15 products in the technology showed that a cross-product analysis would have made the yield excursion substantially more visible during the early weeks.

1.4 Lessons Learned

In this introduction section, we recap the work reported in [46]. In retrospect, we can use the example to illustrate several lessons, as we elaborate below.

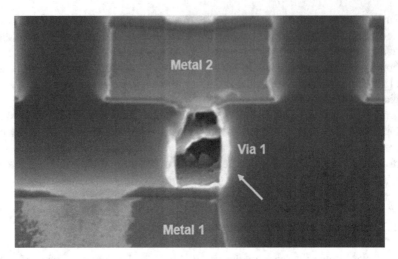

Fig. 3 Results from FA showing the cross section of a partially-filled via [46]

Fig. 4 The "ML" step within an iterative analytics process

1.4.1 Iterative Analysis

In practice, data comes in a stream. For example, in the application example above, the data samples come in as batches of wafers over time. If we consider the analysis performed in [46] including a "machine learning" (ML) step, then Fig. 4 illustrates that the analysis is performed in an iterative fashion.

In one iteration, the analysis is on the current data D. The ML step produces some result which possibly leads to some discussion and action. For example, this action can be "elevating the issue to the FA team" or "trying on more wafer data". In the next iteration, additional data D' may be included in the analysis. It should be noted that between the two iterations, the underlying manufacturing process might be adjusted. As a result, D and D' can be seen as from two different "data generators" (e.g. if we consider the manufacturing process as a data generator for wafer maps).

In view of Fig. 4, seeing the ML box as solving a *supervised learning* problem would not be effective. This is because if D and D' were from two different data generators, trying to learn from D to predict about D' would not be effective. Hence, the analysis supported by the methodology in [46] was *unsupervised*. The most important goal is to identify wafers into a *group* which can exhibit a systematic issue and make it visible to a team of engineers so that they know to elevate the issue to the FA team. In this view, the essence is to solve a *wafer grouping* problem.

1.4.2 Systematic Issue

It is interesting to notice that in the methodology in [46], the focus was on looking for "failure clusters" that could be considered as representing a systematic yield issue. Differentiating the shapes of the clusters was not of much concern. For example, the yield excursion was described by two failure clusters that are "along the wafer edge at 5 o'clock and at 11 o'clock directions". The description comprises two attributes, the location and the direction.

The added value provided by the methodology is that it enables analysis across many products so that the yield issue can become more visible at an earlier time. This is in contrast to the practice where engineers within each product team analyze the wafer maps of their own product. Each team might not notice the significance of a yield excursion. Across the teams, the excursion can become more apparent.

In reality, yield is never a steady number. It fluctuates over time. If we consider an "excursion" as simply a deviation from the norm, then there can be many yield excursions. In practice, a company has limited resources and hence, prioritizing yield excursions is required. With limited resources, only those relatively severe and persistent ones get noticed, elevated, and resolved.

The methodology was proposed with this view. Its goal was trying to make the severity and persistence of a yield excursion more visible. This is perhaps why the description of a failure cluster focuses on its location and direction, without concerning its shape. Differentiating failure clusters in terms of their shapes might improve the resolution for classifying the issues, but might also make a systematic issue less visible due to less number of wafer maps having the sample shape. In practice, improving the resolution of the analysis might be the next level of concern after those more systematic yield issues have been resolved.

1.4.3 Plot Visualization

Because yield excursions are prioritized based on their severity and persistence, it would be desirable to present an analysis result in such a way that the severity and persistence of a yield excursion can be easily seen and evaluated. Figures 1 and 2 are good examples for such a presentation. The analysis result is presented as a summary plot where the concerned aspects of the information are visualized. In practice, plots are often used to summarize analytic findings. These plots can go into a PPT presentation to be discussed in a meeting to make an informed decision.

If plots are among the final products from an analytics effort, when developing an analytics tool, it might be more intuitive to think about its functionality in view of supporting the generation of a summary plot. We will come back to this point in the last section of the chapter.

2 Wafer Map Pattern Recognition

In the field of semiconductor manufacturing, the problem of Wafer Map Pattern Recognition (WMPR) has been studied for decades [57]. The authors in [57] published a comprehensive wafer map dataset called WM-811K. This dataset has 811,457 wafer maps where 172,950 are labeled and the rest are not labeled. The work in [57] includes two learning approaches, one for classifying wafer maps and the other for searching similar wafer maps. The approaches are based on engineering a set of *features* to build a model for pattern recognition and similarity ranking.

Multiple types of features are used, including those based on Radon transform and those based on analyzing geometric properties such as failing die counts, region labeling, line detection, etc. For model building, Support Vector Machine (SVM) [39] is used. The work reported 94.63% classification accuracy on the dataset, comparing to the deep learning approach at the time which achieved 89.64% accuracy.

2.1 The Multi-Class Classification Problem

The WM-811K dataset pre-defines 9 classes. Figure 5 illustrates 8 pattern classes (where yellow dots indicate the die locations of fails). In addition, a "None" class is used to denote those wafer maps containing "no pattern". Figure 6 illustrates some examples of the "None" class.

Using the labeled portion of the WM-811K dataset, it is intuitive to treat the WMPR problem as solving a supervised multi-class classification problem, a supervised learning problem in view of the general ML. For multi-class image classification, we can consider two common approaches as depicted in Fig. 7. With

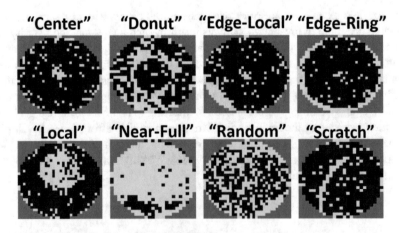

Fig. 5 Eight pattern classes in the WM-811K dataset

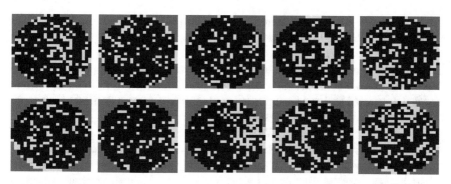

Fig. 6 Examples of the "None" class in the WM-811K dataset

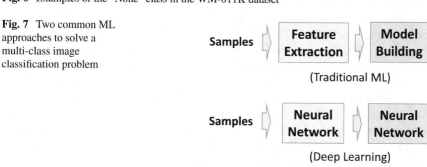

Fig. 7 Two common ML approaches to solve a multi-class image classification problem

a traditional ML approach, one first develops a set of *features* and converts each sample (an image) into a *feature vector*. Then, a model building method (e.g. SVM [39]) operates on the set of feature vectors to learn a model.

With a *deep learning* approach, the feature extraction step is automated by, for example layers of Convolution Neural Network (CNN). The output of the last CNN layer is treated as a feature vector. Then, for model building, one or more FC (Fully-Connected) network layers can be used. Those FC layers correspond to the model building step in the traditional ML.

2.2 Prior Works on the WM-811K Dataset

A number of works had been published using the WM-811K dataset. For example, the work in [10] followed the feature-based approach with a slightly different feature set and aimed to improve the multi-pattern detection accuracy, i.e. a wafer map containing patterns from two or more pre-defined pattern classess. The work in [61] advocated using more discriminant features based on Linear Discriminant Analysis (LDA) such that the model building step did not require a sophisticated method such as SVM. Instead, the work used Fisher discriminant analysis to replace SVM.

Using Radon transform features, the work in [32] proposed a special Decision Tree based ensemble learning method. Decision tree models are generally more interpretable than SVM.

The author in [60] proposed using an Autoencoder (AE) as a pre-training step to learn features before learning a classifier. The author found that to make it work, some pre-selected features were needed so that feature vectors, instead of the wafer map images themselves, were fed into the AE.

To apply deep learning on WM-811K, the author in [62] proposed a 2-stage classification: first to classify between having a pattern and having no pattern (i.e. the "None" class in WM-811K) and if there is a pattern, classify which class it is. Note that both works [60] and [62] acknowledged that the wafer map images were noisy and a de-noising step was required to make the learning work.

WM-811K is an extremely imbalanced dataset where some classes have many more samples than others [57]. This contributes to the inferior deep learning result reported in [57]. The work in [51] proposed a special data augmentation method based on GAN (Generative Adversarial Network) [14]. The network architecture also took the imbalance for learning a class into account, i.e. some classes are harder to learn than others. Instead of using GAN, the authors in [47] used an AE [36] for data augmentation. Moreover, the author found that augmenting the samples with rotation could help. A CNN-based network was then used for training the classifier. In contrast, the authors in [38] used pre-determined methods to augment the dataset while using a deeper CNN architecture for training the classifier.

The authors in [2] approached WM-811K dataset from a different angle. They tried to address the concern that a wafer map to be predicted might contain a new pattern class or a multi-pattern class not defined with the dataset. On those cases, the classifier might not be *applicable*. The work proposed using Selective Learning to determine an applicability of the model, where the deep learning model included the choice to abstain from making a prediction, i.e. the neural network has an integrated option to reject some samples and only make prediction on others.

In view of Fig. 7, we can say that the earlier works [10, 32, 57, 61] follow a traditional ML approach. The work [60] implements a deep learning approach involving a traditional feature extraction step to help. The later works [2, 38, 47, 51, 62] follow a deep learning approach. As pointed out in several works, for learning on this dataset, data augmentation and sample de-noising are two helpful steps.

2.3 A Multi-Class Neural Network Classifier

The work in [63] reports a study on the WM-811K dataset based on training a neural network classifier. The classifier employs the popular VGG-16 architecture [44]. In this section, we recap the results reported in [63] to illustrate the difficulties for training a high-accuracy classifier.

The WM-811K dataset [57] comprises wafer maps with various wafer sizes. The study selects two wafer sizes which have the largest number of wafer maps. Table 1

Table 1 Labeled wafer maps from the WM-811K dataset, used in the study in [63]

Center	Donut	Edge-L	Edge-R	Loc	N-full	Random	Scratch
81	10	402	9	345	16	28	76

Table 2 Confusion matrix (on all 967 wafer maps): ⇒: given label, ⇓: predicted label

	Center	Donut	Edge-L	Edge-R	Loc.	N-Full	Random	Scratch
Center	71	0	0	0	3	0	0	0
Donut	0	10	0	0	1	0	1	2
Edge-L	0	0	387	0	16	0	1	1
Edge-R	0	0	5	9	0	0	0	1
Loc.	10	0	7	0	319	0	0	7
N-full	0	0	1	0	0	16	0	0
Random	0	0	1	0	1	0	26	0
Scratch	0	0	1	0	5	0	0	65
Total mistakes	10	0	15	0	26	0	2	11

summarizes the selected set of labeled samples from the 8 pattern classes for the experiment. In total, there are 967 wafer maps considered. They can be called "in-class" samples. In addition, there are 22,115 wafer maps in the "None" class. From the unlabeled set, there are 19086 wafer maps with the two wafer sizes. In the study, these unlabeled wafer maps were used to evaluate the classifier.

In the dataset, there are substantially more "None" wafer maps than other wafer maps. A common way to deal with this imbalance is to train two models (e.g. see [62]), one to differentiate between the "None" class and the "in-class", and the other to differentiate among the eight pattern classes.

The study in [63] follows this strategy. One binary classifier is trained to separate "in-class" samples from "None" samples. Then, an 8-class classifier is trained to separate samples into 8 pattern classes.

Training the 8-class classifier relies on the samples summarized in Table 1. As seen, this training can still be challenging because some classes have many fewer samples than others. To make the data more balanced, we can involve a data augmentation strategy. A simple strategy to generate more samples for a class is by image rotation. This strategy was used for this dataset before, e.g. [60, 62]). In the study, rotated samples were added to make every class comparable to the number of the "Edge-Local" class (i.e. 402).

For training a classifier, the augmented dataset is split into a *training* dataset and a *validation* dataset. The study uses 2/3 of the samples as training dataset and 1/3 of the samples as validation dataset, randomly selected. With this setup, the work in [63] reported a model with training accuracy 99.24% and validation accuracy 92.5%. The classifier is then applied to the original 967 "in-class" samples. The resulting *confusion matrix* is shown in Table 2 (where the diagonal entries are highlighted).

2.3.1 Manual Review

In total, there are 64 mistakes shown in Table 2. In the study [63], these mistakes were manually inspected to determine the sources of difficulties for predicting them. These sources can be categorized into three types:

- **Label Ambiguity**: Two similar wafer map patterns have two different labels in the dataset.
- **Under-specification**: The wafer map contains a unique pattern. This wafer map is not included in the training dataset.
- **Model Deficiency**: If the mistake is not due to the above two reasons, then it is put into this category.

Among the 64 mistakes, the study reported 30 cases in the label ambiguity category, and 3 cases in the under-specification category. The remaining 31 cases were left in the model deficiency category. In other words, we considered the 33 cases in the first two categories as problems with the dataset itself.

Figure 8 shows several mistakes from the label ambiguity category. Additional examples can be found in [63]. Note that for each wafer map in this category, we found a wafer map with a similar pattern from the predicted class to justify the prediction. For example, if a wafer map was predicted by the model as "Center" but was labeled as "Local", we considered this as label ambiguity if we could find another similar-looking wafer map in the "Center" class to justify the prediction. In other words, we cannot blame the model for the mistake because of the two similar-looking wafer maps with two different class labels in the dataset.

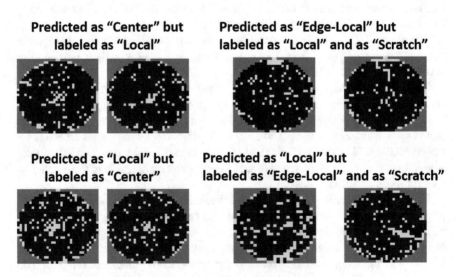

Fig. 8 Examples of mistakes in Table 2, potentially caused by label ambiguity

Fig. 9 Examples of mistakes in Table 2, potentially due to under-specification

Fig. 10 Two-step classification done by the two trained VGG models [63]

The three mistakes in the under-specification category is shown in Fig. 9. Each of these patterns appears only on one wafer map. If the wafer map were not in the training set, the model would never see the pattern. Hence, it is understandable that the model would have made a mistake on them.

2.3.2 Applied to Unlabeled Wafer Maps

The same VGG architecture was used to train a binary classifier model to differentiate the "in-class" wafer maps from the "None" wafer maps [63]. The same image rotation strategy was used to make the training dataset more balanced. The training dataset comprises all 967 original "in-class" wafer maps and their rotated images. It also includes 9K randomly-selected "None" wafer maps. The validation dataset comprises the rest of the "None" wafer maps without any "in-class" wafer maps. The validation dataset is used to evaluate how many wafer maps can be filtered out by the model. Ideally, we would like the model to filter out all wafer maps in the validation dataset. The work in [63] reported a binary classifier model with training accuracy at 95.25%. This model filters out 99.33% of the "None" samples in the validation dataset. However, it also excludes 249 of the "in-class" wafer maps.

The binary classifier and the 8-class classifier were then applied to the 19,086 unlabeled wafer maps in sequence, as illustrated in Fig. 10. After the first step, only 741 wafer maps were left (they were supposed to have an "obvious" pattern). Their classification result is shown in Table 3. After manually reviewing the result, those "questionable" were identified (see [63]). The number of questionable samples are also shown in the table and we consider them as potential mistakes.

Figure 11 shows some examples of those potential mistakes. Additional examples can be found in [63]. An interesting aspect to observe is that although previously

Table 3 Classification on the 741 unlabeled wafer maps

Class	Center	Donut	Edge-L	Edge-R	Loc	N-full	Random	Scratch
Classifier's result	114	37	211	44	233	9	34	59
Potential mistakes	2	24	4	10	37	0	10	49

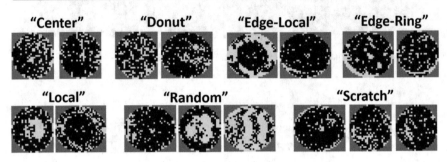

Fig. 11 Examples of questionable classification (the shown labels were reported by the model and were considered questionable in the manual review) [63]

the 8-class VGG classifier has a high validation accuracy at 92.25% on the labeled dataset, its performance on the set of 741 unlabeled wafer maps is much worse. In a sense, the earlier 92.25% accuracy result for the model might be somewhat misleading (considered from a practical application point of view). Also we observe that its performance on different classes can differ significantly.

2.4   Lessons Learned

The two classifiers reported in [63] were by no means optimal. However, optimizing model's classification accuracy was not the goal of the study. The study focused on understanding the sources of the mistakes and suggested that for dealing with some mistakes, it might not be effective to look at the problem from the perspective of model optimization.

2.4.1   Definition of Pattern Classes

A dataset like WM-811K is based on pre-defined pattern classes. This definition is subjective in view of the intended application context. With a fixed class definition, it is assumed that in the application, there is no need to differentiate wafer maps beyond those defined classes. In Sect. 4, we will see that this is not true in some application contexts.

The pattern classes defined in WM-811K can be a rather limited way to differentiate wafer maps. For example, Fig. 12 shows six examples from each of the

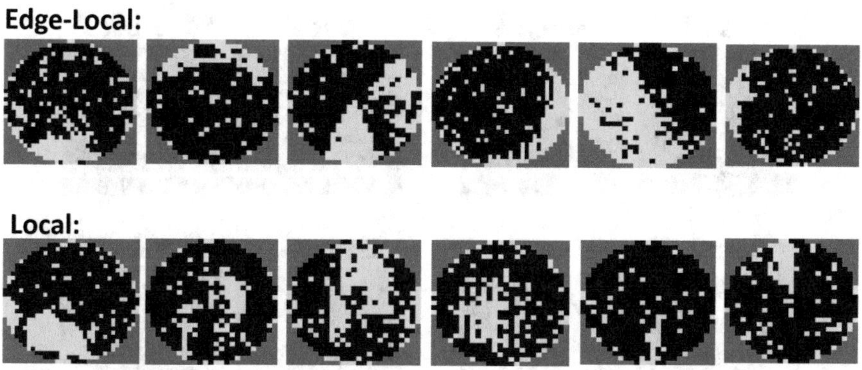

Edge-Local:

Local:

Fig. 12 Examples of pattern variation within a single class in WM-811K dataset

Fig. 13 Six consecutive wafer maps from the same lot

two classes, "Edge-Local" and "Local". As seen, the six wafer maps in each set do not look similar. In one application context, it might make sense to consider them as a single class. In another, it might make sense to separate them into their own classes. In other words, the class definition can largely depend on the application context.

Deciding on a pre-defined set of pattern classes that makes sense in view of an application might not always be easy though. For example, Fig. 13 shows six consecutive wafer maps from one single lot in the WM-811K unlabeled dataset. These six wafers clearly show a systematic trend. However, based on the WM-811K's class definition, the 1st, 3rd, and 5th wafers (from the left) might be called a "Local" pattern. The 2nd and the 4th might be a "Donut". The rightmost one might be called a "Center" pattern or a "Local" pattern (i.e. label ambiguity). From the pattern class definition, they are not the same. However, by the fact that they are consecutive wafers in the same lot, when presented in such a figure, the systematic trend can be exposed. Ideally, we would like to capture all these wafers in a single group. However, with the class definition they would be separated into multiple classes.

2.4.2 The Quality of the Training Dataset

Given a labeled dataset like WM-811K, it is intuitive to treat WMPR as a multi-class classification problem, as many prior works had done (see Sect. 2.1). Then, the

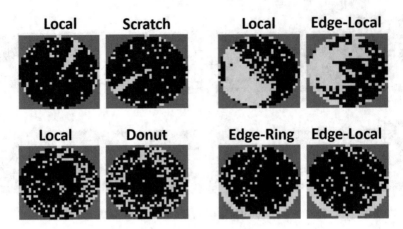

Fig. 14 Examples of possible label ambiguity

primary goal is to obtain a classifier with a prediction accuracy as high as possible. In other words, it is solving an *optimization* problem.

Solving the optimization problem would make more sense if two assumptions are verified first: (1) The labels assigned to the wafer maps in the dataset have little ambiguity; (2) The samples provided in the training dataset is representative to all future samples to be analyzed. For the study in [63] based on the WM-811K dataset, we have seen that these two assumptions are not met.

Regarding the label ambiguity mentioned earlier, Fig. 14 shows a few examples to illustrate the issue. Also, we notice that some wafer maps shown before in Fig. 6 for the "None" class might look like they containing a pattern.

One might think that label ambiguity is an issue that can be mitigated, say by following a more rigorous labeling process and by allowing a model developer to work closely with the practitioner who labels the samples. In practice though, likely the data preparation is limited by cost, available resources, among other factors.

For the second assumption, consider again the three wafer maps shown in Fig. 9 used to illustrate the issue. Earlier, these wafer maps were called "unique". However, it might not be clear what "uniqueness" means precisely.

Intuitively, one can explain "uniqueness" as that the patterns on those wafer maps do not appear on any wafer map in the training dataset. As a result, the training dataset is "under-specified". However, this does not explain why a model fails to predict the labels of those patterns even though they do not appear in the training dataset. After all, a learning model is supposed to be able to *generalize* from the patterns it has seen to other patterns it has not seen. Hence, a pattern not seen in the training dataset does not imply (and should not imply) the model shall fail on predicting the label of the pattern.

If we take the model into account, then "uniqueness" might be more precisely described with two conditions: the pattern is not in the training dataset, and more

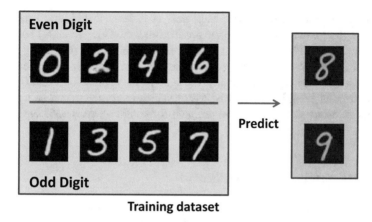

Fig. 15 An analogy to illustrate the difficulty with under-specification

importantly the pattern is beyond the generalization capability of the model. In other words, it is with respect to a given model that the pattern looks like a novel pattern.

It might be easier to illustrate the difficulty of learning with an under-specified dataset using a pattern recognition context that is more intuitive. Figure 15 depicts a simple example based on the popular MNIST dataset [23].

The MNIST dataset contains image samples for 10 digits, 0 ... 9. Typically, a training dataset would have samples from each of the ten digits. Figure 15 poses a different classification problem. Instead of ten classes, the dataset is defined to have only two classes: "even" digit and "odd" digit. The dataset includes training samples from the first eight digits, 0–7. In the evaluation, the trained classifier is asked to correctly predict the label for digit 8 as "even" and digit 9 as "odd".

From the pattern recognition point of view, the classification problem would be hard because the training dataset never includes any sample from digit 8 and digit 9. The learning problem is under-specified. For this example, one can say that the under-specification is caused by the subjectivity on the class definition. In other words, people understand what "even" and "odd" mean, but this understanding is not completely reflected in the training samples.

In view of the iterative analytics process discussed in Sect. 1.4.1, there can be a situation that the current class definition is based on the data D and it is under-specified in view of the additional data D'. Deciding whether or not a class definition for D is under-specified is not easy though. This is because to claim the class definition is under-specified, one might need to show that even the most optimized learning model would have failed to predict some "unique" patterns in D', i.e. under-specification depends on the generalization ability of a model.

In practice, a ML model developer might desire a class definition that facilitates reaching a high-accuracy model. More importantly, sufficient training samples are required for every class. For a practitioner who defines the classes and provides the dataset, these might not be the objectives. For example, a practitioner might want to define the classes based on how each class of patterns would have been used to debug the yield issue and improve the production line. This gap adds another level of complexity in solving WMPR as a multi-class classification problem.

Overall, if issues related to label ambiguity and under-specification can be mitigated or if they are not a major concern in an application context, then in practice solving WMPR as a multi-class classification problem can be a preferred view which offers simplicity in the problem formulation that is well supported by ML tools.

2.4.3 Optimization Vs. Personalization

For the works reviewed above based on the WM-811K dataset, the approaches to the WMPR problem essentially follow a flow like the following:

1. Define the pattern classes.
2. Collect a set of training and test samples.
3. Search for the best model based on the training samples.
4. Check if the accuracy on the test samples is acceptable for deploying the model.

The focus is on step 3 and possibly step 4, i.e. learning the best classifier. In this sense, *optimization* (of model accuracy) is the focus.

In practice, there can be a scenario where a practitioner does not have a well-defined set of pattern classes to begin with. Instead, the practitioner desires a tool to help clarify and define the pattern classes.

For example, suppose wafers come in batches. Each time an engineer receives a batch of wafers $w_1, \ldots, w_n$ and would like to know what patterns in the batch require attention. Note that whether or not a pattern requires attention, can be subjected to the engineer. Hence, it is up to the engineer to decide the importance of a pattern, and separate the pattern from others by calling it a "class". For a tool aiming to provide this flexibility to the user, it would be difficult to assume there exists a well-defined set of pattern classes to start all subsequent analyses.

Consequently, in view of the 1–4 steps, even for step 1 we might not assume that a tool can completely decide the pattern classes for its user. This decision is left to the user. If we take this view, then it is not even sufficient to simply consider WMPR as a traditional unsupervised learning problem where there is still a clear objective for a learning algorithm to optimize the pattern classification. To clarify this point, we state three postulates, and in the rest of the chapter we will discuss wafer map pattern analytics in view of these postulates.

1. Defining pattern classes is part of the problem.
2. There is no single answer to the definition.
3. A user can decide a pattern class of interest.

These postulates essentially state that a wafer map pattern analytics tool does not aim to provide a definite answer to the pattern classification. It can make suggestions and provide information to facilitate a user to attain their own classification. In this view, the focus of the tool development is no longer about optimization, because there is no clear classification objective to drive the optimization. Instead, the focus is on designing a tool that facilitates a user to attain an *understanding* of what is in the data and reach a comfort level to make a decision of their own.

3 Learning A Recognizer For A Pattern Class

Suppose a user decides that a set of wafer maps should represent a pattern class of their own. The user would like to have a model to recognize just the particular pattern class. For example, this model can be used to monitor future wafer maps. The works in [30, 41] were based on this perspective. For a given pattern class, the work in [30] proposes to learn a so-called *concept recognizer*. Such a recognizer can then be used to identify wafer maps that exhibits the pattern. In the work, a pattern class was called a a *concept*. This was because the work built recognizers not only for wafer map but also for other types of plots as well [30].

3.1 Concept Recognition

In [30], the approach is based on Generative Adversarial Networks (GANs) [14] for learning a concept recognizer. GANs are methods to learn a *generative model*. Given a dataset, a generative model synthesizes new samples similar to the training samples. A GAN's architecture consists of two neural networks. The *generator* network **G** is trained to generate samples. The *discriminator* network **D** is trained to differentiate the training samples from the generated samples.

Figure 16 illustrates the GANs design employed in [41]. The discriminator is a simple CNN with one convolution layer and one fully-connected (FC) layer. The generator network basically has a reverse-CNN architecture with more convolution

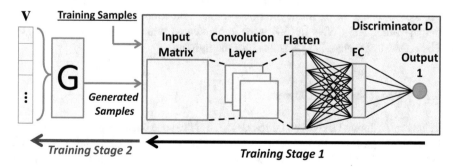

Fig. 16 Illustration of the GANs architecture used in [41]

layers and FC layers. The design uses a simple discriminator and a more complex generator because generation of samples is harder than classification of samples. In other words, the generator is given with more *capacity* than the discriminator.

This idea was inspired by the regularization ideas proposed in [3, 6]. Note that in typical GANs training, the goal is to learn a good generator. The goal here is different. After learning, the discriminator is used as the concept recognizer. Hence, the training focuses on the quality of the discriminator.

To train a discriminator as the concept recognizer, the experiments in [30] show that a minimum of five training samples are required. It should be noted that this minimum requirement depends on the pattern class to be learned. In [30], the five wafer maps used in the experiments were very similar. If we were to learning a recognizer for a class where within-class patterns vary significantly (as those examples shown in Fig. 12), then the required number of samples could increase.

The training process is iterative. Each iteration has two stages of training as shown in Fig. 16. Both can follow the common *stochastic gradient descent* (SGD) approach [13]. In the first stage, the goal is to learn the weights in the discriminator **D** in order to separate the generated samples from the training samples as much as possible. In the second stage, weights in **D** are fixed. SGD is applied to learning the weights in **G**. The gradients calculated on inputs of **D** are further back propagated to the inputs of **G**. The inputs of **G** are drawn from a *latent space* where each **v** is a random vector. The optimization objective for **G** is to generate samples such that at **D**'s output, their values are as close as possible to the values of the training samples, i.e. making it difficult for **D** to differentiate between the generated samples and the training samples.

For training GANs, attention is required to ensure two aspects: the output quality of both neural networks and the convergence of the training iteration. The work in [33] suggests several guidelines to improve the quality.

The work in [30] found that the performance of the CNNs were sensitive to whether or not (1) Batchnorm was used in both generator and discriminator CNNs, and (2) the Leaky ReLU activation function was used for all perceptrons. Implementation of a *Dropout* strategy was also found to be crucial [45]. For convergence, the feature matching technique [37] is crucial. Furthermore, in **D** the Sigmoid function is used to convert the output of the last perceptron into a value between 0 and 1. In **G**, the Hyperbolic Tangent function is used for adjusting the output value. The CNNs in [30] were implemented with Google TensorFlow [1]. The ADAM optimizer [22] was used for the training.

3.1.1 The Robustness Concern

In general, training GANs can be tricky [33, 37] because balancing the convergence between **G** and **D** can be a challenge. For training a recognizer, an additional tricky aspect is deciding when to stop the training. This aspect was studied in detail in [41]. If the goal was to obtain a good generator, the training could proceed until the discriminator fails to separate the training samples from the generated samples. This strategy would not work if the goal is to obtain a discriminator as the recognizer. In [41], deciding when to stop the training is based on so-called *separability*, a measure how well the in-class and out-of-class samples are separated by the discriminator.

3.2 Use of Tensor-Based Methods

Because of all the subtleties in training a GANs-based recognizer, it is difficult to guarantee the performance of a concept recognizer. The robustness concern motivated the use of another method to check the performance of a concept recognizer [53]. A tensor computation based method was developed and refined through a sequence of works [48–50]. Initially, the method was for implementing an online checker for a concept recognizer. In [41], it was further extended as a way to extract training samples for training a GANs-based recognizer. Overall, Fig. 17 shows the use of the two tensor-based methods in [41].

3.2.1 The Basic Ideas

One can think of the primary use of a tensor-based method is for implementing a similarity measure for a pair of wafer maps. Once the similarity between any pair of wafer maps can be calculated, it is easy to invoke a cluster algorithm. For the clustering, the work in [41] chose the Hierarchical DBSCAN algorithm (HDBSCAN) [26]. The minimum cluster size is specified at 5. This means that if the number of similar wafer images is less than 5, it is insufficient to be considered as a new pattern class. The number 5 was chosen because training a recognizer required a minimum of 5 training samples. From each cluster the best 5 samples are selected as the training samples. This selection is based on a *learnability* measure proposed in [48] where the best 5 samples are those with the highest learnability.

The clustering step is shown as the concept extraction step in Fig. 17. The output is a set of *preliminary pattern classes*, each represented by five training samples. Based on each class of samples, another tensor-based method is implemented to

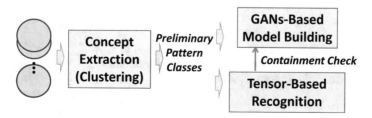

Fig. 17 Two places to apply a tensor-based method

perform recognition, i.e. for a given wafer map, decide which class it should fall into or it should be classified as not in any class.

An easy way to implement the classification would be using a threshold, i.e. if the similarity value between the given wafer map and a class of samples is above the threshold (similar enough), then the wafer map is considered in that class. This simple method is not robust because of its dependency on the threshold selection.

To avoid using a threshold, a *binning* strategy is implemented to decide whether a wafer map should be recognized as in-class or not. Given a set of preliminary pattern classes $C_1, \ldots, C_m$, the strategy is the following: if a wafer map is closest (most similar) to C_i, it is considered recognized by C_i as in-class and by all others as out-of-class. However, this strategy requires a way to allow a wafer map not recognized by any of $C_1, \ldots, C_m$. To do so, the work in [41] proposes to use a separate set of *baseline* recognizers as the "attractors" for those out-of-class samples.

As shown in Fig. 17, the tensor-based recognition is performed in addition to the GANs-based modeling. In [41], the tensor-based recognition is used to check the performance of the corresponding GANs-based model. This check is called a *containment check*.

For a pattern class, let T be the set of wafer maps recognized by the tensor-based recognizer. Let G be the set of wafer maps recognized by the GANs-based recognizer. Because recognition by the tensor-based recognizer is expected to be more conservative, we would expect to see the containment property: $T \subseteq G$. If this property is violated, it is an indicator that the GANs-based recognizer might have an issue. In [41], this containment check is used to check the robustness of the GANs-based recognizers.

3.3 An Application Result

The work in [41] implemented a wafer map pattern recognition flow and experimented with 8300 wafer maps from an automotive product line. The experiment divided the wafer maps into two batches, the first batch with the first 50 wafer lots and the second batch with the remaining wafer lots. The first batch has 1243 wafer maps. Figure 18 summarizes the result.

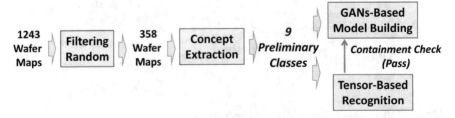

Fig. 18 Result from processing the first 50 wafer lots [41]

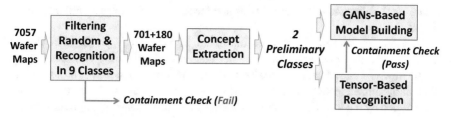

Fig. 19 Result from processing the remaining 287 lots [41]

Before the concept extraction step, a "Filtering Random" step was added. The goal of this step was to filter out wafer maps that contain no obvious pattern (e.g. sparse random failing). The filtering was implemented through a set of rules [41]. The work found that this step could be quite helpful to filter out a large number of wafer maps. This could dramatically simplify the clustering task that followed. Note that this step was not supposed to be perfect, i.e. it was not intended to filter out all no-pattern wafer maps, which would be difficult to do. This step tried to make the problem easier for the next step.

The next step is to perform the clustering scheme as described above to extract a list of preliminary pattern classes and their training samples. After the filtering step, there were 358 wafer maps left. Then, nine preliminary classes were identified. After GANs-based modeling, nine GANs-based recognizers were obtained. The containment check was performed on these recognizers using the 358 wafer maps. No violation of the check was observed.

Figure 19 then shows the result from continuous processing of the remaining 287 wafer lots from the second batch. This batch has 7057 wafer maps. After the filtering step, there were 1178 wafer maps left. Because we already had the 9 recognizers in place, these wafer maps were sent to the recognition step first. In this recognition, it was found that two of the recognizers violated the containment check.

For the 701 unrecognized wafer maps, they were added to the 180 unrecognized wafer maps from the first batch to form a single set and then to be processed in the concept extraction step. The clustering scheme found two new preliminary classess, resulting in building two new recognizers. The containment check of these recognizers did pass. For detail of the experiment, please see [41].

3.4 Lesson Learned

Because training a GANs model is complicated, the robustness of a GANs-based recognizer can be a concern. While the tensor-based containment check was proposed to mitigate the concern, it did not completely address it [41]. The flow presented in [41] was comprehensive. Nevertheless, its use was still limited to the application scope of multi-class classification. In other words, its output is still a set of "best-determined" pattern classes and the set of wafer maps identified in each class.

In view of the three postulates discussed in Sect. 2.4.2, the flow in Fig. 17 addresses the first postulate, but does not address the remaining two. The output of the flow is a classification of wafer maps, and in a sense the flow is still trying to find an optimized answer to the classification problem. However, the classification determined by the flow (i.e. by the tensor-based clustering scheme) might not always be what a person would like the classification to be in an application context.

For the recognition methods discussed in this section, training a recognizer requires five wafer maps. In the next section we will discuss a method that can turn every wafer map into its own recognizer [63, 64]. With this capability, we will then discuss how to build a software App that aims to address all three postulates mentioned in Sect. 2.4.2.

4 Describable Analytics

While optimizing the prediction accuracy of a model might be a common practice from the perspective of a ML technology developer, it is not always intuitive from the viewpoint of a practitioner. Figure 20 illustrates that in some application contexts, there can be a gap between the two.

For the ML tool developer, if a dataset is provided with pattern classes defined, the objective is well defined, i.e. providing a tool that can get to a model that has a high classification accuracy. For the practitioner, the objective can be somewhat vague. While the practitioner might have stated a general outcome to be attained, the meaning of some words used in the statement are up to further interpretation.

For example, what does it mean by a "pattern" can be subjected to the practitioner. The phrase "correlates to" refers to some sort of relationship between the identified pattern and test results from other stages. For example, these other stages can include both E-test (parametric test) and final test. However, what kind of relationship is considered as a correlation is up to further interpretation. Moreover, the wafer maps are produced and tested over time. They are presented as

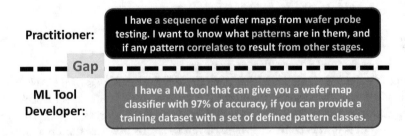

Fig. 20 A gap between a practitioner and a ML tool developer

Fig. 21 A describable analytics App driven by natural language queries, as proposed in [58, 65]

"a sequence". In this context, the practitioner might also be looking for a systematic trend as well.

To mitigate the gap, we can envision an App that allows the practitioner to express an analytic intent in terms of *queries*. Ideally the queries are stated in natural language as input to the App, and the App outputs *summary plots*. Then, the practitioner can visualize the plots and decide if any plot is interesting or not.

Figure 21 depicts this thinking which includes two essential ideas: (1) The summary plot is driven and dictated by user queries. (2) More importantly, the App does not decide on an answer for the user. Instead, the App enables a user to search for plots of their own interest. This search can be carried out at the natural language level, which makes it more intuitive to the user.

In view of the three postulates presented in Sect. 2.4.2, we can see that such an App can address the 2nd and the 3rd postulates. For example, the user can describe a pattern class to include all wafer maps used to generate a plot. This is in contrast to that presented in Fig. 17 before, where the software decides the pattern classes for its user. In this section, we discuss ideas to enable realization of such an App as depicted in Fig. 21. The approach involves a process that starts with a *semantic parser* to translate user queries into instructions executed by an *analytics backend*. The backend analyzes the wafer data and produces summary plots according to the instructions. The work in [58] describes an implementation of the analytics backend. The work in [65] describes an implementation of the natural language frontend. This section summarizes the high-level ideas presented in both works.

4.1 IEA and Concept Recognition

At first glance, we can see that Fig. 21 is proposing an App acting like a *virtual assistant* (e.g. Siri, Alexa). Indeed, the idea to implement a so-called Intelligent Engineering Assistant (IEA) was proposed in [52] before. While Fig. 21 might be related to the idea of IEA, as we will explain below there is a fundamental difference.

Figure 22 depicts the three components in an IEA App. The IEA was proposed as an autonomous system to perform an analytics task. The performance of the task is carried out by the analytics workflow. In this workflow, the decision (e.g. an if-then-else branch) is based on recognition of a *concept*. Hence, to support execution of the workflow, a concept recognition component is needed. This idea inspired the concept recognition work described in [30]. The third component, data processing interface, then serves as an interface to the raw data [52].

Fig. 22 Three components in
an Intelligent Engineering
Assistant (IEA) as proposed
in [52]

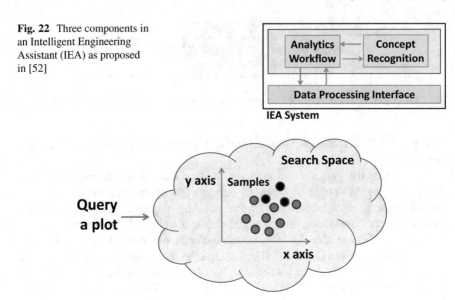

Fig. 23 Plot-based analytics working on a search space of plots

An IEA was supposed to be an end-to-end solution where input is the data and
outputs are the analytics findings, ideally summarized into a PPT presentation (see
Figure 3 in [52]). The App depicted in Fig. 21 is different. In view of the analytics
workflow to conduct an analysis task, the idea presented in Fig. 21 is to leave the
workflow out of the App. In other words, the App assumes that each user has their
own workflow in mind. The App does not dictate that. Instead, the App's job is to
support and facilitate a user to execute a workflow of their own.

4.2   Plot-Based Analytics

The App assumes that in each step of an analytics workflow, the user is asking for a
plot or a set of plots. The user can specify what type of plot as well as what aspect(s)
of the plot are considered interesting. In a sense, the user specifies a *search space of
plots* for the App to help search for an interesting plot based on the user's query.

Figure 23 depicts the idea which can be called *plot-based analytics*. For example,
the work in [43] discusses how to formulate the process for defining a plot-based
search space. The process comprises a sequence of *operators* to specify various
aspects of a plot. That work is different from Fig. 23 though. The goal of that
work was to learn a *process model* from various sequences of operators. In contrast,
Fig. 23 relies on user inputs to specify a search space of plots.

For example, suppose the search space comprises scatter plots that intend
to show a correlation. The x-axis is specified as the measured value from one

measurement in category C1. The y-axis is specified as the measured value from another measurement in category C2. Each dot represents a wafer.

Suppose a group of wafers has been specified. In a plot, the group of wafers are colored differently (e.g. as red dots) from the rest of the wafers (e.g. as blue dots). We are interested in finding a plot such that the red dots show a bias in view of the distribution of the blue dots, as shown in the example in Fig. 23. There can be many plots in the search space when there are many measurements in C1 and C2.

In the example, we see that various aspects of a plot can be specified: the x-axis, the y-axis, what group of wafers to focus on, and what property to check for. In plot-based analytics, the central idea is that a collection of plot types are pre-defined. A user can select a plot type, and specify the various aspects to define a plot-based search space. Then, the App helps a user to identify an interesting plot visually.

Because the plot types are pre-defined, one might think that it is critical for the App to provide a comprehensive list of plot types, anticipating all potential analytic needs. This could be an unrealistic goal to accomplish in one shot. In practice, a more realistic question can be asking how to design the App so that when needed, a new plot type can be conveniently added. Note that addressing this scalability aspect is a different subject beyond the scope of this chapter.

4.3 A "Try-and-See" Analytics Process

Suppose in Fig. 20, what the practitioner meant, includes checking to see if there exists a failure pattern that can be correlated to an E-test parameter. Based on the same test data analyzed in the experiments reported in [58], below we will use an example to illustrate the analytic process.

There are 2905 wafers included in the analysis in this example. There are a number of E-test parameters measured on each wafer across multiple sites. Suppose this number is N. Suppose on each plot, we will use the same parameter across two sites. Suppose the total number of site combinations is k. Then, in total a search space will contain $N \times k$ plots.

In the first try, we would like to see if the pattern "Edge-Local" exhibits a correlation to an E-test parameter. Among the 2905 wafer maps based on the wafer probe test result, there are 179 wafer maps showing an "Edge-Local" pattern. As discussed before, the pattern class definition "Edge-Local" is rough and hence, the patterns on the 179 wafer maps can vary. Figure 24 shows some examples to illustrate the within-class pattern variation.

Figure 25 shows one of the plots based on one E-test parameter denoted as "E1" on site 1 and on site 2. The orange dots represent the 179 wafers. The blue dots represent other wafers. On the right, the approximate locations of the two sites are shown. Because there are N E-test parameters, we would have N such plots. To rank plots, we could simply use the average position of the orange dots and the average position of the blue dots. Call their distance a "bias". Then, the plot can be ranked from the largest bias to the smallest bias, to facilitate visual inspection. Note that in

Fig. 24 Examples to show within-class pattern variation of the "Edge-Local" pattern class

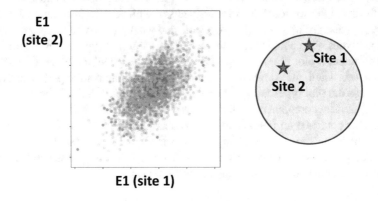

Fig. 25 A plot based on the E-test parameter E1 for the group of "Edge-Local" wafer maps

this search space, the plot based on E1 did not show up on the top. We show it in this figure to facilitate the discussion below.

Suppose among the 179 wafer maps, the App can present us the various patterns shown in Fig. 24 separately where each is presented as a suggested pattern. For example, as a user we can look at the pattern represented by the leftmost wafer map in Fig. 24. We can describe this pattern as a "thin arc at direction from 11 o'clock to 12 o'clock along edge". With this description, we mean the pattern is a "thin arc" (its shape), is "along edge" (its location), and the center of the pattern is somewhere between 11 o'clock and 12 o'clock (its direction).

Suppose the App can interpret the description in such a way that the App extracts a set of wafer maps satisfying the description. Figure 26 shows a set of wafer maps identified by the App according to the pattern description above. These wafer maps are from three lots, namely A, B, and C. Using this group of wafers, the App can re-generate the N plots, rank them, and present the plots for us to look at.

Figure 27 shows the new plot based on the same E-test E1, which is now one of the top-ranked plots. As seen, focusing only on those wafers following the described pattern led us to a plot showing a bias toward the bottom-left. Whether this plot truly means something or not, is subjected to further investigation and discussion. Nonetheless, the plot provides a good pointer for further investigation.

After seeing the plot, suppose we change our description from looking for a "thin arc" to a "cluster fails" and keep the rest of the description the same. The App would again identify the wafer maps satisfying this new description. From the new group of wafer maps, we realized that there were 10 wafer maps from the same Lot C.

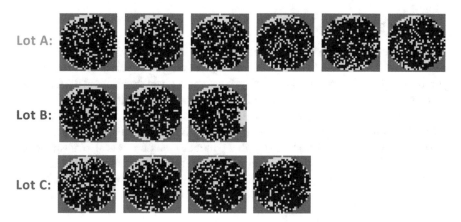

Fig. 26 A selected group of wafer maps based on the described "thin arc" pattern class

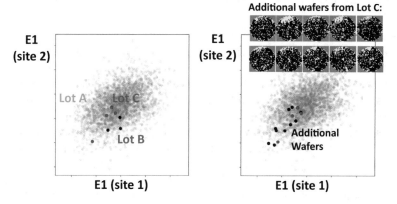

Fig. 27 Left: A top-ranked plot based on the selected group of wafers. Right: The plot based on wafers only from Lot C with additional 10 wafer maps (black dots)

Using the same E-test E1, we made a new plot with these 10 wafer maps together with the 4 wafer maps from Lot C before. Figure 27 shows the resulting plot.

In this second plot, the two groups of wafers come from the same Lot C. One is described as "thin arc" and the other is described as "cluster fails". Notice that their patterns are in the same location (both along edge) and at the same direction (both at direction from 11 o'clock to 12 o'clock). The difference is in their description of the pattern shape. This second plot reminds us an interesting aspect discussed in Sect. 1.1 before: In some application context, the exact shape of the failure pattern is not that important. The location and the direction of the pattern are. In this case, if we want to describe both groups of wafers, we can use a more generic term like "some pattern" to capture a pattern without explicitly constraining its shape.

4.3.1 Usage of the App

The example discussed above illustrates our view toward analytics, that it is seen as an iterative "try-and-see" process. In each try, the App's job basically is to perform three tasks: (1) Extract a group of wafers based on a given description; (2) Generate a set of plots based on the group; (3) Present the plots to facilitate a person to pick an interesting plot, if any. Note that the first task is key to the remaining two which would be relatively easy once the group of wafers is decided. In addition to the three main tasks, when one does not know what pattern to look for, it is also desirable for the App to provide a suggestion what patterns might be available to choose from. In the following, we will focus the discussion on accomplishing the first task and explain how to extract a set of *primitive patterns* to start a "try-and-see" process.

4.4 Three Major Components

The work in [58] and the work in [65] together present an overall approach to enable an implementation of the proposed App. Figure 28 depicts three major components in the implementation. It comprises a natural language frontend and an analytics backend, connected through a *plot-based worldview*. The worldview defines what plot types are supported as well as for each plot, what *plot attributes* can be controlled. For making a plot, it includes a plotting module driven by the frontend and a result database to store analytic results from the backend [65].

The worldview essentially defines an actionable space for the frontend. The core of the frontend is a *semantic parser* [65]. The parser translates a user's input in natural language into a sequence of instructions to the backend for two types of tasks: (1) selecting the *group of wafer maps* to be used in plotting; (2) selecting a plot type and setting the values of the plot attributes.

The backend supports two types of analytics: Minions analytics [63] and NLI analytics [58] (NLI stands for Natural Language Interpreter). They provide two perspectives integrated to support analytics of wafer maps. Each perspective provides a set of *wafer attributes* to describe a wafer map and the two work together to determine the values of those attributes. Given a set of wafer maps, one can think that the analytics results are stored in a database as tables where each row corresponds to a wafer map and each column corresponds to a wafer attribute [65].

Fig. 28 Three major components in our language driven analytics approach

In our implementation, there are multiple tables, each stored with wafer attribute values based on a *primitive pattern* extracted from the Minion analytics [58].

4.4.1 The Minions Approach

The analytics backend is built upon the Minions (MINiture Interactive Offset Networks) approach as its foundation. The approach was presented in [63], which was based on the *manifestation learning* approach presented in [64]. The NLI analytics are carried out on top of Minions' analytic results [58]. Figure 29 illustrates the high-level idea of the Minions approach.

With the approach, a neural network (NN) model is independently learned for each wafer map. This NN model serves as a *recognizer* dedicated for the wafer map. In [63], each recognizer is a neural network model called a Minion. The wafer map used to train a Minion is called its *anchor*. With one Minion for every wafer map, we can then perform mutual recognition, which will result in a *recognition graph*. In this graph, every node is a wafer map. Two nodes have an edge connecting them if their recognizer recognizes each other (i.e. mutual recognition).

Given a recognition graph, we can analyze wafer maps using well-known operations on graphs. For example, clusters of wafers can be attained by finding all Connected Components (CC) where each CC is treated as a separate cluster. Figure 30 shows a CC example. It is interesting to notice that while two wafer maps with a direct connection have a similar pattern, not all wafer maps in the CC have a similar pattern. For example, wafer A is indirectly connected to wafer D through wafers B and C, and wafer A does not have a pattern similar to wafer D. This is because each Minion can recognize a wafer map with certain variations from its anchor. As a result, the more other wafers along an indirect connection between two wafer maps, the more dissimilar the two wafer maps can be.

One major advantage of the Minions approach is that it turns a traditional analysis such as clustering from statistical to graph-based. After a mutual recognition graph is constructed, all analyses can be done with graph-based operations, which can improve traceability and robustness of the analyses.

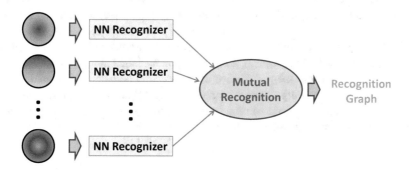

Fig. 29 High-level idea of the Minions approach

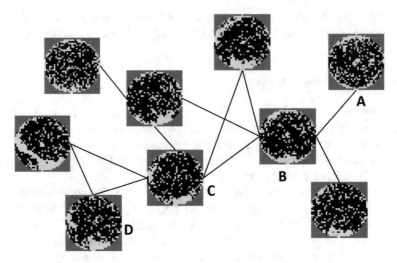

Fig. 30 An example connected component extracted from a Minions recognition graph

4.4.2   A Brief Note on Training a Minion

A Minion is trained with one sample. Training with one sample is generally referred as *one-shot learning* [11, 24]. In Machine Learning, three approaches have been proposed to tackle one-shot learning, as illustrated in Fig. 31.

Two domains are shown, a source domain A and a target domain B. Source domain has many training samples. Target domain does not. The idea of data augmentation is to learn from samples in domain A to generate more samples and augment the dataset in domain B [17, 27, 40]. In *feature augmentation* (e.g. [8, 16, 17, 25]), generated samples are in the feature space rather than in the input space.

In *feature transfer* [59], neural network weights learned from the source domain are transferred to the target domain. *Domain adaptation* [4, 12] is a specialized approach where the transfer is between two domains for performing the same task. In *meta-learning* [18], a learning strategy that was learned in the source domain is transferred into the target domain. A recent study [15], though, shows that meta-learning is not as effective as people had claimed on a variety of tasks.

For learning a Minion model with one wafer map, the work in [63] did not find the three approaches effective. Instead, a fourth approach called *manifestation learning* [64] was adopted. From the perspective of transferring between two domains, this approach transfers an "output vector space" (a *codebook*) from domain A to domain B [64]. In addition, a two-part Minion training scheme is used based on a variant of the Triplet Loss Siamese Network [9].

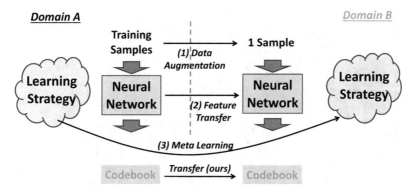

Fig. 31 Different ways to attain one-shot learning

4.4.3 Primitive Pattern and Describable Set

The example in Fig. 30 shows that it might not be a good idea to use a CC to represent a pattern class. In addition to extracting CCs from a recognition graph, we can also extract *maximal cliques*. From our past experiments [58, 63], we had observed that wafer maps within a maximal clique usually looks very similar. Therefore, in [58] maximal cliques are used to provide so-called the *primitive patterns*.

Figure 32 show a CC that contains a maximal clique of size 4. This clique represents a primitive pattern. In the backend, the NLI is used to find a group of wafers, which is extended from a primitive pattern. This group of wafers is called a *describable set* [58]. Note that the NLI finds the describable set based on a user's pattern description. For example, the describable set might be found by following the description: "an arc along edge spread from 6 o'clock to 9 o'clock".

Please refer to [58] for detail of the NLI implementation. In summary, the NLI relies on a formal language model, i.e. a context-free grammar, to capture the scope of what is describable. The *lexicon* of the grammar comprises a list of terms. A software script is implemented for each term to determine what descriptive value should be used. For example, a "REGION" term is used to describe the location of a pattern. The value can be "center", "edge", or "in-between".

Overall, the primitive patterns provided by the Minions approach can be used to suggest a good starting point for carrying out a "try-and-see" process. Then, in each try the user can provide a description for what pattern to be focused on. The NLI then finds the describable sets as the group of wafers used in the plotting. On the recognition graph, a describable set satisfies two conditions [58]: (1) It is based a primitive pattern and includes its maximal clique; (2) Within itself, all wafers are connected. As a result, describable sets are not only limited by the grammar defined in NLI but also constrained by the recognition graph provided by the Minions. Conceptually, given a user query, the job of the frontend is to instruct the backend to look for the describable sets that best satisfy the query. Because the describable sets are constrained, a query can result in no wafer found.

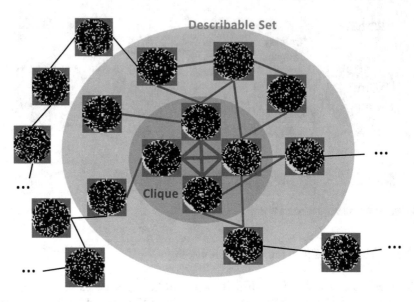

Fig. 32 Illustration of a primitive pattern and one of its describable sets

4.5 The Frontend

The core of the frontend is a semantic parser. *Semantic parsing* is the process that assigns real-world meanings to linguistic inputs [21] (e.g. words). Specifically, in *computational semantics*, formal structures called *meaning representations* are used to link the non-linguistic knowledge (e.g. data stored in database, API function calls, etc.) to linguistic elements such as English words. Semantic parsing is a wide field that has been studied for some time. In machine learning, most works to achieve semantic parsing rely on supervised learning with large amounts of human-created semantic parses [21, 54]. Following such an approach is challenging in our context, because we are unable to obtain large amounts of training data.

Consequently, our work presented in [65] does not follow a supervised learning approach. Instead, it follows the approach called *semantic parsing as paraphrasing* [5]. This approach makes use of triples (*natural query q*, *canonical utterance c*, *meaning representation m*), where the parser maps $q \rightarrow c \rightarrow m$. The work in [5] also observed that mapping $c \rightarrow m$ and vice-versa could be achieved by fixed rules (e.g. as some software scripts), which makes the approach more feasible in our context.

Figure 33 illustrates the two major components in the frontend. Given a query, the constrained parsing component first converts the query into a canonical utterance. The rule-based scripts then convert the utterance into meaning representation which comprises executable instructions for retrieving the requested group of wafer maps. The meaning representation employed in our implementation is called *Question Decomposition Meaning Representation* (QDMR) [56]. QDMR contains natural

Fig. 33 Two major components in the frontend

utterances, which is easier to understand than complex logic forms used in [54]. For detail of the frontend implementation, please refer to [65].

It is worth noting that our implementation in [65] leverages the power of a pretrained language model (LM) [42] to realize the semantic parsing component. With a LM, we can use *few-shot learning* to teach the LM about the acceptable utterances designed in our frontend. Following [42], our implementation uses the state-of-the-art language model GPT-3 [7], as the natural language interface. GPT-3 handles the variations of wordings presented in spoken language. By utilizing GPT-3's powerful *in-context learning*, we can finetune GPT-3 to learn the *translation* from a variation of input query given in natural language to the canonical utterance as defined by us, with a small number of demonstrating examples (i.e. few-shot learning). The translation is essentially a many-to-one mapping function that generates a single unique interpretation of various input statements.

4.6 Lesson Learned

In this section, we discuss a novel approach where analytics of wafer map patterns are driven by natural language queries. The central idea is to view analytics as an iterative "try-and-see" search process. Under this view, we can then envision an App that enables a user to conduct a search of interesting plots using natural language queries, which makes the search more intuitive and efficient. Our implementation includes a frontend that performs constrained semantic parsing and an analytics backend that supports two analytics perspectives (the Minions and the NLI). The frontend and the backend are connected through a plot-based worldview. This section reviews the works reported in [58, 63, 65] where the works in [58, 63] are about the backend and the work in [65] is about the frontend. The Minions approach [63] is the foundation for the backend which adds the NLI on top of it to support the envisioned describable analytics [58].

Note that in our view, presenting analytic results as plots is essential. In a "try-and-see" process, we desire an App to provide analytic results easily understandable by its user. Plots make it easier for a person not only to understand the analytic findings but also to interpret them to others. As seen in the yield excursion example in Sect. 1.1, a truly valuable finding is the one that generates a meaningful action, e.g. elevated to the FA team. An analytics App might be unable to dictate how a finding is interpreted and used by a person. Nevertheless, the App should at least aim to provide findings in a way that makes it easy for a person to decide the next step.

References

1. Abadi, M., et al.: TensorFlow: Large-scale machine learning on heterogeneous systems (2015). Software available from https://www.tensorflow.org/.
2. Alawieh, M.B., Boning, D., Pan, D.Z.: Wafer map defect patterns classification using deep selective learning. ACM/IEEE Design Automation Conference (2020)
3. Arjovsky, M., Chintala, S., Bottou, L.: Wasserstein generative adversarial networks. In: Precup, D., Teh, Y.W. (eds.) Proceedings of the 34th International Conference on Machine Learning, ICML 2017, Sydney, NSW, Australia, 6–11 August 2017. Proceedings of Machine Learning Research, vol. 70, pp. 214–223. PMLR (2017)
4. Ben-David, S., Blitzer, J., Crammer, K., Pereira, F.: Analysis of representations for domain adaptation. Adv. Neural Inf. Proces. Syst. (22), 137–144 (2007)
5. Berant, J., Liang, P.: Semantic parsing via paraphrasing. In: Proceedings of the 52nd Annual Meeting of the Association for Computational Linguistics, vol. 1: Long Papers. Association for Computational Linguistics (2014)
6. Berthelot, D., Schumm, T., Metz, L.: Began: Boundary equilibrium generative adversarial networks. arXiv:1703.10717v4 (2017)
7. Brown, T.B., et al.: Language models are few-shot learners. In: Larochelle, H., et al. (eds.) Advances in Neural Information Processing Systems 33: Annual Conference on Neural Information Processing Systems 2020, NeurIPS 2020, December 6–12, 2020, virtual (2020)
8. Cheny, Z., Fuy, Y., Zhang, Y., Jiang, E.A.: Multi-level semantic feature augmentation for one-shot learning. IEEE Trans. Image Process. 28(9), 4594–4605 (2019)
9. Dong, X., Shen, J.: Triplet loss in siamese network for object tracking. In: European Conference on Computer Vision (2018)
10. Fan, M., Wang, Q., van der Waal, B.: Wafer defect patterns recognition based on optics and multi-label classification. IEEE Advanced Information Management, Communicates, Electronic and Automation Control Conference (IMCEC) (2016)
11. Fink, M.: Object classification from a single example utilizing class relevance metrics. Adv. Neural Inf. Proces. Syst. 449–456 (2005)
12. Ganin, Y., Ustinova, E., Ajakan, H., Germain, E.A.: Domain-adversarial training of neural networks. J. Mach. Learn. Res. 17(1), 1–35 (2016)
13. Goodfellow, I., Benjio, Y., Courville, A.: Deep Learning. The MIT Press, Cambridge (2016)
14. Goodfellow, I.J., Pouget-Abadie, J., Mirza, M., Xu, B., Warde-Farley, D., Ozair, S., Courville, A.C., Bengio, Y.: Generative adversarial networks. Commun. ACM 63(11), 139–144 (2020)
15. Guo, Y., Codella, N.C., Karlinsky, L., Codella, E.A.: A broader study of cross-domain few-shot learning. In: Vedaldi, A., et al. (eds.) ECCV 2020. Lecture Notes in Computer Science, vol. 12372, pp. 124–141. Springer Nature, Cham (2020)
16. Gao, H., E.A.: Low-shot learning via covariance-preserving adversarial augmentation networks. Adv. Neural Inf. Proces. Syst. 983–993 (2018)
17. Hariharan, B., Girshick, R.: Low-shot visual recognition by shrinking and hallucinating features. International Conference on Computer Vision (2017)
18. Hochreiter, S., Younger, A.S., Conwell., P.R.: Learning to learn using gradient descent. International Conference on Artificial Neural Networks, pp. 87–94 (2001)
19. Illyes, S., Baglee, D.: Statistical bin limits: an approach to wafer dispositioning in IC fabrication. IEEE/SEMI Conference on Advanced Semiconductor Manufacturing Workshop, pp. 95–98 (1990)
20. Jeong, Y.S., Kim, S.J., Jeong, M.K.: Automatic identification of defect patterns in semiconductor wafer maps using spatial correlogram and dynamic time warping. IEEE Trans. Semicond. Manuf. 21(4), 625–637 (2008)
21. Jurafsky, D., Martin, J.H.: Speech and Language Processing: An Introduction to Natural Language Processing, Computational Linguistics, and Speech Recognition, 3 draft edn. (2022)

22. Kingma, D.P., Ba, J.: Adam: A method for stochastic optimization. In: Bengio, Y., LeCun, Y., (eds.) 3rd International Conference on Learning Representations, ICLR 2015, San Diego, CA, USA, May 7–9, 2015. Conference Track Proceedings (2015)

23. LeCun, Y., Cortes, C.: MNIST handwritten digit database (2010). http://yann.lecun.com/exdb/mnist/

24. Li, F.F., Fergus, R., Perona, P.: One-shot learning of object categories. IEEE Trans. Pattern Anal. Mach. Intell. **28**(4), 594–611 (2006)

25. Liu, B., Wang, X., Dixit, M., Kwitt, R., Vasconcelos, N.: Feature space transfer for data augmentation. Conference on Computer Vision and Pattern Recognition, pp. 9090–9098 (2018)

26. McInnes, L., Healy, J., Astels, S.: hdbscan: Hierarchical density based clustering. J. Open Source Softw. **2**(11), 205 (2017)

27. Miller, E.G., Matsakis, N.E., Viola, P.A.: Learning from one example through shared densities on transforms. Conference on Computer Vision and Pattern Recognition, pp. 464–471 (2000)

28. Miller, R., Riordan, W.C.: Unit level predicted yield: a method of identifying high defect density die at wafer sort. International Test Conference (2001)

29. Moreno-Lizaranzu, M.J., Cuesta, F.: Improving electronic sensor reliability by robust outlier screening. Sensors (Basel, Switzerland) **13**(10), 13521–13542 (2013)

30. Nero, M., Shan, C., Wang, L.C., Sumikawa, N.: Concept recognition in production yield data analytics. IEEE International Test Conference (2018)

31. Pedregosa, F., et al.: Scikit-learn: machine learning in Python. J. Mach. Learn. Res. **12**, 2825–2830 (2011). https://scikit-learn.org/stable/

32. Piao, M., et al.: Decision tree ensemble-based wafer map failure pattern recognition based on radon transform-based features. IEEE Trans. Semicond. Manuf. **31**(2), 250–257 (2018)

33. Radford, A., Metz, L., Chintala, S.: Unsupervised representation learning with deep convolutional generative adversarial networks. In: Bengio, Y., LeCun, Y., (eds.) 4th International Conference on Learning Representations, ICLR 2016, San Juan, Puerto Rico, May 2–4, 2016. Conference Track Proceedings (2016)

34. Riordan, W., Miller, R., St Pierre, E.: Reliability improvement and burn in optimization through the use of die level predictive modeling. In: Proceedings of the 43rd Annual Reliability Physics Symposium, pp. 435–445. IEEE International (2005)

35. Rosenblatt, M., et al.: Remarks on some nonparametric estimates of a density function. Ann. Math. Stat. **27**(3), 832–837 (1956)

36. Rumelhart, D.E., et al.: Learning internal representations by error propagation. Parallel Distributed Processing: Explorations in the Microstructure of Cognition, vol. 1, pp. 318–362. ACM, New York (1986)

37. Salimans, T., et al.: Improved techniques for training gans. In: Lee, D.D., et al. (eds.) Advances in Neural Information Processing Systems 29: Annual Conference on Neural Information Processing Systems 2016, December 5–10, 2016, Barcelona, Spain, pp. 2226–2234 (2016)

38. Saqlain, M., Abbas, Q., Lee, J.Y.: A deep convolutional neural network for wafer defect identification on an imbalanced dataset in semiconductor manufacturing processes. IEEE Trans. Semicond. Manuf. **33**(3), 436–444 (2020)

39. Schölkopf, B., Smola, A.J.: Learning with Kernels: support vector machines, regularization, optimization, and beyond. The MIT Press, Cambridge (2001)

40. Schwartz, E., Karlinsky, L., Shtok, J., Harary, E.A.: Deltaencoder: An effective sample synthesis method for few-shot object recognition. Advances in NIPS, pp. 2850–2860 (2018)

41. Shan, C., Wahba, A., Wang, L.C., Sumikawa, N.: Deploying a machine learning solution as a surrogate. In: IEEE International Test Conference, pp. 1–10. IEEE (2019)

42. Shin, R., et al.: Constrained language models yield few-shot semantic parsers. In: Moens, M., et al. (eds.) Proceedings of the 2021 Conference on Empirical Methods in Natural Language Processing, EMNLP 2021, Virtual Event / Punta Cana, Dominican Republic, 7–11 November, 2021, pp. 7699–7715. Association for Computational Linguistics (2021)

43. Siatkowski, S., Wang, L.C., Sumikawa, N., Winemberg, L.: Learning the process for correlation analysis. IEEE VLSI Test Symposium (2017)

44. Simonyan, K., Zisserman, A.: Very deep convolutional networks for large-scale image recognition (2015). https://arxiv.org/abs/1409.1556
45. Srivastava, N., Hinton, G., Krizhevsky, A., Sutskever, I., Salakhutdinov, R.: Dropout: a simple way to prevent neural networks from overfitting. J. Mach. Learn. Res. **15**(1), 1929–1958 (2014)
46. Sumikawa, N., Nero, M., Wang, L.C.: Kernel based clustering for quality improvement and excursion detection. IEEE International Test Conference (2017)
47. Tsai, T.H., Lee, Y.C.: A light-weight neural network for wafer map classification based on data augmentation. IEEE Trans. Semicond. Manuf. **33**(4), 663–672 (2020)
48. Wahba, A., Shan, C., Wang, L.C., Sumikawa, N.: Measuring the complexity of learning in concept recognition. In: International Symposium on VLSI Design, Automation and Test, pp. 1–4. IEEE (2019)
49. Wahba, A., Shan, J., Wang, L.C., Sumikawa, N.: Wafer plot classification using neural networks and tensor methods. In: ITC-Asia, pp. 79–84. IEEE (2019)
50. Wahba, A., Wang, L.C., Zhang, Z., Sumikawa, N.: Wafer pattern recognition using tucker decomposition. In: 2019 IEEE 37th VLSI Test Symposium (VTS), pp. 1–6. IEEE (2019)
51. Wang, J., Yang, Z., Zhang, J., Zhang, Q., Chien, W.T.K.: Adabalgan: an improved generative adversarial network with imbalanced learning for wafer defective pattern recognition. IEEE Trans. Semicond. Manuf. **32**(3), 310–319 (2019)
52. Wang, L.C.: An autonomous system view to apply machine learning. In: IEEE International Test Conference (2018).
53. Wang, L.C., Shan, J., Wahba, A.: Facilitating deployment of a wafer-based analytic software using tensor methods: invited paper. In: International Conference on Computer-Aided Design (ICCAD). IEEE/ACM (2019)
54. Wang, Y., et al.: Building a semantic parser overnight. In: Proceedings of the 53rd Annual Meeting of the Association for Computational Linguistics and the 7th International Joint Conference on Natural Language Processing, vol. 1: Long Papers. Association for Computational Linguistics (2015)
55. White, K.P., Kundu, B., Mastrangelo, C.M.: Classification of defect clusters on semiconductor wafers via the hough transformation. IEEE Trans. Semicond. Manuf. **21**(2), 272–278 (2008)
56. Wolfson, T., et al.: Break it down: a question understanding benchmark. Trans. Assoc. Comput. Linguist. **8**, 183–198 (2020)
57. Wu, M.J., Jang, J.S.R., Chen, J.L.: Wafer map failure pattern recognition and similarity ranking for large-scale data sets. IEEE Trans. Semicond. Manuf. **28**(1), 1–12 (2015)
58. Yang, M.J., Zeng, Y.J., Wang, L.C.: Language driven analytics for failure pattern feedforward and feedback. IEEE International Test Conference (2022)
59. Yosinski, J., Clune, J., Bengio, Y., Lipson, H.: How transferable are features in deep neural networks? Adv. Neural Inf. Proces. Syst. **2**, 3320–3328 (2014)
60. Yu, J.: Enhanced stacked denoising autoencoder-based feature learning for recognition of wafer map defects. IEEE Trans. Semicond. Manuf. **32**(4), 613–624 (2019)
61. Yu, J., Lu, X.: Wafer map defect detection and recognition using joint local and nonlocal linear discriminant analysis. IEEE Trans. Semicond. Manuf. **29**(1), 33–43 (2016)
62. Yu, N., Xu, Q., Wang, H.: Wafer defect pattern recognition and analysis based on convolutional neural network. IEEE Trans. Semicond. Manuf. **32**(4), 566–573 (2019)
63. Zeng, Y.J., Wang, L.C., Shan, C.J.: Miniature interactive offset networks (minions) for wafer map classification. In: IEEE International Test Conference, pp. 190–199. IEEE (2021)
64. Zeng, Y.J., Wang, L.C., Shan, C.J., Sumikawa, N.: Learning a wafer feature with one training sample. In: IEEE International Test Conference, pp. 1–10. IEEE (2020)
65. Zeng, Y.J., Yang, M.J., Wang, L.C.: Wafer map pattern analytics driven by natural language queries. IEEE International Test Conference in Asia (2022)

Summary and Conclusions

The interest in Machine Learning (ML) for use in various domains is expanding as the available amount of data increases exponentially with time. Machine learning proposes an abundance of techniques to extricate knowledge from data that can be rendered into purposeful objectives. Moreover, ML along with computer vision has augmented many domains where medical diagnostic, statistical data analysis and algorithms, scientific research, etc. are included. Such practices have already been done in the arena of smartphone applications, computer appliances, online websites, cybersecurity, etc.

Utilizing the power of ML and applying it to automate and optimize semiconductor manufacturing process and associated data analysis are now a hot hub of research for both academia and industry in recent years. A wide variety of ML algorithms and models have been developed recently. Depending on whether the label is provided or not for the training data, ML algorithms are categorized into supervised or unsupervised learning. On the other hand, from the model perspective, there exist discriminative models and generative models. Given the large number of well-labelled historical data collected from the existing manufacturing process, supervised discriminative models typically learn from the experience to accelerate future manufacturing and design efficiency. Moreover, supervised generative models are usually employed to further explore the design space, and facilitate or even replace manual design.

ML techniques are employed in many domains with great success because of their ability to build powerful models from data. One domain includes electrical and computer engineering, and the subfield of CAD, where ML promises to fill the gaps that exist in heuristic algorithms, and open new possibilities. Employing ML techniques allows designers to raise the abstraction level by focusing on the objective itself and leave the technical details on how to reach the objective to the ML model. The work in [1] categorizes how ML may be used, and how it is now used for design-time reduction and run-time optimization. A comprehensive state of the art on ML for CAD is hence presented, in all areas in CAD that are well-explored

© The Author(s), under exclusive license to Springer Nature Switzerland AG 2023

P. Girard et al. (eds.), *Machine Learning Support for Fault Diagnosis of System-on-Chip*, https://doi.org/10.1007/978-3-031-19639-3

and underexplored with ML, as well as trends in the employed ML algorithms. From this survey, it is shown that a wide spectrum of ML techniques already exists for CAD.

Another prominent domain that uses ML is in the test and reliability aspects of ICs. Defects are inevitable during semiconductor manufacturing, making it critical to screen them out during production test flow to prevent them from becoming failures in customer applications. With technology scaling approaching atomic levels, IC test and diagnosis of complex System-on-Chips (SoCs) become an overwhelming challenge due to more complexity and eventually less accessibility to internal nodes of these systems. In addition, sustaining the reliability of transistors as well as circuits at such extreme feature sizes, for the entire projected lifetime, also become profoundly difficult. This holds even more when it comes to emerging technologies that go beyond convectional CMOS where the underlying physics are not yet fully understood. In this context, one now finds ML is now embedded in many EDA tools from major companies dedicated to test, design-for-test, design-for-manufacturability, design-for-reliability, yield prediction and analysis, diagnosis, etc.

The topics covered in this book deal with the use of ML techniques in the context of IC fault diagnosis. The reader has been introduced first to the basic concepts on test and diagnosis, including failure analysis, defect modeling, fault simulation, yield improvement, etc. Existing conventional methods for fault diagnosis have also been introduced in the Chap. 2. Then, machine learning and its applications in test have been discussed in the Chap. 3. In order to improve fault diagnosis, various solutions that use ML have been developed in the last decade. The next part of the book is dedicated to the description of these solutions at various hierarchical levels of application and in different contexts: logic diagnosis and fault classification, cell-aware diagnosis, diagnosis of analog circuits, board-level functional fault diagnosis and wafer-level failure characterization.

Learning-based fault diagnosis is an active area of research and development that has steadily moved from research labs to industrial practice. This book has detailed both the basic and the advanced techniques in the field. It is anticipated that with the growing need for better diagnosis performance (resolution and accuracy), the techniques presented in this book will continue to be adopted. With the ongoing advances in technology, circuits and architectures, more innovation for learning-based fault diagnosis will happen; in this respect, we believe that this book will serve as an inspiration for future research and development in the field.

Reference

1. Rapp, M., Amrouch, H., Lin, Y., Yu, B., Pan, D.Z., Wolf, M., Henkel, J.: MLCAD: A Survey of Research in Machine Learning for CAD. IEEE Trans. Comput-Aided Des. Integr. Circuits Syst. (2021). https://doi.org/10.1109/TCAD.2021.3124762

Index

© The Author(s), under exclusive license to Springer Nature Switzerland AG 2023
P. Girard et al. (eds.), *Machine Learning Support for Fault Diagnosis of System-on-Chip*, https://doi.org/10.1007/978-3-031-19639-3

315

Printed in the United States
by Baker & Taylor Publisher Services